ABHANDLUNGEN
DER AKADEMIE DER WISSENSCHAFTEN DER DDR
Abteilung Mathematik – Naturwissenschaften – Technik
Jahrgang 1988 · Nr. 2 N

Characterization, production and application of food flavours

Proceedings
of the
2nd Wartburg Aroma Symposium 1987

Organized by
Central Institute of Nutrition Potsdam-Rehbrücke/GDR
Academy of Sciences of the GDR
Eisenach/GDR, November 16th–19th, 1987

Edited by
Manfred Rothe

AKADEMIE-VERLAG BERLIN 1988

Herausgegeben im Auftrage des Präsidenten
der Akademie der Wissenschaften der DDR
von Vizepräsident Prof. Dr. Heinz Stiller

Die Beiträge dieser Abhandlung wurden nach den Originalmanuskripten der Autoren reproduziert.

ISBN 3-05-500594-5
ISSN 0138-1059

Erschienen im Akademie-Verlag Berlin, DDR-1086 Berlin, Leipziger Straße 3–4

Lizenznummer: 202 · 100/374/88
Printed in the German Democratic Republic
Gesamtherstellung: IV/2/14 VEB Druckerei »G. W. Leibniz«, 4450 Gräfenhainichen · 7147
LSV 3695
Bestellnummer: 763 904 7 (2001/88/2N)
05200

Preface

Under the view of the consumer of foodstuffs flavour by far holds the first place among the attributes participating to food quality. Consumption statistics of highly developed countries clearly indicate this fact as well as the trend to more and more specialized and aroma-rich foods and meals. Whether this trend may be caused by the consumer's behaviour itself or originates in the supply of food shops and restaurants with a broad and still increasing variety of special foods and delicacies is a question of secondary importance.

As one of the consequences there is an increasing demand for food flavourings with respect to amount and variety. Because of limits in the production and use of natural flavour resources new ways have to be developed and introduced in the flavour industry. This actual task, however, must be based on more basic knowledge and practical experience as important tools for the expanding flavour industry.

Discussions between the specialists of different countries can stimulate the progress and must be profitable for all experts working or interested in this rapidly developing field. With the organisation of the 2nd Wartburg Aroma Symposium the Central Institute of Nutrition Potsdam-Rehbrücke tries to contribute to this process by the discussion of actual flavour problems and the personal exchange of opinions. We hope that such activities indirectly will support the interest of the consumer, too, as flavour belongs to those physiological sensations which may produce pleasure, satisfaction and contentment in human life.

With pleasure we noticed the resonance of our invitiation to this meeting by the guest lecturers as well as by a lot of expert groups within the socialistic countries. More than 50 % of the participants came from abroad thus giving a good chance for an effective discussion of flavour problems from a different point of view.

I want to thank the Academy of Sciences of the GDR as well as the director of the Central Institute of Nutrition Potsdam-Rehbrücke, **Prof. H. Schmandke**, and his deputy, **Prof. R. Macholz** for supporting our organisation work. As editor of these proceedings I am much indebted to lector **Karl Abel** from the Akademie-Verlag Berlin for an always good and effective atmosphere when discussing questions connected with the publication. Finally I am very much obliged to my colleagues **Dr. H.-P. Kruse** and **Edith Weise** for help and assistance in solving many detailed organisation problems as well as for correspondence and writing of manuscripts.

December 1987 Manfred Rothe

Contents

ACTUAL TRENDS IN FOOD FLAVOURING

M. Rothe, H. Ruttloff, R. Schrödter[1] and F.-J. Strüber[2]

Various aspects have stimulated our decision to focus this symposium somewhat to food flavouring problems. First of all there are three anniversaries in the flavouring field. In these days it is exactly 100 years ago that Gustav Theodor Fechner died who first discussed the relationship between stimulus concentration and stimulus response. This year a symposium in Leipzig remembered his well-known activities in the field of physics and physiology.

Also exactly 100 years ago Wallach established the so-called 'isoprene rule' which gives us a simplified concept of the pathways of terpene biosynthesis: most terpenoids can be hypothetically constructed by a 'head-to-tail' joining of isoprene units. Though we know today that mevalonic acid is the real intermediate of terpene synthesis this one century old hypothesis has created much advance in the terpene field, and that means in the large area of essential oils which had and still have a central place in food flavouring.

Finally it is now 100 years ago that the threshold value of an odorant was first determined by Emil Fischer and his co-workers. Today the story looks rather curious for us: The determination was realized in such a way that known amounts of the odorant were sprayed into a room of known volume. After some time necessary to get an equal distribution single persons were asked to enter the room and to answer the question whether they could find an odour or not. Using different people, different concentrations and repeated tests it was possible to define the threshold values of thiophenol and chlorophenol, respectively, in air as the diluting medium.

Most congresses in the flavour field held in the last two decades did not deal with flavouring problems directly. This may be the domaine of the flavour industry. In basic aroma research the activities are focussed first to the analytical side of flavour and aroma. Here the progress reached in the last 25 years indeed was fascinating. Despite of high prices of modern gas chromatographic and spectrometric equip-

[1] Zentralinstitut für Ernährung Potsdam-Rehbrücke, Akademie der Wissenschaften der DDR, Arthur-Scheunert-Allee 114/116, Bergholz-Rehbrücke, DDR-1505

[2] VEB Chemisches Werk Miltitz, Miltitz, DDR-7154

ments and of their steady improvement more and more flavour research groups make use of these modern techniques.

Table 1. Trends in the use of modern chromatographic and spectrometric methods in aroma research
(Data taken from literature documentation of the Central Institute of Nutrition Potsdam-Rehbrücke)

	1956/60	1965/69	1970/75	1981/83	1984/86
tested number of aroma publications	300	1000	1100	1000	1000
classical methods* and reviews (%)	95	77	63	52	41
GC methods (%)	4	14	22	27	27
GC/MS methods** (%)	1	9	14	18	25
HPLC methods (%)	÷	÷	÷	3	7

* = sensory, enzymatic, TLC, PC and colorimetric methods
** = including GC combination with other spectrometric methods
÷ = < 1%

We can register and confirm this trend from Table 1 with material of our own documentation 'Nahrung und Ernährung' edited by our institute [1]. Only such articles were taken into consideration and included in the table where flavour and flavouring problems are in the center of the discussion. Some actual aspects of instrumental aroma research will be discussed in the course of the symposium.

The use of modern analytical tools has enlarged our knowledge also in another field of interest which is very important for the flavouring practice. That is the high and still increasing number of identified odorous volatiles which on one hand causes progress in knowledge and arises on the other hand new problems simultaneously. We have to thank our colleagues in the Netherlands for their early started work of collecting all data on identified aroma components in the literature [11, 12].

Fig. 1 indicates the trend. Nearly all food flavours tend to be recognized as more and more complex from the compositional point of view. Till today a total of approximately 5000 volatiles with odorant character has been reached. A maximum value of 10 000 is considered

[5]. It seems to depend, however, on the sensitivity of our instrumental techniques, especially.

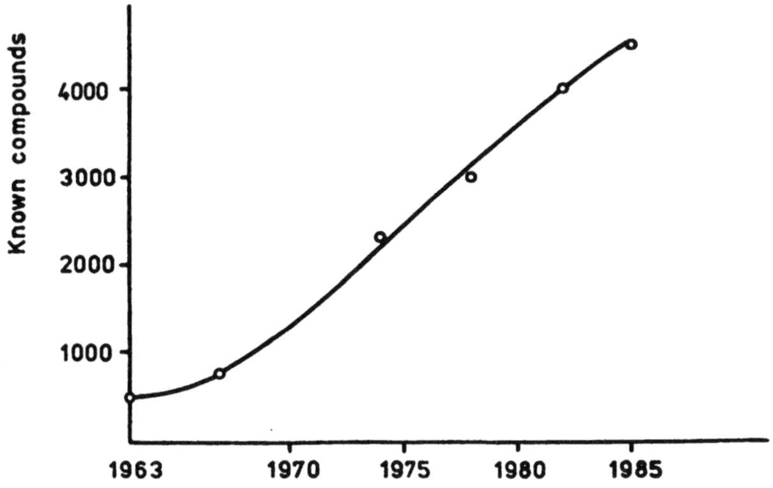

Fig. 1. Number of volatile compounds reported to occur in foods (H. Maarse, 1985)

The flavour producing industry makes use of these lists of naturally occurring food volatiles. In most European food laws such aroma components are defined as 'nature-identical'. Their application often is facilitated. Thus in several countries nature-identical aroma components may be applied for flavouring purposes independent on their origin.

The critical discussion of such food law and declaration problems is another modern trend in aroma research, though there are still differences between the different countries. Because of the increase in health risks in modern life the consumer today is very sensitive against an uncontrolled application of new additives. Here the lists of volatiles identified in food are of high value within the flavouring practice.

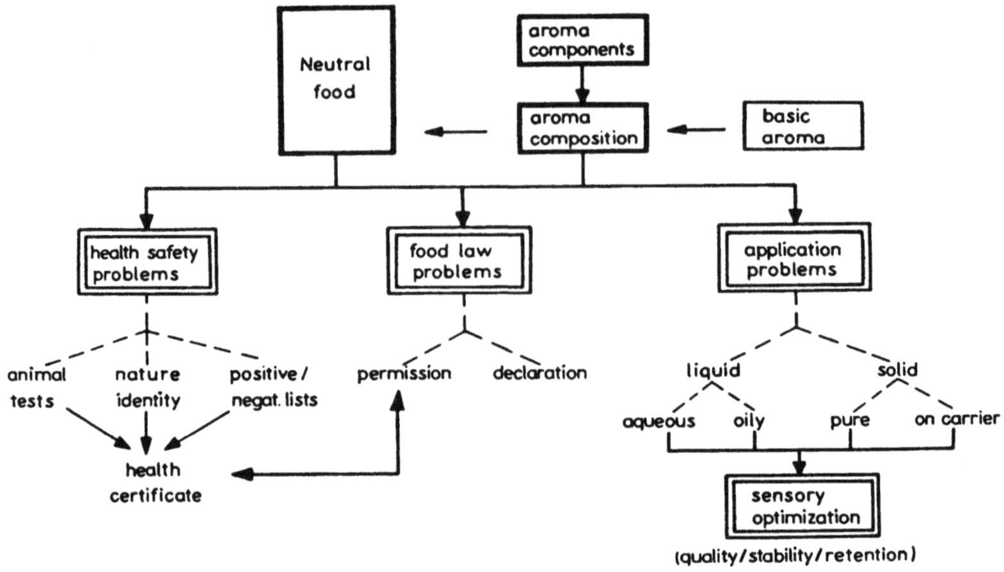

Fig. 2. Main problems in food flavouring

Another main problem for the flavour industry originates from the high complexity of natural food aromas. The higher our knowledge about this very complex composition the more single aroma components must be available and have to be combined for aroma compositions of optimum quality. When 30 years ago e. g. a margarine flavouring may have been composed by between 10 and 30 single components this value actually has been doubled. Considering this situation several conclusions become obvious: there is need for a more and more increased palette of aroma components to be produced by chemical processes or isolated from natural material. On the other hand the more complicated the aroma composition the better should be the sensory methodology to find the optimum composition which is best adapted to the consumer's interest. Improved and new sensory methods are a useful tool for solving these problems by the flavour industry. When considering the central interest for the flavourist we will also deal with some of these problems during the symposium.

The various possibilities and pathways of modern aroma production are summarized in Fig. 3. According to this scheme 4 principles can be used:

1) isolation from natural aromatic raw materials (by extraction of distillation procedures)

2) chemical synthesis of aroma compounds

3) biotechnological processes, that means the application of enzymes, microorganisms or plant cell cultures resulting in most cases in aroma complexes

4) thermic processes, that means the use of aroma-forming thermic procedures as the nonenzymatic browning reaction or other chemical reactions also resulting in complex aroma mixtures.

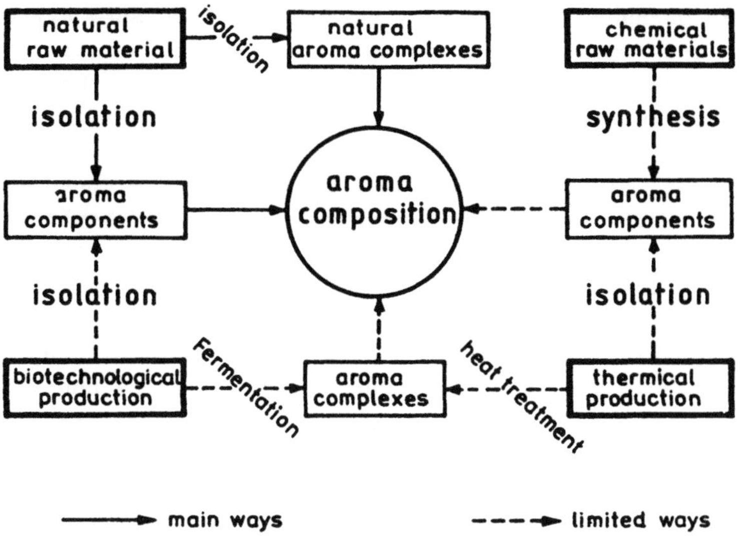

Fig. 3. Production principles for food aroma compositions

The indicated limits in Fig. 3 originate in economic or food law problems especially. Each of these different possibilities for the production of aromas and flavourings has its own large variety of specific problems.

The same is true for the application of flavourings. Within a contribution like this it is not possible to discuss all the problems indicated schematically in Fig. 2 and 3. Thus a few examples are selected as demonstration objects.

When considering the actual health aspects of aromatization natural aroma concentrates seem to be the preferred category. There are, however, also problems which restrict production as well as application of naturally derived aroma concentrates and flavourings. On the one hand the continuously increasing consumption of aroma-rich foods and beverages in highly developed countries can limit the availability thus resulting in growing or instable prices on the world market. Another disadvantage of natural aroma concentrates often is a rather low aroma efficiency compared with most of the synthesized aroma components. Thus the combination of natural and synthetic aroma components represents a preferred way to overcome this problem. Such compositions can combine the characteristics of high aroma intensity (from the synthetic top aroma or key component, respectively) with a broad background (from the natural part).

A typical example actually practicized by the flavour industry of some countries is yoghurt. This product becomes more and more attractive for the consumer with steadily increasing consumption rates. Here the normal rather neutral and one-sided aroma character of the basic product can be intensified and improved not only by natural fruits but more and better by their combination with synthetic nature-identical aroma compounds. Similarly the actual interest of the consumer for exotic flavour notes within milk products, sweets and beverages in many cases is the result of an attractive flavouring on the base of natural and synthetic components.

Flavouring legislation, however, must be considered. Although there are differences in the food law situation of different countries with respect to flavouring additives we can register as a general trend that nature-identity today is one of the most important criteria.

As to aroma compounds the term 'nature-identity' means not only the raw chemical structure but also fine structure and conformation. In some cases there are great differences in aroma quality and/or aroma efficiency between isomers. A well-known practical example is menthol which as to its annual consumption now holds the second place of all aroma components. Fig. 4 demonstrates the isomeric forms of this important aroma component together with some differences in their sensory properties [3].

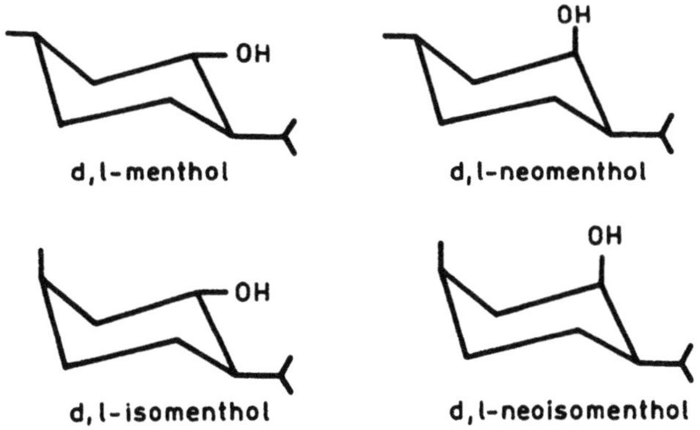

	threshold (ppm)*		
	taste	cooling	bitter
l-menthol	0.4	0.8	15
d-menthol	0.3	3	25
l-isomenthol	0.6	30	50
d-isomenthol	0.7	7	20
l-neomenthol	0.65	25	30
d-neomenthol	0.5	3	25
l-neoisomenthol	1.0	7	65
d-neoisomenthol	0.2	65	65

*R. EMBERGER and R HOPP 1985

Fig. 4. Structure and threshold value of 8 menthol isomers

The partial synthesis of menthol (Fig. 5) which starts from m-cresol is realized today in industrial scale and has to respect these sensory differences. l-Menthol as the most efficient aroma compound must be separated out of the complex of 8 isomeric menthols. The pathway demonstrated by R. Emberger and R. Hopp (1985) thus includes the separation of the racemic d,l-menthol by a special selective crystallization process [3].

In many cases the chemical synthesis of aroma components is easier and much more economic compared with the time- and work-consuming isolation steps of natural aroma compounds. If nature-identity of a component is proved it is an important as well as an interesting task for the flavour industry to find out the best and most economic synthesis pathway. As an example Fig. 6 shows the partial synthesis of a sulphur-containing terpene aroma compound which was found to act as one of the key components in grapefruit. The elegant synthesis which starts from

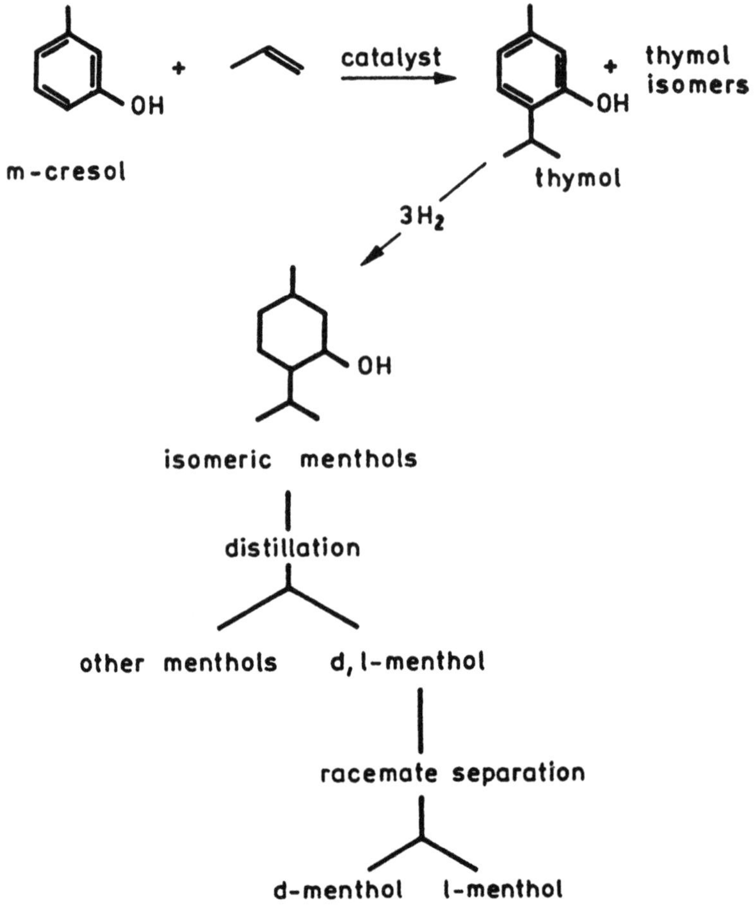

Fig. 5. Chemical synthesis of l-menthol
(R. Emberger and R. Hopp 1985)

limonene as an available raw material was realized by G. Ohloff and co-workers [2].

Such syntheses are of interest in those cases of aroma components especially where a high aroma efficiency of the compound has been proved. There are high differences as to this property between aroma components which can be demonstrated by means of their threshold values. Table 2 summarizes threshold values of some aroma components which today are actual for flavouring purposes [4, 6, 8]. The values were obtained in water as a medium. Obviously there are differences of magnitudes. They are of practical importance on account of their influence on dosage and costs. It may surprise somewhat that mainly

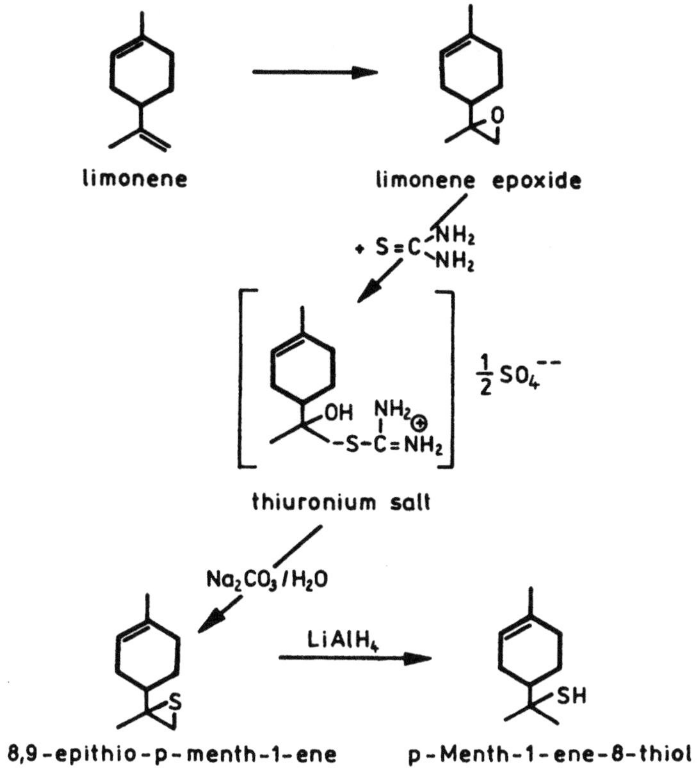

Fig. 6. Chemical synthesis of p-menth-1-ene-8-thiol
(E. Demoli, P. Enggist and G. Ohloff 1982)

produced and applied aroma components like vanillin, menthol, diacetyl or maltol are not found on the top of the list which is ranked according to the threshold concentration or aroma efficiency of the components, respectively. Threshold determination has its own problems we are going to discuss within this symposium, too.

Table 2. Detection threshold of some important food aroma components (water as diluting agent; 20 °C; in ppm)

Aroma component	Threshold concentration
[3,3'-(2,2'-Dimethyl)-bifuryl] disulphide	0,000 000 02
p-Menth-1-ene-8-thiol	0,000 000 02
2-Methoxy-3-isobutyl-pyrazine	0,000 002
Methyl thiol	0,000 02
4-Hydroxy-2,5-dimethyl-3(2H)-furanone	0,000 04
cis-3-Hexenal	0,000 3
Ethyl 2-methylbutyrate	0,000 1
(+)-Nootkatone	0,001
Diacetyl	0,004
Vanillin	0,02
l-Menthol	0,4
(-)-Nootkatone	1
Maltol	35
Pyrazine	300

B i o t e c h n o l o g i c a l (fermentative) and t h e r m i c (chemical) formation of aroma concentrates are rather new possibilities of aroma production. Characteristic for both principles is the technical simulation of aroma producing mechanisms. As an advantage of biotechnological and thermic aroma production the formation of complete aroma complexes is discussed which may be applied without further composition with other aroma compounds. Disadvantages are the relatively low aroma concentration requiring high dosages which arise the flavouring costs. Thus for both types of aroma forming processes we can find a lot of interesting examples in laboratory scale, but - up to now - only a few for practical use in the industry.

As an example for some of the problems integrated in biotechnological aroma production Fig. 7 indicates the development of methyl ketone formation during industrial fermentation of a two phase aqueous-lipid system by means of a strain of Penicillium roqueforti [7]. Obviously there are two phases within the aroma formation curve which may limit a rapid and high yield in methyl ketones. First, just after the start of the process we can observe a lag phase with rather low increase of the microorganism as well as of the formation and availability of precursor substances. This limitation may be overcome by chemical, microbial or technical means. Much more critical is the asymptotic

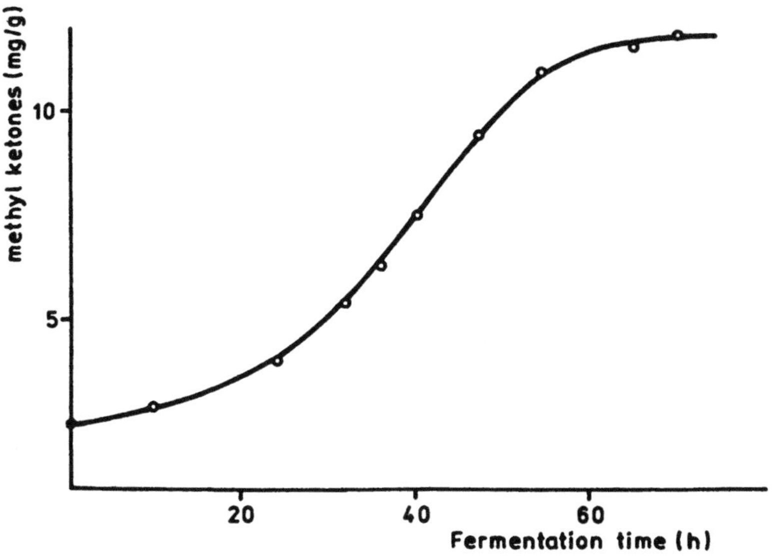

Fig. 7. Formation of methyl ketones during biotechnological conversion of butter fat by Penicillium roqueforti

approximation of the aroma formation curve to a maximum value. In our example the yield in methyl ketones is limited to 1.2 % representing a rather low value compared with the average yields of chemical syntheses.

The influence factors on such equilibria are complex and often not easy to understand. Generally this effect may be a consequence of the fact that aroma formation in most biological materials is a side-process only which cannot be intensified without disturbing the complicated multienzymatic systems.

As a practical consequence of such low aroma amounts, however, concentration processes are required which demand extra equipment and which may be time- and work-consuming. Thus production costs of aroma concentrates generated in a biotechnological way often are high.

T h e r m i c aroma formation today is used mainly for the production of meat-like flavourings [10]. In principle, however, other types of Maillard flavours like cocoa or bread aroma can also be produced in this way. Fig. 8 shows the most important types of meat flavour concentrates as well as main precursors and additives. Nonenzymatic browning reactions are in the center of the aroma formation. In the case of chicken flavour unsaturated lipids act as precursors via peroxides as reactants with free amino acids, in the case of smoked meat products phenolic compounds are important additives.

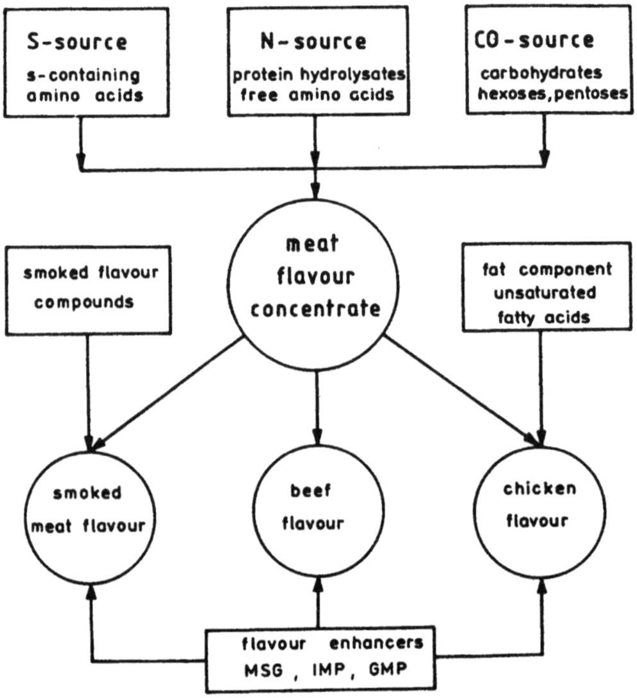

Fig. 8. Principle ways in the production of meat flavourings

Sulphur-containing amino acids or other sulphur compounds play an important role as precursors for the formation of meat flavours. The number of aliphatic and heterocyclic sulphur-containing aroma compounds identified in meat and meat-like flavourings is one of the highest found in foods. This topic will be discussed in the course of the symposium, too.

The proof of health safety is an important problem for biotechnologically as well as for thermically produced aroma complexes. The complexity of such flavourings may be an advantage for flavour quality, however, that complexity is a problem under a toxicological point of view. Before practical application health certificates based on animal tests often are necessary which are rather expensive and time-wasting.

Some topics of flavour application will also be discussed within the symposium program. At a first view the practical use of aroma concentrates and flavourings may be understood as a dosage problem only. The actual trend to more and better flavourized products also depends, however, on more knowledge about the basic laws of flavour reception as well as about the acceptability by the consumer. As an introduction to this large field two examples shall be discussed here.

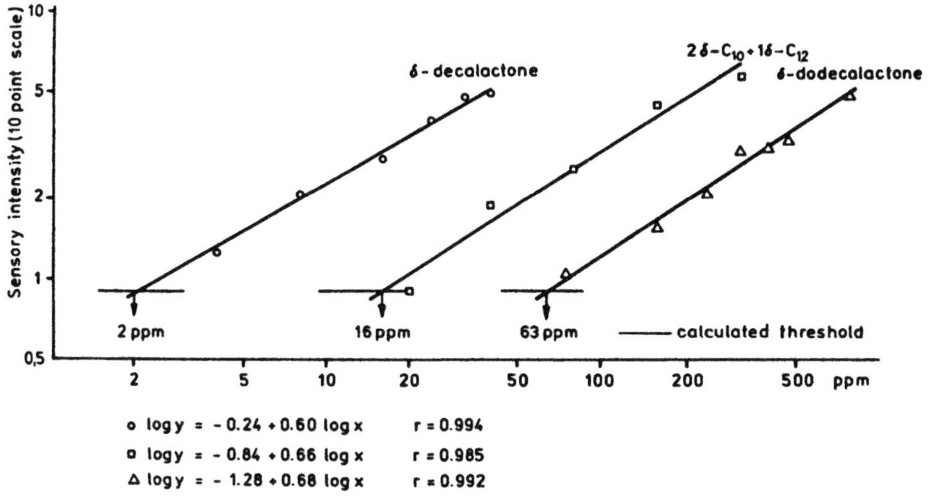

Fig. 9. Sensory characteristic of two δ-lactones and of their mixture in energy-reduced butter

An optimum flavouring composition often needs knowledge about interrelation effects between different aroma components. In the case of similar chemical substances this interrelation may be additive, however, potentiating or subtractive (masking) effects are also possible. In Fig. 9 the aroma intensity of two δ-lactones in low flavoured energy-reduced butter is characterized quantitatively in a double logarithmic diagram. Data were obtained by means of a trained panel of 10 people using an unstructured scaling procedure. Within the range interesting under practical flavouring conditions the relationship is linear. δ-decalactone was found much more effective compared with δ-dodecalactone with an (extrapolated) threshold value of 2 ppm and 63 ppm, respectively. Such differences as found between the two lactones have consequences for the dosage and obviously must lead to economic calculations.

When the two lactones are combined in a 2 : 1 ratio the resulting line is found between the lines of the pure compounds and near the calculated line. This indicates - together with the identical slope of all 3 lines - that the interrelation is additive. In the mentioned example combination or exchange of the two lactones seems to be without quality problems when the different aroma efficiency is considered.

Flavouring is also a concentration or dosage problem, respectively. The optimum amount of the additive depends on the aroma components present as well as on the medium. The exact determination of the optimum dosage is a critical problem in the case of new types of flavourings, especially. Consumer tests with a high number of people may be required. In Fig. 10 the results of such tests with a Blue cheese aroma concentrate in different dosages are summarized. For each of these dosage tests between 150 and 250 people were asked for their opinion as to the quality of a cheese spread offered on rolls. There were 3 possibilities for answering, that means for defining the quality as

- excellent or good
- still acceptable or
- insufficient.

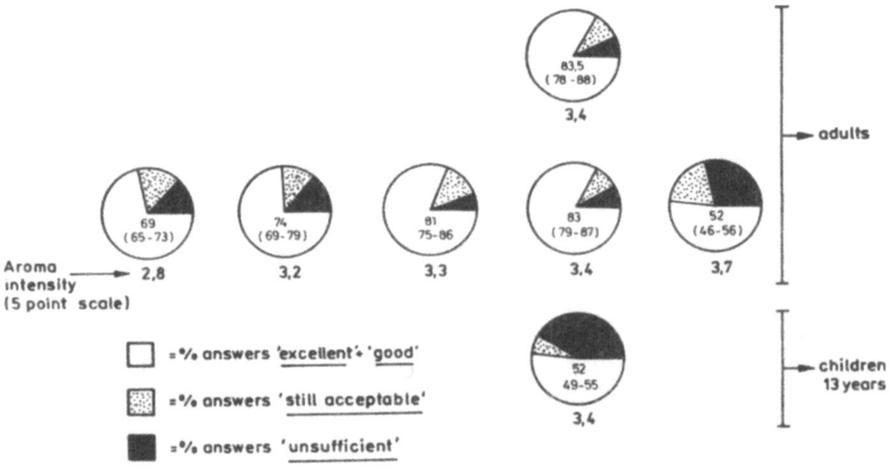

Fig. 10. Influence of aroma intensity (dosage) on the acceptability of a flavourized cheese spread presented on rolls (150 - 250 untrained persons per test; each test with different people; percentage and confidence area)

The results are indicated in circle segments showing the percentage of all answers. Obviously in the area of low aroma intensity increasing amounts of the additive improve the acceptability up to a maximum dosage. As a consequence of the high number of test people the reproducibility of this consumer test was rather high: two different test groups revealed a very similar answer spectrum at the optimum level (aroma intensity = 3.4).

Above the optimum flavouring dosage the hedonic value decreases rapidly. As to our experience this turning of the acceptability curve is not so sharp with natural, more complex flavourings compared with pure aroma components or artificial flavourings which sometimes show a very strong decrease within a narrow concentration range just above the turning point of the acceptability curve. In any case of food flavouring, however, the optimum application dosage has to be determined exactly and by means of methods which are well appropriated for testing the consumer's behaviour.

Fig. 10 touches quite another aroma problem, too. We often have to consider that different groups of our population may have a varying opinion as to quality and hedonic value of aroma-rich foods and beverages. An example is one of the consumer tests in Fig. 10 which was carried out with school children in the age of 13 years. The optimum level of the Blue cheese aroma was added as it was found according to the tests with adults. In comparison with all further tests the result was quite different. Obviously children are less adapted to special and new flavours and thus show a much more negative behaviour. They also tend to decide more clearly in a positive or negative way which is indicated by a rather low percentage in the 'still acceptable' answers.

Summarizing modern food flavouring includes a lot of different aspects:

- the isolation and identification of natural food aromas,
- the weighting of the numerous numbers of components with the aim to find key components,
- synthesis and technical production of interesting aroma components and aroma complexes,
- application problems like optimum dosage, carriers or binding effects,
- consumer behaviour, food law, health and nutritional aspects.

Within this limited lecture it was only possible to introduce to some of these problems. If accepting it as a frame everybody will easily be in the position to find the relationship between the various lectures within the symposium program. They all together can be understood as a further small step for increasing our general knowledge in food flavour and flavouring. Let us hope that at the end of the symposium and our discussions most of us will come to the conclusion that there are some new aspects and ideas which will promote the progress in aroma research and food flavouring.

References

[1] Bibliographie 'Nahrung und Ernährung des Menschen', Reihe F.O. Lebensmittelwissenschaft, AdW der DDR, Zentralinst. Ernährung Potsdam-Rehbrücke 1971 - 1984 and collection of literature 1956 - 1970

[2] Demoli, E., P. Enggist and G. Ohloff; 1-p-menthen-8-thiol: a powerful flavour impact constituent of grapefruit juice (Citrus paradisi MacFayden), Helv. chim. Acta 65 (1982), 1785 - 1794

[3] Emberger, R. and R. Hopp: Synthesis and sensory characterization of menthol enantiomers and their derivatives for the use in nature-identical peppermint oils, in R. G. Berger, S. Nitz and P. Schreier, Topics in flavour research, H. Eichhorn-Verlag, D-8050 Marzling-Hangenham 25, 1985, p. 201 - 218

[4] Gora, J. and W. Brud: Progress in synthesis of sensory important trace components of essential oils and natural flavours, Nahrung 27 (1983), 413 - 428

[5] Rijkens, F. and H. Boelens: The future of flavour research, in H. Maarse and P. J. Groenen, Proceed. Internat. Symp. on Aroma Research, Zeist/The Netherlands 1975, Centre Agric. Publish. and Doc., Wageningen 1975, p. 203 - 220

[6] Rothe, M.: Introduction to aroma research, Handbuch der Aromaforschung, Akademie-Verlag, Berlin 1988, p. 64

[7] Rothe, M., H. Ruttloff, H. Herrmann and W. Engst: Problems of technical production and characterization of Blue cheese aroma concentrate, Nahrung 30 (1986), 791 - 797

[8] Rothe, M., G. Wölm, L. Tunger and H.-J. Siebert: Schwellenkonzentrationen von Aromastoffen und ihre Nutzung zur Auswertung von Aromaanalysen, Nahrung 16 (1972), 483 - 495

[9] Schreier, P.: Some aspects of biocatalysis in natural products chemistry, in R. G. Berger, S. Nitz and P. Schreier, Topics in flavour research, H. Eichhorn-Verlag, D-8050 Marzling-Hangenham 25, 1985, p. 353 - 369

[10] Schrödter, R., J. Schliemann and G. Wölm: Lipid oxidation products and their importance in the formation of meat aroma, Nahrung 30 (1986), 799 - 808

[11] van Straten, S. and H. Maarse: Volatile compounds in food, 5th edition, TNO-CIVO Food Analysis Inst., Zeist/The Netherlands 1983, and supplements

[12] Weurman, C.: Lists of volatile compounds in food, Report R 1687, TNO-CIVO Food Analysis Inst., Zeist/The Netherlands 1963

ON-LINE COUPLED HRGC TECHNIQUES FOR FLAVOUR ANALYSIS

P. Schreier

Lehrstuhl für Lebensmittelchemie, Universität Würzburg, Am Hubland, 8700 Würzburg, FRG

Introduction

Approximately 5000 volatile chemicals have been identified in food (1,2) and there is little doubt that others exist and will be ultimately detected. These complex chemical mixtures, responsible for food aroma, can number as high as 1000 for an individual food and originate via a variety of processes (3,4).

In order to analyze such complex mixtures of substances often occurring only at trace levels, gas chromatography (GC) has been introduced already in an early stage of flavour research. Flavour chemists were among the pioneers of GC and also, more recently, capillary GC (HRGC) (5,6). This early work has been reviewed by Teranishi (7).

High resolution gas chromatography (HRGC)

The major advances of the past 10 years or so which have led to the present generation of high efficiency, inert and thermostable capillary columns may be summarized in a few sentences:

- development of high quality glass columns,
- beginning of a new era in HRGC with the introduction of inert, easy to handle flexible fused silica columns,
- development of immobilized stationary phases with improved thermostability which are not damaged by large injections of solvents.

These developments have been treated in recent texts (8,9). Reviews on separation techniques with specific emphasis on flavours include those of

Jennings (10-12), Teranishi (7) and Shibamoto (13).

Nowadays, the analysis of flavour volatiles by HRGC and on-line HRGC techniques places quite stringent demands on the performance of the separating column which should possess the following properties:

- high resolution power,
- adequate sample capacity,
- minimal stationary phase "bleed",
- maximum degree of chemical inertness,
- ready compatibility with different methods of sample introduction.

The ideal column is one which should optimize the attainment of all these features - and does not exist. At present, wall-coated open tubular columns of intermediate internal diameters and stationary phase thicknesses, prepared using modern bonded phase technology, appear to offer the best compromise of separating power, sample capacity and compatibility witn different injection techniques, for general purpose use in HRGC or on-line coupled HRGC techniques.

High resolution gas chromatography - mass spectrometry (HRGC-MS)

High resolution gas chromatography in combination with mass spectrometry (HRGC-MS) was and remains the most powerful technique available to the chemist for the separation and identification of complex mixtures of volatile compounds because it yields the maximum amount of structure information for the smallest amount of material in the shortest time.

The coupling of capillary columns with mass spectrometers was problematic for many years, but these problems have been largely solved and linking of high resolution capillary columns to modern mass spectrometers is possible without compromising the performance of either instrument. The interface region must, of course, be properly deactivated and heated to minimize sample losses by adsorption, condensation or decomposition. Two main methods are in widespread use, i.e. (i) direct connection and (ii) the open-split interface.

(i) Ion source pumps with very high pumping capacity are available on modern mass spectrometers which can readily handle helium flow rates maintaining source housing pressures of 10^{-5} to 10^{-6} bar. Thus, the entire effluent from the column can be transferred directly to the ion source. Loss of column efficiency between 10-20 % has been estimated (14,15). The only disadvantage of direct coupling, especially when dealing with very complex samples, is the difficulty in relating chromatograms, previously obtained under optimized conditions using a GC detector, with those produced by the reconstructed gas chromatogram on total ion monitoring in HRGC-MS runs. If the analysis speed is increased in the latter, the elution temperatures of compounds are different from those in the HRGC run, and this may have implications for the retention times of certain compounds (16).

(ii) For the open-split interface, introduced by Henneberg et al. (17,18), a number of all-glas devices have been developed (19-22). The main advantages of the open-split interface are as follows:

- well-constructed devices have no adverse effect on the efficiency of the column;
- chromatograms obtained by HRGC-MS are directly comparable with separate HRGC runs because the column outlet is at atmospheric pressure; this is essential in chemical ionization (CI) applications where the TIC trace may differ considerably from that recorded in electron impact (EI) mode;
- column change can be quickly performed since no vacuum sealing is involved;
- the bulk of large solvent peaks may be diverted to waste thereby maintaining the vacuum and protecting the source from contamination.

There is another point of discussion which has to be stressed in the context with HRGC-MS, i.e. the mass spectrometric techniques. In Table 1, some ionization methods other than the usually applied EI technique are outlined. With most of these techniques, only molecular or quasi-molecular ions are formed. Thus, in low-resolution mode the molecular masses or in high-resolution mode the composition of molecular ions are detectable by intense signals. In flavour research, the use of most of these techniques is rather limited. Only the CI technique (23) has found some interest in the investigation of volatiles (24-28). The other methods shown in Table 1 are

Table 1. Ionization techniques

Type	Ionization		Compounds
Chemical Ionization (CI)	+/-		Volatile
Field Ionization (FD)	+	gas phase	To be vaporized
Desorption Direct CI (DCI)	+/-	gas/cond.phase	
Field Desorption (FD)	+(-)		Less or nonvolatile
Fast Atom Bombardment (FAB)	+/-		
Secondary Ion MS (SIMS)	+/-	cond. phase	High-molec.;high-polar
Laser Desorption	+/-		Thermolabile
^{252}Cf-Plasma-Desorption	+/-		e.g., polysaccharides

generally used for the study of non-volatile material, i.e. they can be important for investigating flavour precursors (29).

Negative ion CI has been little used in the analysis of flavour volatiles. Both positive and negative modes of ionization can yield complementary information. An illustration of the value of negative ion CI is shown in Fig. 1. As demonstrated by Bruins (30), e.g., the monoterpenoid ester iso-

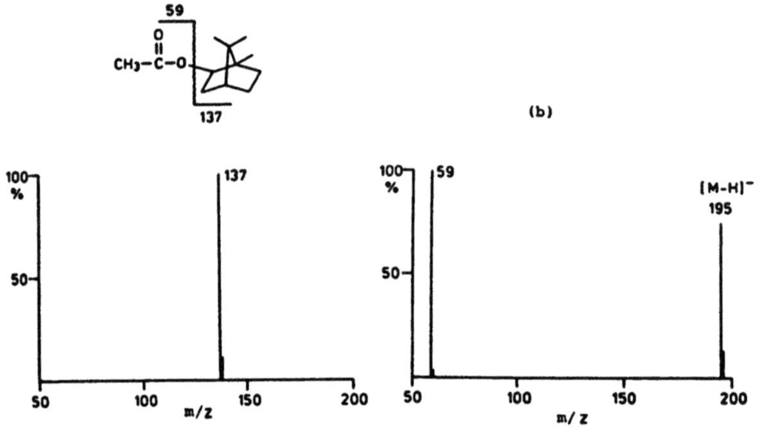

Fig. 1 CI spectra of isobornyl acetate. (a) positive ions, reactant gas isobutane; (b) negative ions, reactant ion OH^- (30).

bornyl acetate was very effectively cleaved in positive ion CI with isobutane, but yielded an intense $M-H^-$ and acetate ion in the negative ion spectrum using OH^- as the reactant.

As to the registration techniques, selected ion monitoring (SIM) has also to be mentioned. In this technique, the mass spectrometer is equipped with a so-called peak selector which instructs the instrument to record either single (SID) or multiple (MID) ions characteristic of the compound being searched for to provide a series of fragmentograms for separated peaks during HRGC run. The main advantage of SIM ist that it enables the mass spectrometer to be used as an extremely sensitive and specific GC detector. By monitoring the ions of one or just a few specific masses instead of the whole spectrum, a 1000-fold increase in sensitivity, i.e. reaching the picogram area, can be attained. SIM has been applied extensively in many fields of flavour research both to detect and quantify trace components separated by HRGC (31-34).

In the analysis of complex mixtures using HRGC-MS, it is commonly observed that two or more compounds may still be eluted simultaneously and it is important to know what can be done to extract useful information in such cases. Remarkable results may sometimes be achieved using the technique of "mass chromatography" with a computerized data system. This allows a reconstruction of the ion current profiles of selected ions as a function of spectrum index number from continuously collected data. The complete mass spectra may be accessed from the disk at any time after data acquisition. This technique should not be confused with the just mentioned SIM (SID, MID), where the data of preselected ions are only recorded and no mass spectra are available. An example of what may be achieved by "mass chromatography" in a favourable case, was illustrated by means of the monoterpene hydrocarbon fraction of a lavender oil sample by Thomas et al. (35). As shown in Fig. 2, the shaded area of a peak separated by HRGC was scanned first for m/z 93, present in all monoterpene hydrocarbons enabling the shoulder on the peak to be clarified observed. A mass spectrum of this shoulder, after subtraction of the background from either side, permitted the identification of 3-carene. Similarly, m/z 119 maximized at two positions, the mass spectrum of the first of these indicating the presence of p-cymene, while the spectrum of the main peak identified it as ocimene, due

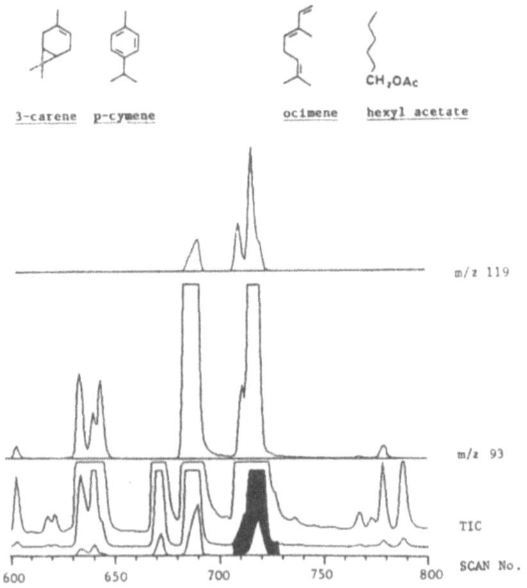

Fig. 2 Mass chromatograms at m/z 93 and 119 of unresolved components (shaded area of TIC trace) in monoterpene fraction of lavender oil (35).

to the fragments of m/z 79 and 80. By examining every mass spectrum over the peak several new fragment ions, m/z 56 and 61, were obtained in the tail. "Mass chromatography" of these fragmentations delivered a GC trace from which a spectrum was obtained; after suitable background substraction hexyl acetate was identified.

Concluding the chapter of HRGC-MS one should also spend a few words about high resolution MS. The application of high resolution MS to yield precise elemental composition of components separated by HRGC has been fairly limited to date (36), but is likely to be more widely exploited in the future due to continuing improvements in the performances of magnetic sector mass spectrometers and the availability of data systems with very high rates of digitization. Impressive results have been recently reported for HRGC in combination with a dual-cell Fourier Transform mass spectrometer (37).

Multidimensional gas chromatography (MDGC)

As mentioned before, the acquisition of clean unambigious spectroscopic information of sensorially important minor components of aroma isolates is frequently compromized by insufficient power and sample capacity even when using modern high resolution capillary columns. This may arise from compounds coeluted with major components or emerging on the leading or tailing edges of large peaks. Off-line prefractionation and trapping techniques are useful, but are also tedious and time consuming. In addition, enrichment of minor components by trapping of multiple fractions from several HRGC runs may result in sample losses and introduction of artefacts. However, if it is achieved to introduce on-line into a second column only the substances of interest, the problem can be easily resolved. Multidimensional gas chromatography (MDGC) allows this and a number of other important functions to be carried out with high efficiency.

An effective MDGC system consisting of two columns - a precolumn and an analytical column - may be used to perform the following operations:

- foreflushing; permits solvent peaks to be vented from the precolumn thereby protecting a HRGC column from overloading or contamination;
- backflushing; allows unwanted compounds of low volatility to be backflushed from the precolumn before entering the analytical column;
- heartcutting; permits selected fractions to be directed from the precolumn to the analytical column, usually with the application of intermediate trapping. The technique may be used to transfer "narrow cut" fractions between two columns of different polarity for improved resolution on the second. In addition, effective enrichment of trace components may be achieved by cold-trapping multiple cuts of a given fraction resulting from several injections on the precolumn.

These operations can be implemented by use of the pressure balance technique introduced by Deans (38) and later developed for HRGC by Schomburg et al. (39,40). The particular advantage of the Deans switching technique is that no valves or moving parts are either in the sample flow path or the high temperature zone. The regulation of line pressures at selected times is controlled by remote solenoid switching valves. Combinations of HRGC/HRGC or packed GC/HRGC columns may be used; commonly the second column is a HRGC column. Depending on the problem the precolumn may be a packed column, a medium resolution wide-bore capillary column or a high resolution capillary column of similar dimensions but different polarity to the analytical column.

The scheme of a commercially available sophisticated MDGC system comprising a double oven for independent temperatures of the two columns is shown in Fig. 3 (41). Switching of two HRGC columns is achieved by modification of Deans switching, the so-called "live column switching technique" developed

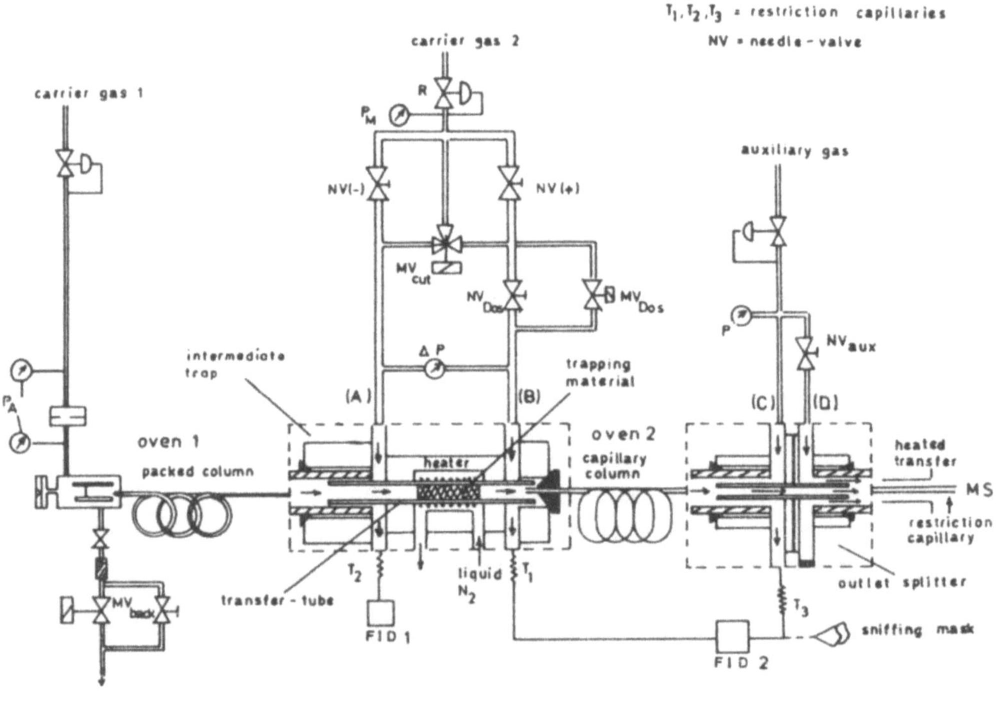

Fig. 3 System for total transfer from packed to capillary columns with intermediate trapping and device for parallel MS/sniffing or MS/FID registration (41).

by Schomburg et al. (42,43). The coupling of packed and capillary columns is effectively accomplished using the so-called "total transfer system", in which selected cuts from a packed column are trapped before transfer into a capillary column (44). This is difficult to achieve with conventional MDGC systems due to large differences in the flow rates of the two column types. Because of the high sample capacity of packed columns the "total transfer system" is especially attractive for the enrichment of trace components in aroma isolates prior to HRGC-MS analysis.

The high power of the system has been demonstrated by Nitz and Jülich (41) (Fig.4). Studying cauliflower extract by GC/sniffing test, the authors de-

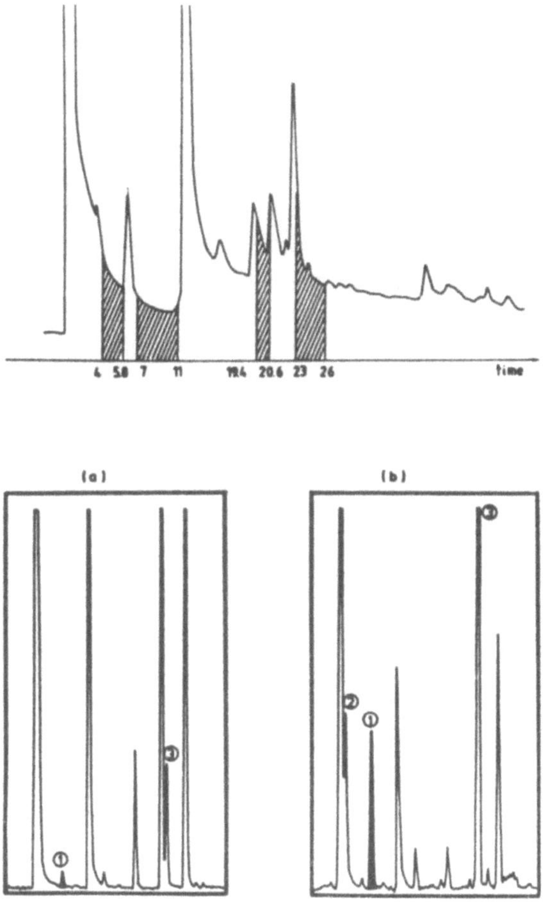

Fig. 4 Top: Chromatogram of cooked cauliflower extract obtained on the packed precolumn (cutting areas are indicated), 1 m x 2 mm i.d. glass column, filled with 5 % OV 101 on Chromosorb W (70-100 mesh). 70-280°C, 5° /min. Bottom: TIC of definite cuts of cooked cauliflower extracts. (a) cut from 19.4 to 26.0 min. (b) cut from 19.4 to 20.6 and 23 to 26 min, respectively.

tected a zone with an intense cooked cauliflower odour. The HRGC-MS of the sample gave a mass spectrum of low purity, as in the same region other minor components were eluted. Therefore, cutting as described above was carried out (Fig. 4, top). Chromatogram (a) (Fig. 4, bottom left) shows the TIC when a cut was performed between 19.6 to 26.0 min. Simultaneous sniffing during MS acquisition made sure that component 1 was the compound to be looked for. Chromatogram (b) (Fig. 4, bottom right) was obtained performing cuts between 19.4 to 20.6 and 23 to 26 min, respectively. A clean mass spectrum could now be recorded. In addition, a second substance 2 which was masked by a major component, could be detected. At the same time, it was possible to enrich component 3 and eliminate the substance eluting immediately before.

Recently, Nitz et al. (45) have demonstrated additional examples by means of combined cryogenic headspace enrichment and MDGC analysis. Fig. 5 shows a scheme of the system used. The extraction of volatile compounds has been achieved by stripping the headspace vial with a constant helium flow followed by cryogenic concentration in a cold trap packed with 5 % OV 101 on Chromosorb. Total purging volume could be adjusted by time-programming a solenoid valve. During this sample period breakthrough of substance into the first column was avoided by actuating another solenoid valve, thus maintaining the preselected column pressure and establishing a minor gas flow through a restriction capillary in the afore-mentioned trap. The sampling period has been interrupted by closing the valves. The gas flow in the trap has been reversed by actuating a solenoid valve and enriched volatiles were introduced into column 1 by thermal desorption. After admittance into column, the original column pressure was reestablished by again operating the valve. The group of peaks of interest was transferred into column 2 by means of the above-mentioned "live-T" switching device. After final separation the substances were directed to the MS and/or sniffing mask.

This system has been applied for a number of flavour problems, from which an impressive example is shown in Fig. 6, i.e the resolution of an "off-flavour" problem in coffee beans, the so-called "peasy" beans. Fig. 6a summarizes the afore-mentioned analytical steps, i.e. the enriched coffee headspace and the heartcuts at the retention times corresponding to the

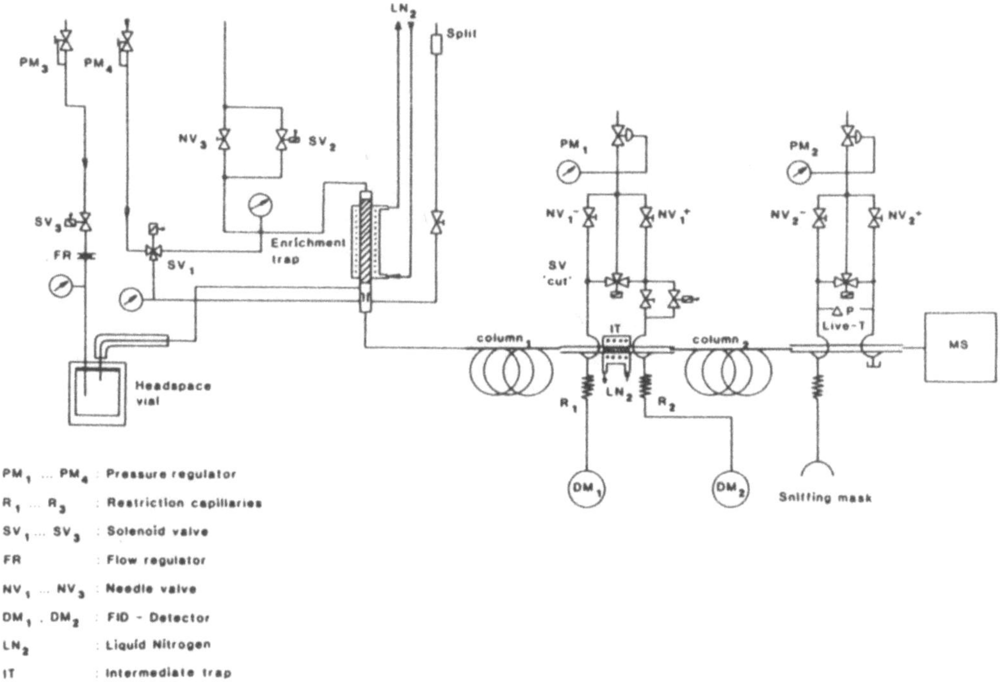

Fig. 5 Scheme of cryogenic headspace enrichment system combined with MDGC (45).

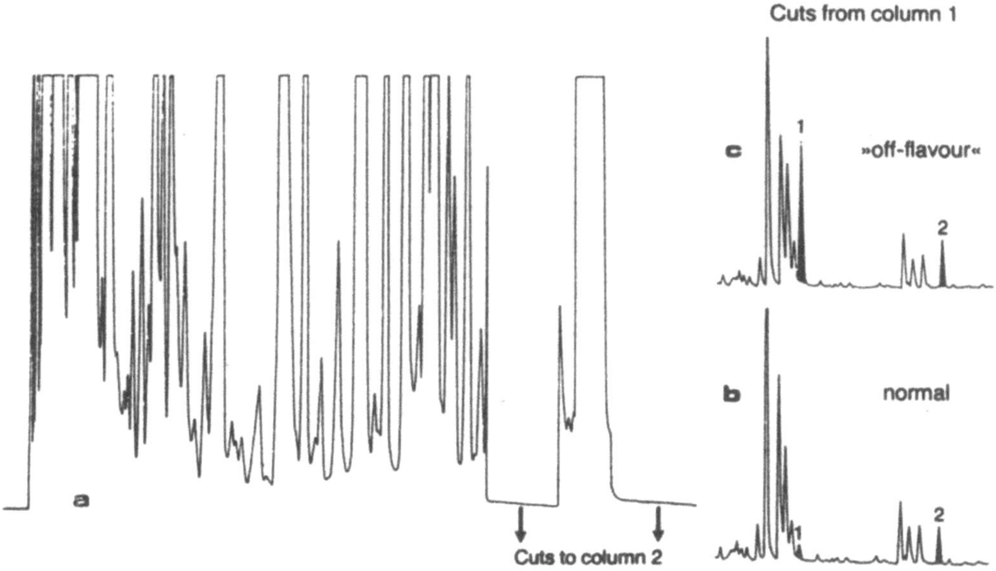

Fig. 6 Headspace of coffee (sample: 3 beans !) using MDGC. Column 1: CW 20 M; column 2: SE 54. 5 min/80°C. 1: 2-methoxy-2-isopropylpyrazine (odour: "peasy"); 2: 2-methoxy-3-isobutylpyrazine (odour: "bell pepper") (45).

elution of the "peasy" compounds. The chromatograms of cuts from column 1 (CW 20 M) on column 2 (SE 54) for three normal and three defective coffee beans are shown in Figs. 6b and 6c, respectively. While 2-methoxy-3-isobutylpyrazine was present in nearly equal amounts in both samples, 2-methoxy-3-isopropylpyrazine was markedly increased in the defective beans and had to be considered as responsible for the "off-flavour". Overloading of the first column was in this case of secondary importance, since the main separation was performed in column 2 with a preselected cut.

Of course, these were very sophisticated examples for the use of MDGC in flavour analysis. There is also an alternative, simpler and cheaper, but limited method for carrying out MDGC, i.e. using an in-line six port rotary switching valve. A very low dead volume valve suitable for connecting fused silica columns and operable to 300°C has been introduced recently. Several applications using a double oven chromatograph have been demonstrated by Jennings (11).

High resolution gas chromatography-Fourier Transform infrared spectroscopy (HRGC-FTIR)

In the chapter on HRGC-MS the disadvantage of mass spectrometry has not been mentioned; however, it is well-known that MS information is often insufficient, when isomers, in particular geometric isomers have to be distinguished. Therefore, in the past, various attempts have been made to elucidate additional spectroscopic information. In this context, IR spectroscopy has to be mentioned which has - from the beginning - a distinct position in flavour research. Each substance possesses its more or less unique IR fingerprint absorption. The capability to deliver unambiguous identification of a compound makes IR well-suited as an identification method coupled with GC.

In 1967, first approaches were made to apply a Fourier Transform IR (FTIR) as a detector combined with a GC (46). Already these early GC-IR experiments showed the feasibility of the technique, although packed columns had to be used. Since then, spectrometer hardware has developed considerably and more powerful computers and quite flexible software were available. Nowadays, FTIR instrumentation is âvailable fast enough to acquire several full mid-IR spectra even over a narrow capillary peak (47,48).

Fig. 7 shows a scheme of a HRGC-FTIR system as used in our laboratory. In a typical measurement, the effluent from the HRGC column is passed down a heated transfer line and continuously to a separately heated IR measuring cell, the so-called "light-pipe". At the "light-pipe" entrance accelerating gas is added parallel with the effluent to maintain linear gas velocity to the whole interface. At the end of the "light-pipe" a fused silica transfer line is employed to guide the effluent into a monitor detector (FID). Thus, IR spectra, IR chromatograms and a conventional FID trace are obtained simultaneously during each single GC-IR run. It has to be stressed that IR detection is non-destructive, i.e. the effluent after having passed the "light-pipe" is still available for other detection techniques.

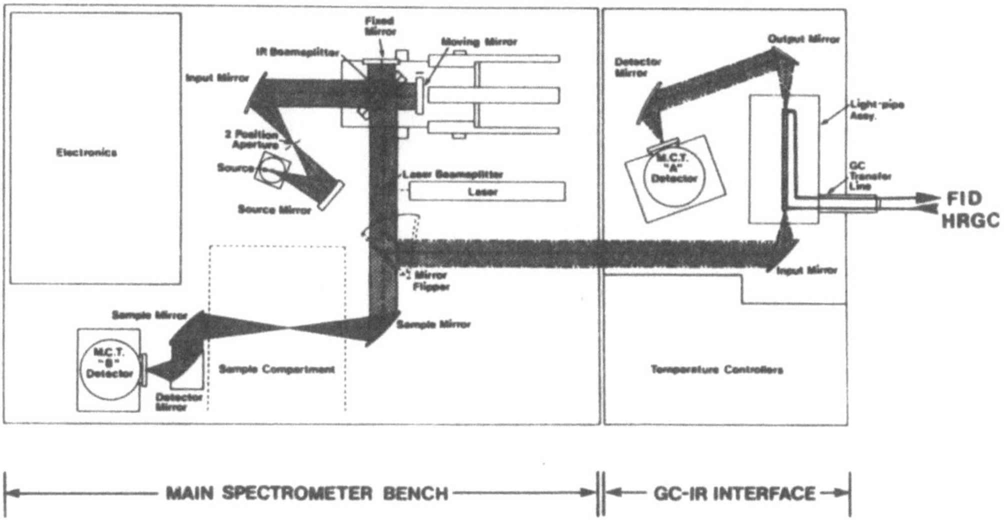

Fig. 7 Scheme of a HRGC-FTIR system (Nicolet 20 SXB)

In Figs. 8-10, the importance of HRGC-FTIR technique for the elucidation of geometric isomers is demonstrated by means of an example taken from our studies on Papaya (Carica papaya, L.) fruit flavour (49). Fig. 8 shows the TIC trace recorded from a HRGC-MS run of a silica gel fraction separated from a Papaya (C. papaya) fruit flavour isolate. The peaks with linear

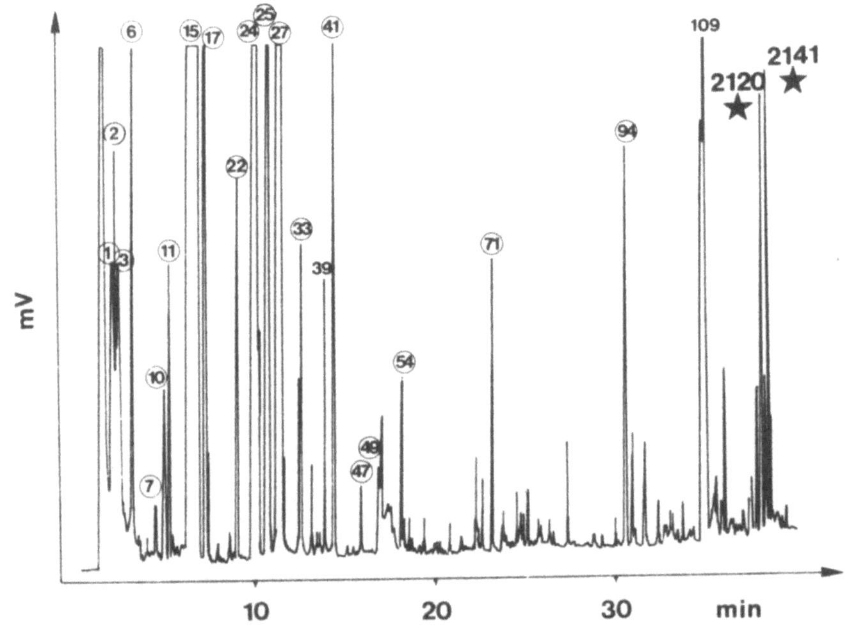

Fig. 8 TIC trace from a HRGC run of a silica gel fraction separated from Papaya (C. papaya, L.) fruit flavour isolate.

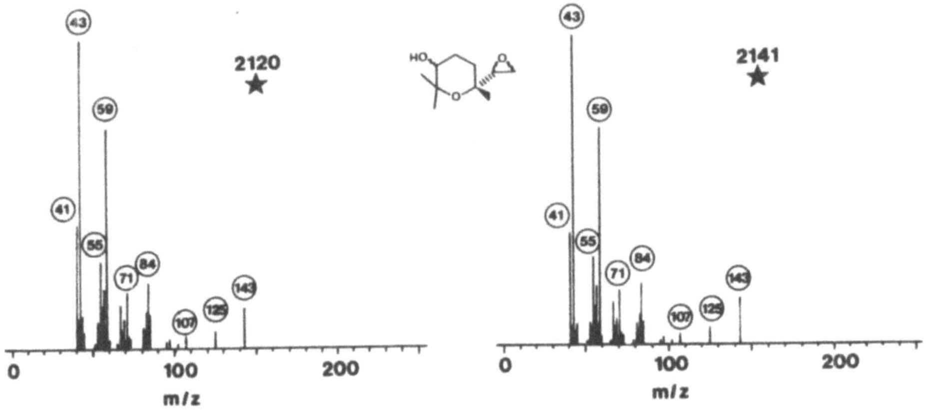

Fig. 9 EIMS (70 eV) spectra of isomeric pyranoid epoxylinalool oxides with R_t (DB-Wax) 2120 and 2141.

retention times (R_t) (DB-Wax) of 2120 and 2141 could by identified as isomeric epoxy derivatives of pyranoid linalool oxides. As shown in Fig. 9 these compounds showed identical mass spectra and their geometric configuration was not known. As outlined in Fig. 10 both diastereoisomers could be clearly differentiated by means of their vapour phase FTIR spectra. Only for the compound with R_t 2120 an intramolecular H-bonding has been found at 3490 cm^{-1} indicating Z-configuration, while this absorption did not appear in the FTIR spectrum of the E-isomer (R_t 2141).

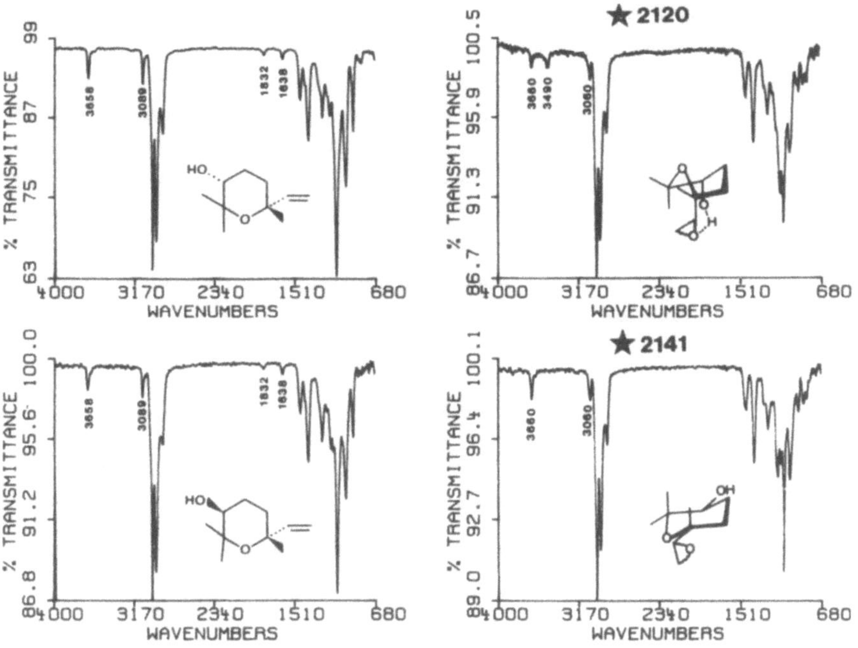

Fig. 10 Vapour phase FTIR spectra of pyranoid linalool oxides (left) and their corresponding epoxy derivatives (right) (R_t 2120: Z; R_t 2141: E)

After this "sunshine" example, the disadvantage of HRGC-FTIR has to be mentioned, too. Problems arise due to the low sensitivity of the method - when compared with mass spectrometry. On a routine basis, HRGC-FTIR analysis still requires, strongly depending on the IR absorption, an amount of approximately 10-100 ng/peak in order to obtain a sufficient vapour phase FTIR spectrum. In principle, there are two possibilities to overcome this problem. One possibility is to use "matrix isolation" (50), a quite different system in which the separated compounds are collected in an argon matrix using a cryogenic system, i.e. an expensive device, or to improve the chromatographic part of the system. A chance to improve the chromatography may be provided by use of MDGC, as recently proposed by our group (51).

Finally, it has to be stressed that one of the advantages of HRGC-FTIR consists of the non-destructive character of detection. Thus, HRGC-FTIR analysis of real samples offers the opportunity to supplement MS in on-line procedure (52,53). Recently, a HRGC-FTIR-MS combination has been realized by our group in which a FTIR spectrometer was linked in parallel with a quadrupole mass spectrometer for concurrent analysis of HRGC eluents from a WCOT column (54) (Fig. 11). Programmed temperature vaporization (PTV) was used for sample injection on a 30 m x 0.32 mm i.d. J & W DB-5 bonded phase WCOT column. The post-column effluent was divided at an open-split system between the FTIR and mass spectrometer (approximately 150:1) with restriction to the MS provided by a narrow bore fused silica column.

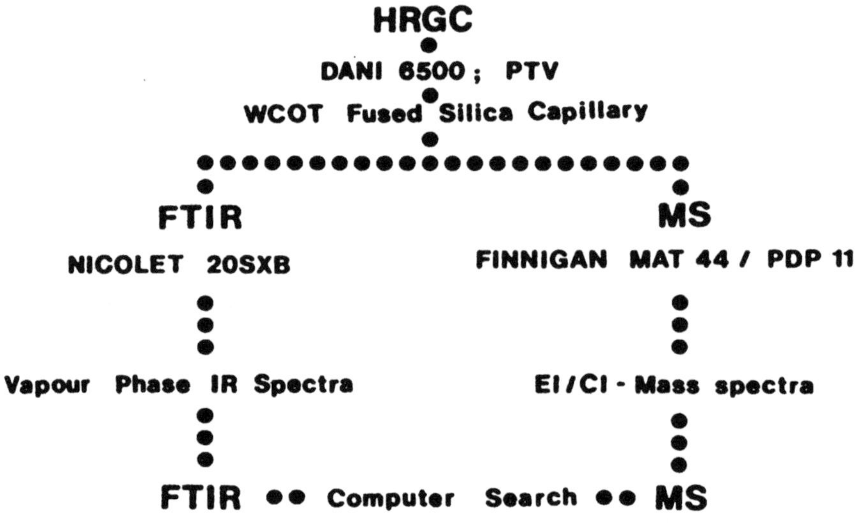

Fig. 11 Block diagram of HRGC-FTIR-MS combination (54).

Conclusions

Discussing the state-of-the art of on-line coupled HRGC techniques for flavour research, it can be concluded:

HRGC-MS

remains the most powerful technique available to the chemist for the separation and identification of volatile components; it will retain this position in spite of recent advances in complementary techniques such as HRGC-FTIR.

Future developments: SFC/MS

MDGC

has tremendous potential for greatly facilitating the identification of volatile constituents of complex mixtures.

Future developments: Coupling of achiral-chiral columns
MDGC-MS; MDGC-FTIR; MDGC-FTIR-MS

HRGC-FTIR

delivers on-line complementary spectroscopic information about an unknwon volatile constituent of a complex mixture. Problems of low sensitivity can be solved by use of MDGC-FTIR or "matrix isolation".

Future developments: Improvement of computers and software
Improvement of "light-pipe"
Development of vapour phase FTIR data banks.

References

1. Van Straten, S. and Maarse, H. (1983). Volatile compounds in food. Qualitative data, CIVO-TNO, Zeist.
2. Maarse, H. and Visscher, C.A. (1987). Volatile compounds in food. Qualitative data. Suppl. 4. CIVO-TNO, Zeist.
3. Schreier, P. (1984). Chromatographic studies of biogenesis of plant volatiles. Hüthig, Heidelberg.
4. Birch, G.G. and Lindley, M.G. (Eds.) (1986). Developments in food flavours, Elsevier Appl. Sci., London, New York.
5. Teranishi, R., Nimmo, C.C. and Corse, J. (1960). Anal. Chem. 32, 1384-1386.
6. Flath, R.A., Black, D.R., Guadagni, D.G., McFadden, W.H. and Schultz, T.H. (1967). J. Agric. Food Chem. 15, 29-35.
7 Teranishi, R. (1979). Int. Lab. 9, 18-24.
8. Lee, M.L., Kong, R.C., Wooley, C.L. and Bradshaw, J.S. (1984). J. Chromatogr. Sci. 22, 136-142.
9. Schomburg, G.G (1987). Gaschromatographie, 2. Aufl., VCH, WEinheim.
10. Jennings, W.G. (1980). In: The analysis and control of less desirable flavours in foods and beverages (G. Charalambous, ed.), Academic Press, New York. pp. 3-16.
11. Jennings, W.G. (1984). J. Chromatogr. Sci. 22, 129-135.
12. Jennings, W. and Takeoka, G. (1984). In: Analysis of volatiles (P. Schreier, ed.), W. de Gruyter, Berlin. pp. 65-75.
13. Shibamoto, T. (1984). In: Analysis of volatiles (P. Schreier, ed.), W. de Gruyter, Berlin. pp. 233-250.
14. Vangaever, F., Sandra, P. and Verzele, M. (1979). Chromatographia 12, 153-154.
15. Cramers, C.A., Scherpenzeel, G.J. and Leclerq, P.A. (1981). J. Chromatogr. 12, 153-154.
16. Grob, K. and Grob, G. (1983). Chromatographia 17, 481-485.
17. Henneberg, D., Henrichs, U. and Schomburg, G. (1975). Chromatographia 8, 449-451.
18. Henneberg, D., Henrichs, U. and Schomburg, G. (1975). J. Chromatogr. 112, 343-352.
19. Henneberg, D., Henrichs, U., Husmann, H. and Schomburg, G. (1978). J.

Chromatog. 167, 137-147.
20. Schmid, P.P., Müller, M.D. and Simon, W. (1979). HRC & CC 2, 225-227.
21. Hurley, R.B. (1980). HRC & CC 3, 147-149.
22. Wetzel, E., Kuster, T. and Curtius, H.C. (1982). J. Chromatogr. 239, 197-214.
23. Harrison, A. (1983). Chemical ionization mass spectrometry, CRC Press, Boca Raton.
24. Rapp, U., Hoehn, M., Kapitzke, C. and Dielmann, G. (1982). In: Recent developments in food analysis (W. Baltes, P.B. Czedik, D. Eysenberg and W. Pfannhauser, eds.), Verlag Chemie, Weinheim. pp. 161-168.
25. Cobb, C.S., Bursey, M.M. and Rice, A.C. (1978). J. Food Sci. 43, 1822-1825.
26. May, H.V. and Williams, A.A. (1981). In: Progress in flavour research (J. Adda, ed.), Elsevier, Amsterdam. pp.583-590.
27. Lange, G. and Schultze, W. (1988). In: BIOFLAVOUR '87 (P. Schreier, ed.), W. de Gruyter, Berlin, in press.
28. Lange, G. and Schultze, W. (1988). In: BIOFLAVOUR '87 (P. Schreier, ed.), W. de Gruyter, Berlin, in press.
29. C. Salles, H. Essaied, P. Chalier, J.C. Jallageas and J. Crouzet (1988). In: BIOFLAVOUR '87 (P. Schreier, ed.), W. de Gruyter, Berlin, in press.
30. Bruins, A.P. (1985). Proc. 10th Int. Mass Spectr. Conf. Abstr. Swansea.
31. Ten Noever de Brauw, M.C. and Ingen, C. (1981). In: Isolation, separation and identification of volatile compounds in aroma research (H. Maarse and R. Belz, eds.), Akademie-Verlag, Berlin. pp. 155-225.
32. Whitfield, F.B. and Shaw, K.J. (1984). In: Progress in flavour research (J. Adda, ed.), Elsevier, Amsterdam. pp. 221-238.
33. Liddle, P.A.P. and Bossard, A. (1984). In: Progress in flavour research (J. Adda, ed.), Elsevier, Amsterdam. pp. 467-476.
34. Hirvi, T. and Honkanen, E. (1984). In: Proc. ALKO Symp. on Flavour research of alcoholic beverages (L. Nykänen and P. Lehtonen, eds.), Foundation for biotechnological and industrial fermentation research, Vol. 3. Helsinki. pp.275-278.
35. Thomas, A.F., Willhalm, B. and Flament, I. (1984). In: Chromatography and mass spectrometry in nutrition science and food safety (A. Frigerio and H. Milon, eds.), Elsevier, Amsterdam. pp.47-65.
36. Lange, G. and Schultze, W. (1984). In: Analysis of volatiles (P.

Schreier, ed.), W. de Gruyter, Berlin. pp. 307-322.
37. Johlmann, C.L., Lande jr., D.A. and Wilkins, C.L. (1985). Anal. Chem. 57, 1040-1044.
38. Deans, D. (1968). Chromatographia 1, 18-22 (1968).
39. Schomburg, G., Husmann, H. and Weeke, F. (1975). J. Chromatogr. 112, 205-217.
40. Schomburg, G., Dielmann, R., Husmann, H. and Weeke, F. (1976). J. Chromatogr. 122, 55-72.
41. Nitz, S. and Jülich, E. (1984). In: Analysis of volatiles (P. Schreier, ed.), W. de Gruyter, Berlin. pp. 151-170.
42. Schomburg, G., Weeke, F., Müller, F. and Oreans, M. (1982). Chromatographia 16, 87-91.
43. Schomburg, G., Bastian, E., Behlau, H., Weeke, F., Oreans, M. and Müller, F. (1983). In: Proc. 5th Int. Symp. on Capillary Chromatography (R. Kaiser, ed.), Elsevier, Amsterdam. pp.381-395.
44. Oreans, M. Müller, F. and Leonhard, D. (1983). J. Chromatogr. 279, 357-364.
45. Nitz, S., Kollmannsberger, H. and Drawert, F. (1988). In: BIOFLAVOUR '87 (P. Schreier, ed.), W. de Gruyter, Berlin, in press.
46. Low, M.J.D. and Freeman, S.K. (1967). Anal Chem. 39, 194-197.
47. Herres, W. (1984). In: Analysis of volatiles (P. Schreier, ed.). W. de Gruyter, Berlin. pp. 183-217.
48. Smith, S.L. (1987). In: Capillary gas chromatography in essential oil analysis (P. Sandra and C. Bicchi, eds.), Hüthig, Heidelberg. pp. 367-396.
49. Schreier, P. and Winterhalter, P. (1986). In: Biogeneration of Aromas (T.H. Parliment and R. Croteau, eds.), ACS Symp. Ser. 317, ACS, Washington, pp. 85-98.
50. Reedy, G.T., Ettinger, D.G. and Schneider, J.F. (1985). Anal. Chem. 57, 1602-1605.
51. Schreier, P. and Idstein, H. (1985). Z. Lebensm. Unters.-Forsch. 181, 183-187.
52. Crawford, R.W., Hirschfeld, T., Sanford, R.H. and Wong, C.M. (1982). Anal. Chem. 54, 817-820.
53. Shafer, K.H., Hayes, T.L., Brasch, J.W. and Jakobsen, R.J. (1984). Anal. Chem. 56, 237-241.
54. O. Fröhlich, C. Kahre and P. Schreier (1986). Fres. Z. Anal. Chem. 324, 314.

HALOGENATED PHENOLS AND CHLOROANISOLES: OCCURRENCE, FORMATION AND PREVENTION

H. Maarse, L.M. Nijssen and S.A.G.F. Angelino[1)]

1. INTRODUCTION

Halogenated phenols (especially chlorophenols) and chloroanisoles give rise to off-flavours in many food products. Although it is to be expected that this phenomenon occurs all over the world it is amazing that only few people are aware of the enormous damage done by the contamination of food products with these compounds. This is probably due to the fact that the identity of the compounds bringing about an off-flavour of this type is not always recognized. Chloro- and bromophenols and particularly chloroanisoles have very low odour recognition thresholds and often lead to off-flavours at extremely low concentrations.
When an off-flavour appears to have been caused by a halogenated phenol or chloroanisole it is still very difficult to find the origin of these compounds. The use of chlorophenols as fungicides is so widespread that many materials are more or less contaminated with these compounds. In many cases chloroanisoles are formed by microbial methylation of chlorophenols. It is even so that there need not be any direct contact between materials contaminated with chlorophenols or -anisoles and food products to cause an off-flavour. Contamination occurs by air, especially when these compounds readily dissolve in the food product.
Examples of off-flavours originating from these compounds are given in the next section. To trace the origin of these compounds knowledge of the way in which they are formed is a prerequisite. Section 3 deals with this information; especially all possible routes leading to the formation of 2,4,6-trichloroanisole, the chloroanisole isomer with the lowest odour detection threshold value, are given. The techniques used to detect these compounds and to determine their concentration and their odour detection threshold value are described in section 4.
In section 5 ways are given to prevent off-flavours of this kind. As stated before (25) the best method would be a statutory regulation, stating that chlorophenols are not to be used as such in materials that could come into contact with food. Even if such measures would be taken it is to be expected that chlorophenols and chloroanisoles will still be present in many materials in the near future. To avoid contamination of a food product with chloroanisoles at concentrations of approximately the detection threshold value during transport, the maximum allowable concentrations

* TNO Division of Nutrition and Food Research, CIVO Food Analysis Institute, Zeist, The Netherlands

of chlorophenols in packing materials have to be determined. An experimental set-up is given enabling the calculation of these maximum concentrations.

2. OCCURRENCE IN FOOD PRODUCTS

2.1 Introduction

Only in a few countries reports on the occurrence of chlorophenols and particularly chloroanisoles in food products are available .
In the early sixties TNO investigators identified chloroanisoles causing a musty taint in eggs and broilers (12). This was the beginning of a story that has not yet resulted in a happy end (25). Research groups in the United Kingdom (e.g. 10,36) and particularly in Australia (e.g. 40) have also played an important part in the study of mustiness of the chloroanisole type. This type of mustiness should and can be clearly distinguished from off-flavours caused by isoborneol or geosmin.

2.2 Halogenated phenols

Whitfield (40) listed in his extensive and well documented review on off-flavours nine problems caused by so called disinfectant off-flavours. In all instances the taint was caused by chlorophenols (Table 1). Four were the result of accidental contamination of food by chlorophenol-based herbicides and biocides during the processing and handling of the product. Chlorophenol contamination of food also occurs as a result of in-plant chlorination of phenol. The major sources of chlorine are the treated water supply and chlorine-based sterilizing agents. The phenol(s) can be derived from plastic and resin fittings, paint, printing inks, packaging materials, boiler water additives and the food itself.
We have also identified some halogenated phenols (Table 1). One of these was 2-bromophenol in paper sacks and herring. The odour of this compound was described as iodinelike and phenolic.

2.3 Chloroanisoles

In Table 2 examples are given of the occurrence of chloroanisoles in various food products. These have been taken from Whitfield's paper (40) and from our own work in this field. Like Whitfield we found that the off-flavour problems we studied frequently concern chloroanisoles.

Table 1 - Off-flavours caused by halogenated phenols

Product contaminated	Compounds involved	Origin	Ref
Poultry	2,4,6-Trichlorophenol 2,3,4,6-Tetrachlorophenol	Cage litter containing wood shavings contaminated with wood preservatives	40
	6-Chloro-2-methylphenol 2,4,6-Trichlorophenol	Chlorophenol-based biocides used to sterilise cages	
Canned carbonated beverages	6-Chloro-2-methylphenol	Empty drink cans contaminated with airborne agro-chemical pollutants	40
Canned vegetables	2,4,6-Trichlorophenol 2,3,4,6-Tetrachlorophenol Pentachlorophenol	Process water contaminated with chlorophenol-based algaecide	40
Bulk chocolate	2,4,6-Trichlorophenol 2,3,4,6-Tetrachlorophenol Pentachlorophenol	Wooden pallets treated with a chlorophenol-based wood preservative	40
Modified flour	2,4-Dichlorophenol 2,4,6-Trichlorophenol	Wash water contaminated with in-plant chlorination products	40
Reconstituted fruit juice	2,4-Dichlorophenol 2,6-Dichlorophenol	Process water contaminated with in-plant chlorination products	
Canned fruit	2,4,6-Trichlorophenol	In-plant reaction involving steam contaminated with phenol, and process water containing chlorine	40
Cocoa powder	2,4-Dichlorophenol 2,4,6-Trichlorophenol 2,3,4,6-Tetrachlorophenol	Multi-walled paper sacks contaminated with fungicide and waste-paper residues	40
Packaged biscuits	2,4,6-Trichlorophenol 2,3,4,6-Tetrachlorophenol	Adhesive containing chlorophenols as a preservative	40
White beans	2,4-Dichlorophenol	Unknown	32
Beer	2,4-Dichlorophenol	Unknown	32
Peanuts	2,4-Dichlorophenol	Unknown	32
Shrimps	2,4-Dichlorophenol	Unknown	32
Peanut butter	2,4-Dichlorophenol	Unknown	32
Ice cream	4-Chloro-2-methylphenol	Unknown	32
Paper sacks	4-Chloro-2-methylphenol	Unknown	32
Paper sacks	2-Bromophenol	Formed during production of paper out of bromo and phenol	32
Herring	2-Bromophenol	Contamination of sea water	5

Table 2 - Off-flavours caused by chloroanisoles

Foods contaminated	Compounds involved	Origin	Ref.
Eggs and broilers	2,4,6-Trichloroanisole	Cage litter containing wood shavings contaminated with chlorophenols	10,12
Wine	2,4,6-Trichloroanisole	Cork oak treated with pentachlorophenol and/or corks initially contaminated with chlorophenols formed during chloride bleaching process	40
Brandy *	2,4,6-Trichloroanisole		25
Packaged flour	2,4,6-Trichloroanisole 2,3,4,6-Tetrachloroanisole	Jute sacks initially contaminated with chlorophenols present either as preservative formed by an in-plant chlorination reaction	40
Packaged rice	2,4,6-Trichloroanisole 2,3,4,6- and/or 2,3,5,6,-Tetrachloroanisole Pentachloroanisole		
Cocoa powder	2,6-Dichloroanisole 2,4-Dichloroanisole 2,4,6-Trichloroanisole 2,3,4,6-Tetrachloroanisole	Multi-walled paper sacks initially contaminated with chlorophenol residues	32,40
Chilled meat	2,4,6-Trichloroanisole	Wooden insulation in cavity walls of chiller cold store initially contaminated with chlorophenol reaction products or residues	40
Packaged dried fruit	2,4,6-Trichloroanisoles 2,3,4,6-Tetrachloroanisoles	Fibreboard packaging material initially contaminated with chlorophenol residues	40
Cheese	2,4,6-Trichloroanisoles 2,3,4,6-and/or 2,3,5,6,-Tetrachloroanisoles	Wooden shelves initially contaminated with chlorophenols	25
Gin	2,4,6-Trichloroanisole	Packaging material of the empty bottles initially contaminated with chlorophenols	32
Potatoes	2,4,6-Trichloroanisole	Wooden shelves treated with pentachlorophenol	32
Pasty	2,4,6-Trichloroanisole	Soil in neighbouring flowershop contaminated with chloroanisole	32

* The off-flavour is not described as musty, but as walnut-like, heavy, it was followed by a dry mouth feel. Even more remarkable was the fact, that the off-flavour was not noticed immediately but 5-10 minutes after consumption of the brandy.

3. PATHWAYS FOR ANISOLE FORMATION

3.1 Introduction

In nature a subtle balance between the biosynthesis and biotransformation of halo-organic compounds existed for an incalculable era. The integrated activities of plants, animals and in particular microbes, resulted in a state of equilibrium in which no excessive amounts of these components occurred. Their natural function had (and has) presumably to be found in the biochemical synthesis of certain metabolites and the production of compounds with enhanced desirable biological activity (29). Nowadays, because of man's activity, a host of halogenated compounds - many of which are recalcitrant to (bio)degradation - are being thrust into the natural ecosystem. There are well over 100 different halogenated compounds in commercial use today, e.g. as insecticides and fungicides (30). Were it not for microbes, these compounds would be rapidly polluting the earth surface. Many naturally occurring microorganisms have been isolated that degrade halogenated compounds to some degree. These processes together with chemically induced degradation will give rise to a vast number of halo-organic metabolites, which may in part escape from their degradative pathways, or are largely excreted in the environment.

When conversion processes of synthetic and/or degradative nature of halo-organic compounds take place in raw materials or in close contact with the production, transportation and/or storage of products meant for human consumption, they can lead to highly undesirable effects. A striking example of this kind is the formation of chloroanisoles from chlorophenols. It is of utmost economic importance to prevent off-flavours caused by chloroanisoles (see section 5).

To enable well-suited approaches in preventing the occurrence of these phenols and anisoles one must first establish "all" possible pathways which can lead to the formation of chloroanisoles. Therefore, in the next sections general features of (bio)chemical reactions will be presented that may play a role in chloroanisole formation. This is illustrated for the formation of 2,4,6-trichloroanisole, an anisole with the lowest threshold value of this group of compounds (see Table 4). The given schemes will be documented as far as possible by examples from literature. In addition, attention is called to the fact that chemical processes which may occur under industrial conditions, e.g. the synthesis of precursors as impurities in fungicide preparations during production, will not be taken into consideration.The generation of precursors can roughly be divided into two clusters of reactions, viz. one of predominantly synthetic nature and one which merely comprises degradation reactions. In some aspects this subdivision is somewhat artificial and does not hold. In spite of this we consider the latter classification to fit in with our discussion.

3.2 Chloroanisole formation by synthesis

If one starts from compounds as precursors possessing one of both functionalities basically present in chloroanisoles, viz. a hydroxyl or a chloro group, an overview of pathways can be presented as given in Scheme 1. We certainly admit that omissions have been made which render the scheme incomplete, but they do not basically affect its conception.

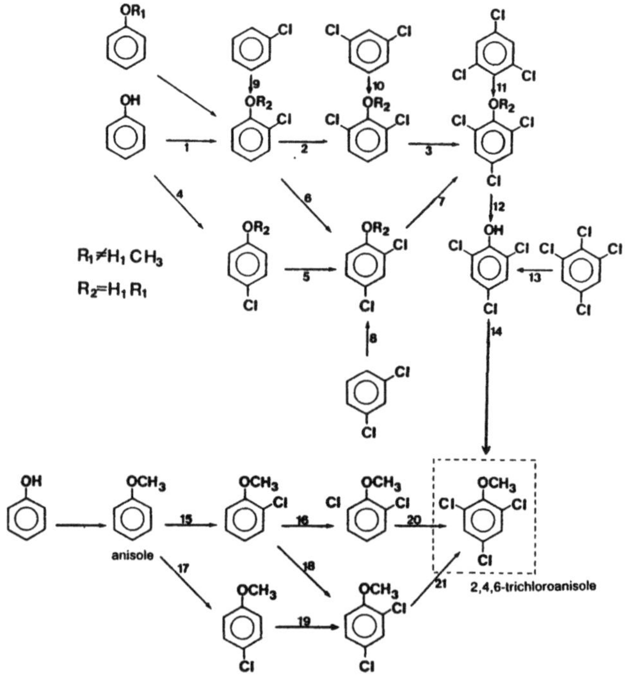

Scheme 1. Synthetic routes from phenol and chlorobenzenes to 2,4,6-trichloroanisole by halogenation and/or hydroxylation reactions

Reactions 1-7: Phenol can rather easily be (poly)chlorinated by reaction with chlorine, for instance present in heavily chlorinated mains water. These reactions do occur completely at the ortho and para positions to the hydroxyl group in the benzene ring and can easily proceed up to di- and trisubstituted chlorophenols (41). The ortho and para directive effect on chlorination in the benzene ring is not only limited to the hydroxyl group, but can also be exerted by substituted analogues (-OR,), like for instance the methoxyl group in anisole (reactions 15-20). The rate of initial chlorination and of subsequent chlorination steps however may be diminished.

An example for reaction 7 could be the chlorination of the agrichemical 2,4-dichlorophenoxyacetic acid (2,4-D) upon contact with aqueous chlorine.

Reactions 8-11: Ballschmiter et al. (4) demonstrated the hydroxylation of mono-, di-, tri-, and tetrachlorobenzenes in special mixed cultures of soil bacteria. Hydroxylation of pentachlorobenzene to its corresponding phenol by lindane-degrading fungi was observed by Engst and co-workers (13).

Reaction 12: In case R is not H, as for instance in 2,4-D and 2,4,5-trichlorophenoxyacetic acid (2,4,5-T), degradation into the corresponding phenol will take place before further conversion occurs (16, 21, 24).

Reaction 13: Another potential route to metabolize chlorobenzenes is the hydrolytic dehalogenation mechanism. This reaction proceeds by nucleophilic substitution replacing the halogen atom with a hydroxyl ion (30). Although for the sake of convenience, this type of reaction has been included in the synthetic scheme, it fits into the degradative scheme as well.

Reaction 14: Biomethylation has been proven to be one of the mechanisms for the detoxification of chlorinated phenols by microorganisms. Especially highly chlorinated phenols, such as pentachlorophenol, which are difficult to degrade, are methylated by fungi to make them less toxic (8,9,28). Also tri- and tetrachlorophenols can - at least in part - be transformed into the corresponding anisoles (30). The methylation reaction is presumably catalysed by so-called methyltransferases, which are very likely membrane-bound enzymes and show a rather high substrate specificity (1). In case of fumigates like methylbromide being used for disinfestation, e.g. of containers, possible chemically induced methylation is feasible as well.

Reactions 15-21: As mentioned before, anisole itself, which may be present in the matrix originally or is formed out of phenol in an early stage, will be subjected to chemical chlorination under relatively mild conditions analogous to phenol as well, although corresponding reaction rates may differ. Biochemical chlorination was found by Jerina et al. (20), who showed the enzymatic conversion of anisole into ortho-chloroanisole and its para isomer by the action of chloroperoxidase. This type of enzyme h s been detected in many fungi species (31). No evidence is available of these enzyme catalyzed reactions leading to the di- or possibly trisubstituted chloroanisole.

3.3 Chloroanisole formation by degradation

In Scheme 2 (bio)degradative routes are presented which lead from various man-made halo-organic chemicals to 2,4,6-trichloroanisole. Compounds like lindane, pentachlorophenol (PCP) and its sodium salt have already been used as an insecticide and fungicide respectively, for a long time. Moreover, it is well-known from literature (15) that many different micro-organisms are able to accomplish the transformation of for instance chlorophenols to chloroanisoles.

Reactions 22-24: These reactions are examples of the dehydrohalogenation, which proceed by the spontaneous removal of the halogen and the adjacent hydrogen on the neighbouring carbon. Dehydrohalogenation is not always due to direct enzymatic dehalogenation, but may result from spontaneous decomposition of an enzyme produced intermediate (30). It is hard to say whether this conversion from lindane into 2,4,6-trichlorobenzene will occur to a large extent, since it presumably has to compete with other degradation pathways for this compound (38).

Reaction 25: Hydroxylation of 1,3,5-trichlorobenzene has been demonstrated with soil bacteria as was described before (4). Of course 2,4,6-trichloroanisole can enter this scheme also by hydrolytic dehalogenation of 1,3,4,5-tetrachlorobenzene (Scheme 1, reaction 13).

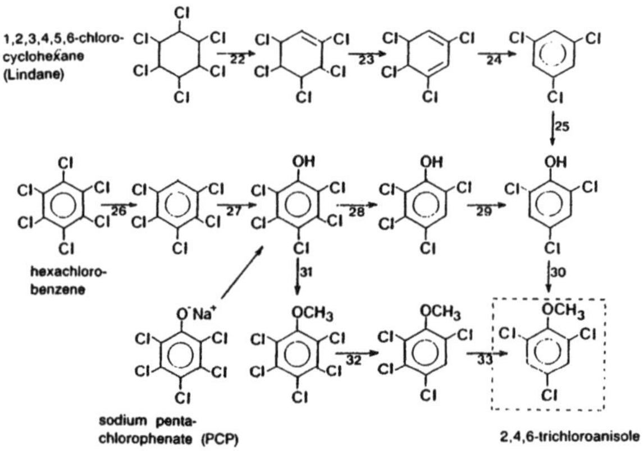

Scheme 2. Degradative routes from various halo-organic chemicals to 2,4,6-trichloroanisole

Reaction 26: Reductive dehalogenation is one of the major pathways for enzymatic degradation of halogenated compounds as well (30). Evidence for this breakdown has been given by Engst et al. (13).

Reactions 27-29: This sequence of reactions comprises first a hydroxylation step analogous to reaction 25, followed by two subsequent reductive dehalogenation steps (30). All compounds could be detected by Engst et al. (13) in their study on the metabolism of hexachlorobenzene by lindane-degrading fungi. In analogy this pathway was also observed for mammalian microsomal oxidation of hexachlorobenzene (33).

Direct application of pentachlorophenol or its sodium salt somewhere along the production line will naturally introduce a major precursor for the formation of anisoles by its own presence as well as by concomitant impurities, such as 2,3,4,6-tetrachlorophenol and 2,4,6-trichlorophenol (41).

Reactions 30 and 31: Methylation has been discussed before in Scheme 1, reaction 14.

Reactions 32 and 33: Analogous to reactions 28 and 29, reductive dehalogenation of pentachloroanisole can lead to the selected trichloroanisole (30).

3.4 Evaluation of (bio)chemical options for anisole formation

In a complex matrix the potential (bio)chemical reactions presented in Schemes 1 and 2 for the formation of 2,4,6-trichloroanisole will not occur exclusively. Competing reactions will lead to other less off-flavour effective anisoles and other transformations of substituted phenols (15,38) will also influence their occurrence. For instance, the degradation of pentachlorophenol does not necessarily result in the formation of pentachloroanisole as by-product, since often a complete conversion of the aromatic ring can be obtained with this kind of compounds (6,30,34). Our exact knowledge of the conditions which favour or will inhibit dechlorination by the various mechanisms is still rather poor.
Another complicating factor is that in the route from plant growth at one end to the product directly or indirectly used for human consumption at the other end, there are many entrances for potential precursors of chloroanisoles (41).
Moreover, reactions as presented in Schemes 1 and 2 that will not proceed concomitantly, may succeed through respective stages of plant growth, processing, storage and transportation. Added to the extremely low flavour threshold values, this makes it very hard to determine which reaction steps will be rate-limiting in the anisole formation process. Only elimination of these steps would actually be successful. Eventually, elimination of one pathway may be by-passed by others.

4. INSTRUMENTAL AND SENSORY METHODS

4.1 Indirect approach

A general approach to off-flavour studies is to first describe the odour character of the off-flavour. Depending on its intensity this can be done informally by two or three persons or with a panel of 12 persons. If thus a clear indication of the presence of an off-flavour is obtained an extract of the product has to be made and analysed by gas chromatography. The compound(s) causing the off-flavour cannot be found by comparing the chromatogram of the product with the off-flavour and that of a reference product. It can only be done by sensory analysis. Therefore the end of the GC column is connected to a splitter which leads part of the effluent to an FID-detector and the other part directly out of the oven to a sniffing port. A person familiar with the character of the off-flavour can determine in this way in which fraction of the chromatogram the component(s) causing the off-flavour elutes. By means of analysis of this fraction on a gas chromatograph coupled to a mass spectrometer (GC-MS) the relevant compounds can be identified.

4.2 Direct approach

Persons familiar with the odour of halogenated phenols and chloroanisoles will in most cases readily recognize an off-flavour caused by these compounds.
Then one can start immediately an appropiate extraction procedure or a combined steam distillation and extraction according to Likens and Nickerson (26). When phenols have to be extracted the aqueous phase has to be acidified. The halogenated phenols and anisoles in the extracts can be detected at low levels by GC-MS using selective ion monitoring or by GC using electron capture detectors.
GC analysis of very small amounts of the highly polar halogenated phenols is only possible as their derivatives (e.g. acetates). The derivatisation step also acts as clean up of the extracts. Extracts containing chloroanisoles can be cleaned up by using disposable silicagel columns. Detection limits are given in Tables 3 and 4.

4.3 Threshold measurements

Important properties of odorous volatile compounds are the odour and flavour threshold values. Threshold values are defined as the concentration which is just perceived by 50 % of the assessors. Knowledge of the threshold values is indispensable in order:

a. to find out whether the instrumental detection limit is sufficiently low, preferably at least as low as the threshold value,
b. to get an idea about maximum allowable concentrations of compounds causing off-odours and/or off-flavours in foods, raw materials and related products such as packaging materials.

The outcome of threshold value measurements is strongly dependent on the experimental conditions. Interlaboratory results can easily differ by a factor 10 and even more. Sensory experiments with compounds as halogenated phenols and anisoles, which have very low threshold values, certainly require careful consideration of all relevant aspects. This starts with the preparation of highly diluted standard solutions. Some critical experimental conditions are:

a. the purity of the compounds (as threshold values of different compounds can differ more than a factor 1000, an impurity of 0.1 % might strongly influence the threshold value).
b. stimulus volume (and in case of odour the amount of air) and temperature of the stimulus.
c. the experience of the panel members with the odour and flavour character of the compounds involved (even if detection thresholds are being measured).
d. psychophysical method (for instance, number and sequence of stimuli; time between stimuli; whether or not forced-choice procedure) and type of threshold (recognition thresholds are higher than detection ones).
e. environmental influences (for instance, extraneous odours).
f. calculation method (i.e. assumptions about the mathematical model upon which the calculation should be based).

Naturally, the threshold values are strongly dependent on the matrix containing the compounds. In principle, the threshold value has to be measured for each food product. Model systems can be a great help in predicting the influence of the matrix. Threshold values obtained in model systems can be checked with those found in air. In that case knowledge of the partition coefficient is essential.
In our laboratory odour detection threshold values have been determined of several chlorophenols and -anisoles in three model systems (water, edible oil and water/ethanol mixtures) and in air.
A panel of 15 members - employees of the institute - familiar with the compounds and methods involved assessed per session 5 pairs of stimuli. Each pair consisted of one blank and one diluted sample. The five different solutions - each having twice the concentration of the preceding one - were assessed in random order (constant stimulus method). The code-marks consisting of the letters X and Y and the spatial positions - left and right - of the samples within each pair were randomized too. The panel members were asked to point out the stimulus containing the diluted compound and were forced to make a choice. Stimuli consisted of 50 ml samples in 400 ml polystyrene beakers covered with a lid. For each threshold value calculation the measurements were repeated once, resulting in 30 responses per concentration. After a correction for chance the 50 % point was calculated based on a log-normal model (5).
For the determination of the odour detection threshold values in air an olfactometer was used. This is an apparatus which is capable of producing airflows with a wide range of concentrations of the relevant volatile compound. This airflow is presented, together with two odourless airflows, to a panel of 8 members. After smelling they are asked to indicate which of the three is the odorous one.
To check the purity of the compounds, standard solutions were first injected splitless on a GC-column. The end of the column was connected to a splitter which led part of the effluent to a FID-detector and the other part directly out of the

Table 3 - Odour threshold values of halogenated phenols in air and water

Halogenated phenol isomer	Air mg/m^3	Ref.	Water mg/kg	Ref.
2-Chloro	0.019	22	0.002	7
	0.0005	23	0.003	43
			0.01	19
3-Chloro	-		0.05	19
			0.1-0.2	27
4-Chloro	0.001	23	0.25	7
			1.24	3
			0.06	19
2,3-Dichloro	-		0.03	11
2,4-Dichloro	0.00027	35	0.002	7
			0.21	3
			0.04	19
			0.00065-0.0065	7
			0.0003	11
2,5-Dichloro	-		0.03	19
2,6-Dichloro	-		0.003	7, 27
			0.2	11
2,3,6-Trichloro	-		0.3	11
2,4,5-Trichloro	-		0.2	11
			0.011-0.333	27
2,4,6-Trichloro	0.021	23	0.3	19
	0.001	2		
	0.0016	2		
	0.00016	35		
2,3,4,6-Tetrachloro	-		0.6	11
			0.915-47	27
Pentachloro	-		0.857-12	27
			1.6	11
2-Bromo	-		0.1	5
3-Bromo	-		50	5
4-Bromo	-		160	5

detection limits 0.001-0.1 mg/kg

Table 4 - Odour detection threshold values and detection limits (mg/l or mg/kg) of chloroanisoles in different media.

	2,4-DCA	2,6-DCA	2,3,6-TCA	2,4,6-TCA	2,3,4,6-TCA	PCA
water	$4x10^{-4}$ (17)	$4x10^{-5}$(17)	$3x10^{10}$ (10)	$3x10^{-8}$ (10)	$4x10^{-6}$ (10)	$4x10^{-3}$ (10)
	$1.9x10^{-3}$ (32)		$7.2x10^{-6}$ (18)	$7.6x10^{-8}$ (32)	$1.1x10^{-6}$ (32)	$4.1x10^{-4}$ (32)
				$2x10^{-5}$ (42*)	$0.2x10^{-3}$ (42*)	$3.2x10^{-3}$ (42*)
ethanol/water						
- 20 %	-	-	-	$5.6x10^{-6}$ (32)	$1.4x10^{-5}$ (32)	-
- 40 %	-	-	-	$3.6x10^{-5}$ (32)	-	-
beer	-	-	-	$7x10^{-6}$ (32*)	-	-
wine	-	-	-	$1x10^{-5}$ (39*)	-	-
edible oil	-	-	-	$7x10^{-3}$ (32)	-	-
egg yolk	-	-	0.12 (14*)	$2.4x10^{-3}$ (14*)	$2.7x10^{-2}$ (14*)	2.8 (14*)
dried fruit	-	-	-	$0.12x10^{-3}$ (42*)	$1x10^{-3}$ (42*)	$33x10^{-3}$ (42*)
				$0.45x10^{-3}$ (42*)		
fruit buns	-	-	-	$0.21x10^{-3}$ (42*)	$1.9x10^{-3}$ (42*)	$126x10^{-3}$ (42*)
plain buns	-	-	-	$1.4x10^{-3}$ (42*)	$5.8x10^{-3}$ (42*)	$183x10^{-3}$ (42*)
air (mg/m^3)	$6.6x10^{-3}$ (32)	-	-	$3.7x10^{-6}$ (32)	$6.7x10^{-5}$ (32)	$2.5x10^{-4}$ (32)
detection limit**	$10^{-6}-10^{-4}$ (25)	$10^{-6}-10^{-4}$(25)	$10^{-6}-10^{-4}$ (25)	$10^{-6}-10^{-4}$ (25)	$10^{-6}-10^{-4}$ (25)	$10^{-6}-10^{-4}$ (25)

* flavour threshold value

** dependent on product/medium

between brackets = lit. references

oven. During elution a 45 l Teflon FEP bag was connected to the splitter outlet. The concentration of the pure compound in the bag was calculated from the concentration of the standard solution, the injection volume and the split ratio at the end of the column.
A survey of the threshold values obtained so far and data found in literature are given in Tables 3 and 4. As could be expected from the lipohilic nature of the chloroanisoles their odour threshold value in water is lower than in edible oil.

Like Griffiths (17) and Whitfield (42) we found that chloroanisoles can cause severe adaptation. This adaptation is disturbing when determining the odour threshold in water/ethanol mixtures. In our investigation application of the constant stimulus method (= random order of concentrations) often resulted in less correct responses for higher concentrations. When the method of limits was applied (increasing concentrations) consistent results were obtained.
For the determination of taste threshold values Whitfield et al. (42) presented only one pair of samples to the panel per day. One contained no chloroanisole and was tasted first, the other contained a given level of chloroanisoles. Presentation of more samples per session would give erroneous results because of adaptation effects. We found that the determination of odour threshold values does not require these severe precautions.
Data on vapour pressures of the chloroanisoles above the model systems for calculation of the partition coefficients are not available at the moment. Experiments are still in progress and the results will be reported later.

5. PREVENTION

Several measures should be taken to prevent contamination of a food product with halogenated phenols and chloroanisoles.
A. During food processing reaction of phenols with halogens should be avoided.
A critical step in processing is the cleaning with phenol containing water and chlorinated cleaning agents. Examples of in-plant formation of chlorinated phenols are given by Whitfield (40)

B. Food must be protected against contamination by the environment.

It is known that for instance plastic packaging materials do not offer complete protection against chloroanisoles. Whether replacing of the present packaging materials by air tight materials is feasible is most of all an economical problem.

C. Mould growth should be prevented so that chlorophenols are not converted into chloroanisoles.

In practice it is very difficult to prevent contamination of packaging materials with moulds. Mould growth can be prevented by maintaining a low water activity. However, in closed containers water condensation may occur. Changing the design of containers as proposed by Sharp (37) may overcome this problem.

D. If food products are stored or transported in containers together with materials containing chlorophenols it should be required that the concentrations of these compounds in these materials are below maximum allowable levels.

Cardboard and wooden pallets almost always contain chlorophenols, which can be converted to chloroanisoles by mould growth. It is very important to control the levels of these compounds in such materials. Effective control is possible only when the levels are known at which these compounds give rise to off-flavours in the packed products. These levels can be determined as follows:
Suppose that a food product is stored in a container for several weeks. In the container there are also wooden pallets and cardboard boxes. First the odour and taste threshold values of 2,4,6-trichloroanisole (TCA), being the isomer with the lowest threshold value, have to be measured in the relevant food product using the method described in the previous section.
Then the diffusion rate of the compound from the surrounding atmosphere through the packaging material into the food product has to be determined in a storage experiment. This can be done by placing the food product in an air tight container, e.g. a tin can. A known amount of TCA is injected in the can through the lid and at predetermined time intervals its concentration in the food product is measured. This must be done with several TCA concentrations in the can. The storage time at a certain concentration of TCA in air at which the level of TCA in the food product comes above the threshold value can be read from a graph.
To be able to estimate the maximum allowable level of chlorophenols in the packaging material, the following assumptions are made:

- the total amount of chlorophenol is converted into TCA within a relatively short period.
- the total amount of TCA is transferred into and evenly distributed in the surrounding atmosphere.
- the container is leak tight.

Furthermore the total amounts of packaging material and food product in the container should be known.

A typical example of such an experiment is shown in Figure 1. For three concentrations of TCA in the atmosphere the concentration of TCA in the food product is given as a function of the storage time. Suppose that the odour threshold value of TCA in the food product is 0.01 μg/kg. The level of TCA in the atmosphere must then be maintained below 1 μg/l as can be read from Figure 1. The empty space in the container is about 50,000 l so the total amount of chlorophenols is allowed to be 50 mg. If the container is filled with 250 kg wood from pallets, the concentration of chlorophenols in the wood must be below 0.2 mg/kg. Concentrations higher than this level are often found. In the given example the diffusion rate is rather low.

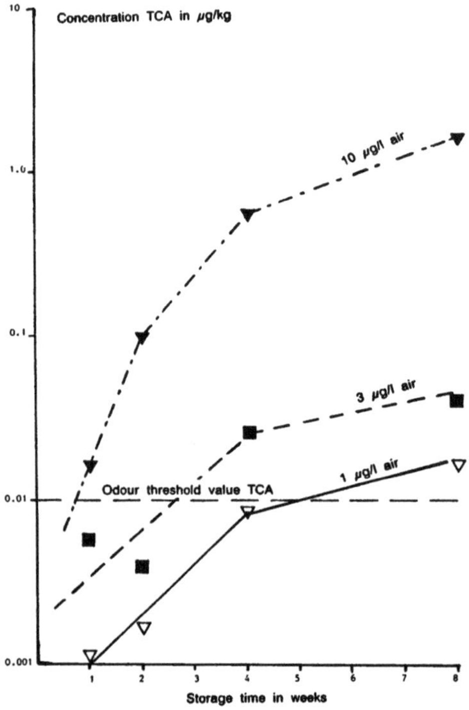

Figure 1 - Concentration of TCA in a product in relation to the concentration of TCA in the surrounding air during a storage experiment

6. CONCLUSIONS

Study of the literature and the results of our own investigations lead to the following conclusions:

a) It is to be expected that halogenated phenols and particularly chloroanisoles will continue to give rise to off-flavour problems in the near future.
b) Several pathways can lead to the formation of a chloroanisole isomer. To find the source of a contamination of chloroanisoles the contaminated products and related materials have to be analysed for the presence of chloroanisole and chlorophenol isomers, and if necessary for 2,4-D; 2,4,- T and HCH.

c) One of the important factors in the occurrence of chloroanisoles is the contamination of packaging material (cardboard, wooden pallets etc.) with chlorophenols.

d) The occurrence of chloroanisoles could be limited if maximum allowable levels of chlorophenols in relation to various applications would be known and used in practice.

e) An experimental set-up to measure and calculate these levels is operational.

7. LITERATURE

1. Axelrod, J. and J. Daly: Phenol-o-methyltransferase. Biochim. Biophys. Acta. 159 (1968) 472-478.

2. Backman, E.L.: Experimentella undersökningar öfver luktsinnets fysiologie. Upsala Laekarefoeren Foerh. 22 (1917) 319-470.

3. Baker, R.A.: Threshold odors of organic chemicals. J.Am. Water Works Assoc. 55 (1963) 913-916.

4. Ballschmiter, K., C. Unglert and P. Heizmann: Bildung von Chlorphenolen durch mikrobielle Umwandlung von Chlorbenzolen. Angew. Chem. 89 (1977) 680-681.

5. Bemelmans, J.M.H. and H.J.A. den Braber: Investigations of an iodine-like taste in herrings from the Baltic sea. Wat. Sci. Technol. 15 (1983) 105-113.

6. Brown, E.J. et al.: Formation of pentachlorophenol as the major product of microsomal oxidation of hexachlorobenzene.

7. Burttschell, R.H. et al.: Chlorine derivatives of phenol causing taste and odor. J. Am. Water Works Assoc. 51 (1959) 205-214.

8. Cserjesi, A.J. and E.L. Johnson: Methylation of pentachlorophenol by Trichoderma virgatum. Can. J. Microbiol 18 (1972) 45-49.

9. Cserjesi, A.J.: Detoxification of chlorinated phenols. Int. Biodetn. Bull. 8 (1972) 135-138.

10. Curtis, R.F. et al.: 2,3,4,6-Tetrachloroanisole association with musty taint in chickens and microbiological formation. Nature 235 (1972) 223-224.

11. Dietz, F. and J. Traud: Geruchs- und Geschmacks-Schwellen-Konzentrationen von Phenolkörpern. Gas-Wasserfach: Wasser/Abwasser 119 (1978) 318-325.

12. Engel, C., A.P. de Groot and C. Weurman: Tetrachloroanisole: a source of musty taint in eggs and broilers. Science 154 (1966) 270-271.

13. Engst, R., R.M. Macholz and M. Kujawa: Identifizierung von Metaboliten und der Abbauweg des Hexachlorbenzols in einer Schimmelpilzkultuur. Nahrung 19 (1975) 603-606.

14. Frijters, J.E.R. and J.M.H. Bemelmans: Flavour sensitivity for chloroanisoles in coagulated egg yolk. J. Food Sci. 42 (1977) 1122-1123.

15. Gee, J.M. and J.L. Peel: Metabolism of 2,3,4,6-tetrachlorophenol by micro-organisms form broiler house litter. J. Gen. Microbiol. 85 (1974) 237-243.

16. Gibson, S.A. and J.M. Suflita: Extrapolation of biodegradation results to groundwater aquifers: reductive dehalogenation of aromatic compounds. Appl. Environ. Microbiol. 52 (1986) 681-688.

17. Griffiths, N.M.: Sensory properties of the chloro-anisoles. Chem. Sences Flavor 1 (1974) 187-195.

18. Guadagni, D.G. and R.G. Buttery: Odor threshold of 2,3,6-trichloroanisole in water. J. Food Sci. 43 (1978) 1346-1347.

19. Hagel, P.: Rijksinstituut voor visserijonderzoek, Ymuiden. Rapport nr. CA 80-05 (1980).

20. Jerina, D., G. Guroff and J. Daly: Enzymic and nonenzymic hydroxylation and chlorination of p-deutero-anisole. Arch. Biochem. Biophys., 124 (1968) 612-615.

21. Karns, J.S., S. Duttagupta and A.M. Chakrabarty: Regulation of 2,4,5-trichlorophenoxyacetic acid and chlorophenol metabolism in Pseudomonas cepacia AC 1100. Appl. Environ. Microbiol. 46 (1983) 1182-1186.

22. Katz, S.H. and E.J. Talbert: Intensities of odors and irritating effects of warning agents for inflammable and poisonous gases. U.S. Bureau of Mines, Technical Report no. 480 (1930).

23. Kendall, D.A., G. Leonardus and E.R. Rubin: Parameters affecting the determination of odor thresholds. Report C-68988 Arthur D. Little Inc. Cambridge, U.S.A. (1968).

24. Loos, M.A., J.M. Bollag and M. Alexander: Phenoxyacetate Herbicide detoxication by bacterial enzymes. J. Agric. Food Chem. 15 (1967) 858-860.

25. Maarse, H., L.M. Nijssen and J. Jetten: Chloroanisoles, a continuing story, in: "Topics in flavour research". eds. Berger, R.G., S. Nitz and P. Schreier, H. Eichhorn, Hangenham, (1985), 241-250.

26. Maarse, H. and R.E. Kepner: Changes in composition of volatile terpenes in Douglas fir needles during maturation. J. Agric. Food Chem. 18 (1970) 1095-1101.

27. Mckee, J.E. and H.W. Wolf: Water quality criteria. The resources of California, State water quality control board (1963).

28. Murthy, N.B.K., D.D. Kaufman and G.F. Fries: Degradation of pentachlorophenol (PCP) in aerobic and anaerobic soil. J. Environ. Sci. Health, B 14 (1) (1979) 1-14.

29. Neidleman, S.L. and J. Geigert: Biohalogeneration: Principles, Basic Roles and Applications. Ellis Horwood Ltd, Chichester, (1986), Chapter 5.

30. Neidleman, S.L. and J. Geigert: Biohalogeneration: Principles, Basic Roles and Applications. Ellis Horwood Ltd, Chichester, (1986), Chapter 8.

31. Neidleman, S.L. and J. Geigert: Biohalogeneration: Principles, Basic Roles and Applications. Ellis Horwood Ltd, Chichester, (1986), Chapter 2.

32. Nijssen, L.M.: unpublished results.

33. Ommen, B. van et al.: Formation of pentachlorophenol as the major product of microsomal oxidation of hexachlorobenzene. Biochem. Biophys. Res. Commun. 126 (1985) 25-32.

34. Pignatello, J.J. et al.: Response of the microflora in outdoor experimental streams to pentachlorophenol: environmental factors. Can. J. Microbiol. 32 (1986) 38-46.

35. Punter, P.H.: Measurement of human olfactory thresholds for several groups of structurally related compunds. Chem. Senses 1 (1983) 215-235.

36. Saxby, M.J. and S. Wragg: Contamination of cocoa liquor by chlorophenols. Chem. Ind. (London) (1985), 310-311.

37. Sharp, A.K. and J.E. van S. Greve: Trial shipment of cocoa, copra and coffee from Papua New Guinea to Europe to evaluate the use of ventilated containers. February-May 1981 "Cocof 81". CSIRO Aust. Div. Food. Res. Techn. Pap. no. 48 (1986) 1-20.

38. Steiert, J.G. and R.L. Crawford: Microbial degradation of chlorinated phenols. Tr. Biotechn. 3 (1985) 300-305.

39. Tanner, H., C. Zanier and H.R. Buser: 2,4,6-Trichloroanisol: eine dominierende Komponente des Korkgeschmackes, Schweiz, Z. Obst. Weinbau 117 (1981) 97-103.

40. Whitfield, F.B.: Food off-flavours: cause and effect, in "Developments in food flavours", eds. Birch, G.G. and M.G. Lindley, Elsevier Applied Science, London and New York, (1986), 249-273.

41. Whitfield, F.B.: Some flavours which industry could well do without. Case studies of industrial problems. CSIRO Food Res. Quart., 43 (1983) 96-106.

42. Whitfield, F.B., R.L. McBride and T.H.L. Nguyen: Flavour perception of chloroanisoles in water and selected processed foods. J. Sci. Food Agric. 40 (1987) 357-365.

43. Zoeteman, B.C.J. and G.J. Piet: Drinkwater is nog geen water drinken. H_2O 6 (1973) 179-189.

'EFFECTS OF TYPES I, II AND III ANTIOXIDANTS ON PHOSPHOLIPID OXIDATION IN A MEAT MODEL FOR WARMED OVER FLAVOUR (WOF)'* (Plate a).

J.P. Roozen [1]

According to Labuza there are three categories of antioxidants:

I : free radical stoppers, primarily phenolic type compounds;

II : free radical production preventers: chelating agents;

III: environmental factors such as redox compounds (Plate b).

Freeze-dried, washed ground meat was mixed with lecithin and antioxidants dissolved in water, heated and then stored in the refrigerator in order to study phospholipid oxidation as related to warmed over flavour (Plate c).

The analysis of thiobarbituric acid reactive substances (TBARS) was based on the distillation method of Tarladgis et al. Type I and II antioxidants decreased the amount TBARS (Plate d), while most Type III's gave an increase, depending on their concentration (Plate e).

Volatiles were analyzed by a gas chromatograph equipped with an external closed inlet device. Similar effects on phospholipid oxidation, as with TBARS, were detected for representatives of each Type of antioxidant, i.e. butylated hydroxy anisole (BHA), ethylenediaminetetraacetic acid (EDTA), sodium bisulfite, cysteine (Plate f) and water activity regulators (Plate g).

The conclusions are presented in Plate h.

*FOOD CHEMISTRY 24 (1987) 167-185.

---

[1] Agricultural University, Department of Food Science, Biotechnion, De Dreyen 12, 6703 BC Wageningen, The Netherlands.

WARMED OVER FLAVOUR

THE RAPID ONSET OF

RANCIDITY IN COOKED MEAT

DURING REFRIGERATED STORAGE:

-PEROXIDATION OF PHOSPHOLIPIDS

-DEGRADATION OF HYDROPEROXIDES

ANTIOXIDANTS
(LABUZA, 1971)

TYPE I: FREE RADICAL STOPPERS;
DONATE H TO A RADICAL

TYPE II: FREE RADICAL PRODUCTION
PREVENTERS (CHELATING AGENTS)

TYPE III: ENVIRONMENTAL FACTORS
(REDOX & WATER ACTIVITY)

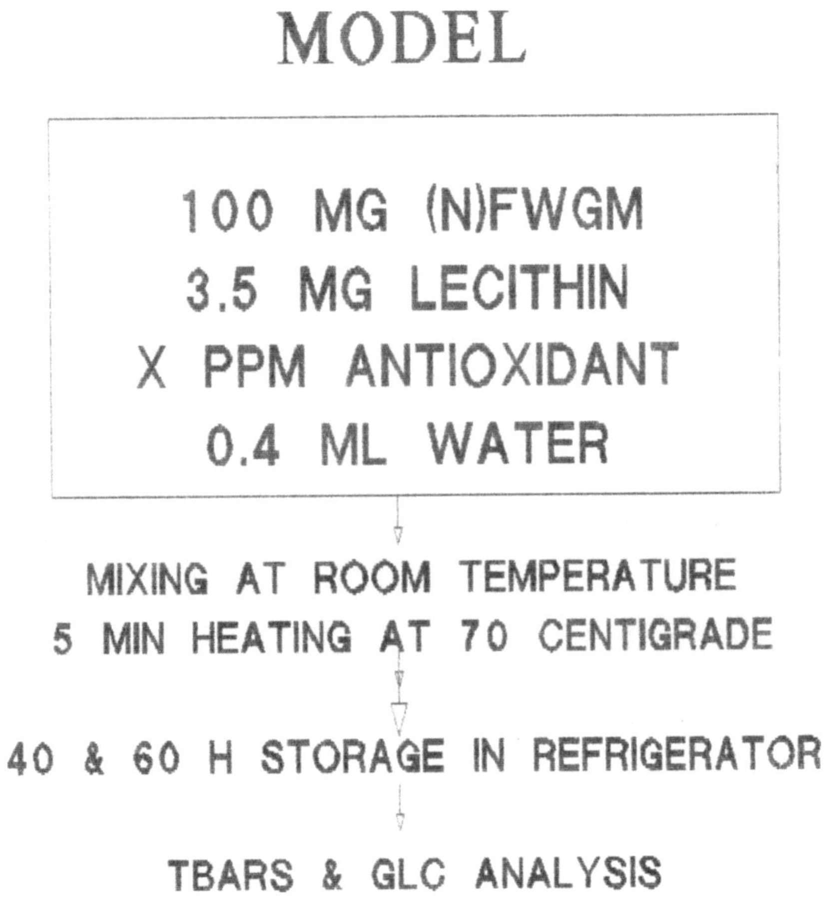
MODEL
100 MG (N)FWGM
3.5 MG LECITHIN
X PPM ANTIOXIDANT
0.4 ML WATER
MIXING AT ROOM TEMPERATURE
5 MIN HEATING AT 70 CENTIGRADE
40 & 60 H STORAGE IN REFRIGERATOR
TBARS & GLC ANALYSIS

EFFECT OF ANTIOXIDANTS IN MEAT MODEL

THIOBARBITURIC ACID REACTIVE SUBSTANCES

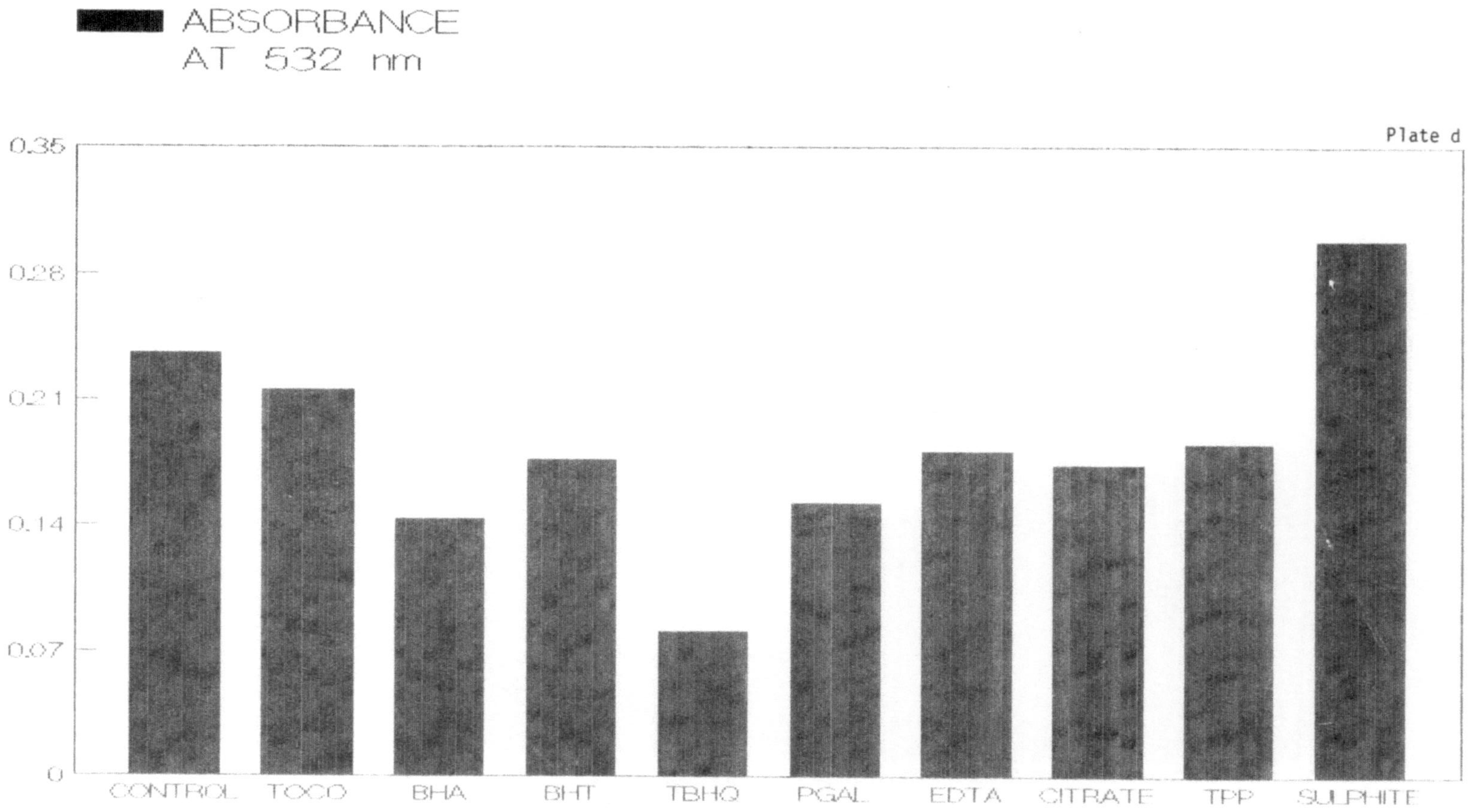

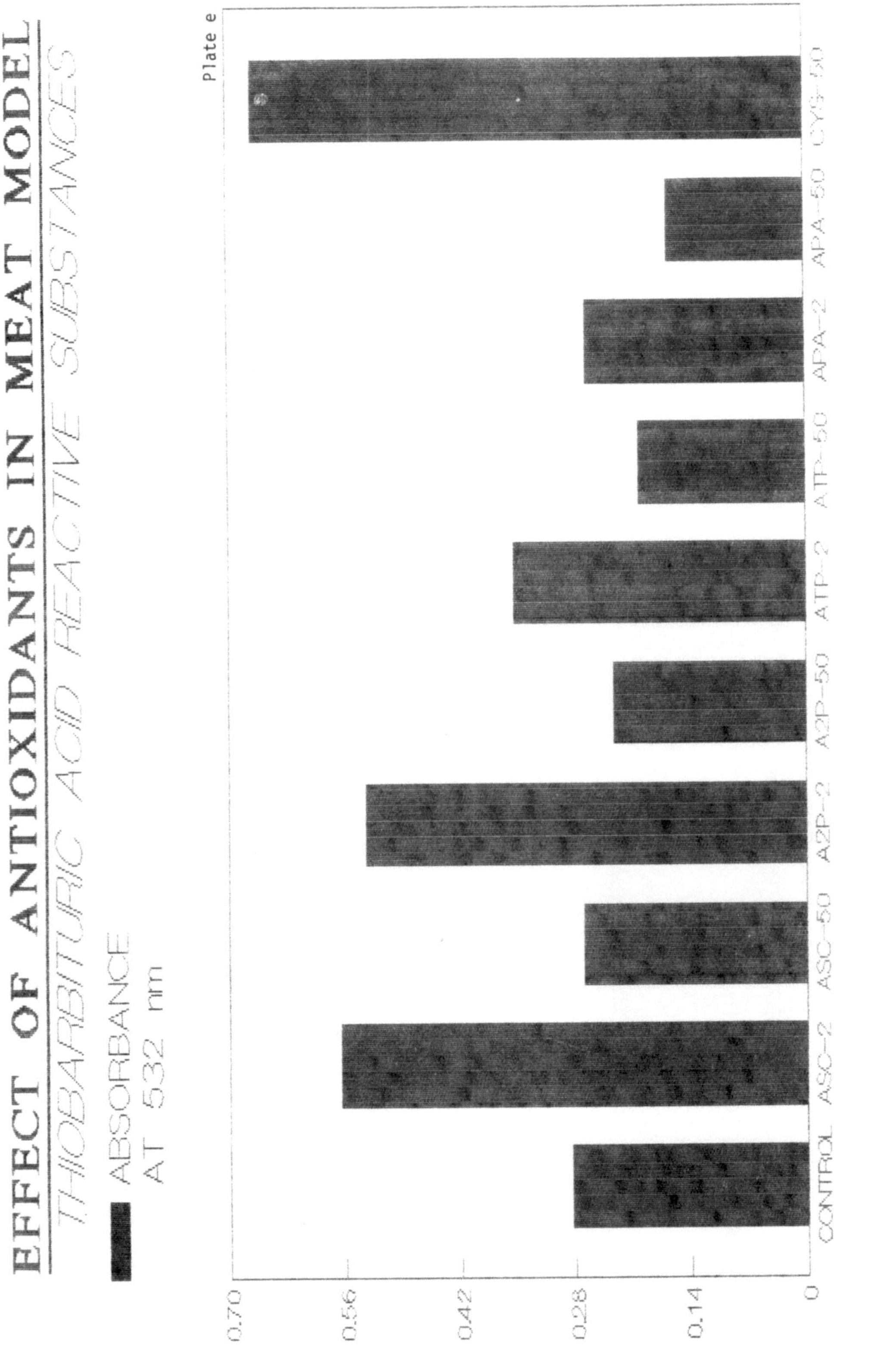

EFFECT OF ANTIOXIDANTS IN MEAT MODEL
THIOBARBITURIC ACID REACTIVE SUBSTANCES
ABSORBANCE
AT 532 nm
Plate e
0.70
0.56
0.42
0.28
0.14
0
CONTROL
ASC-2
ASC-50
A2P-2
A2P-50
ATP-2
ATP-50
APA-2
APA-50
CYS-50

EFFECT OF ANTIOXIDANTS IN MEAT MODEL

GLC AREAS OF MAJOR PEAKS (COMP. COUNTS)

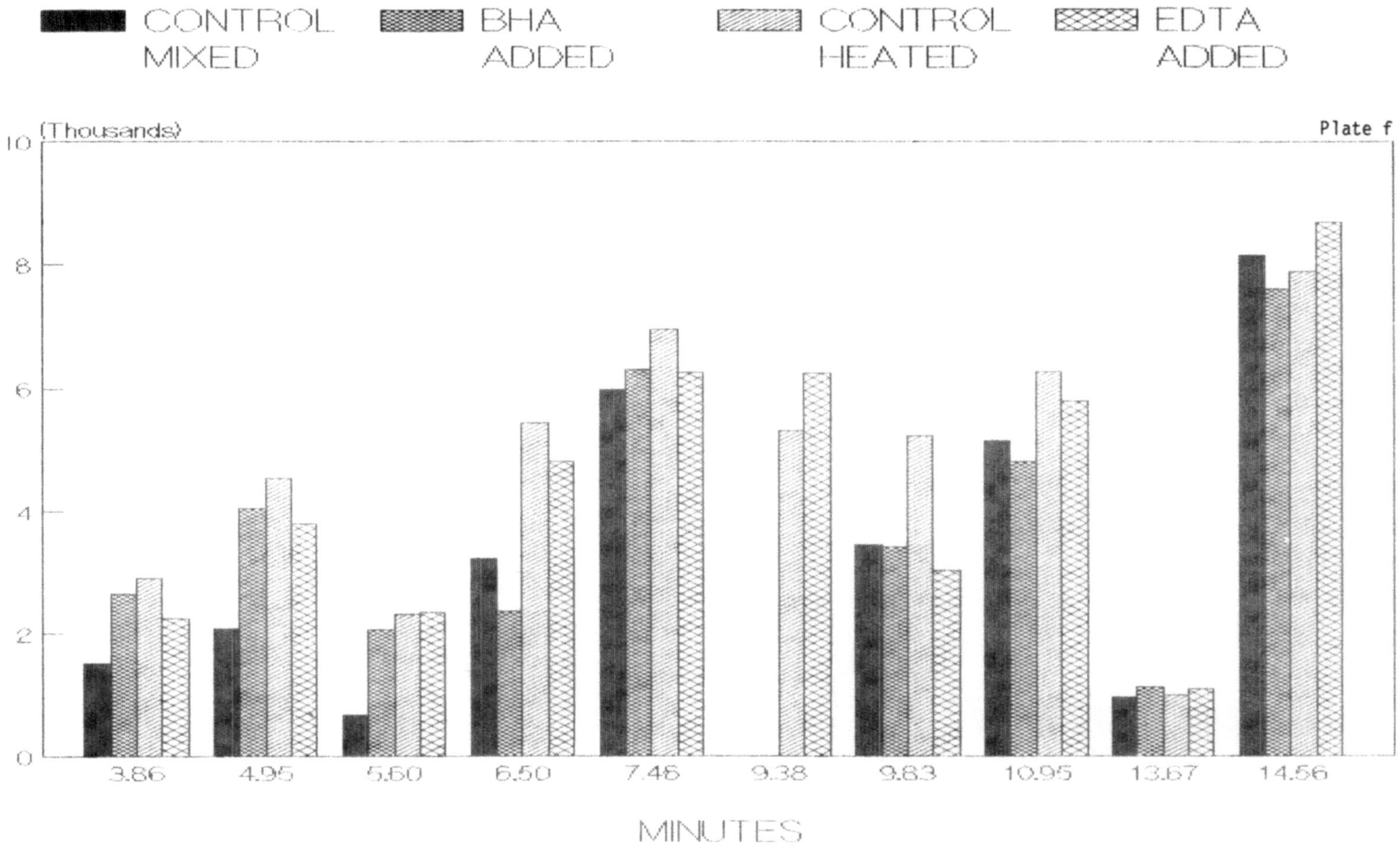

EFFECT OF rH IN DRIED MEAT MODELS

TOTAL AMOUNTS OF VOLATILES FORMED

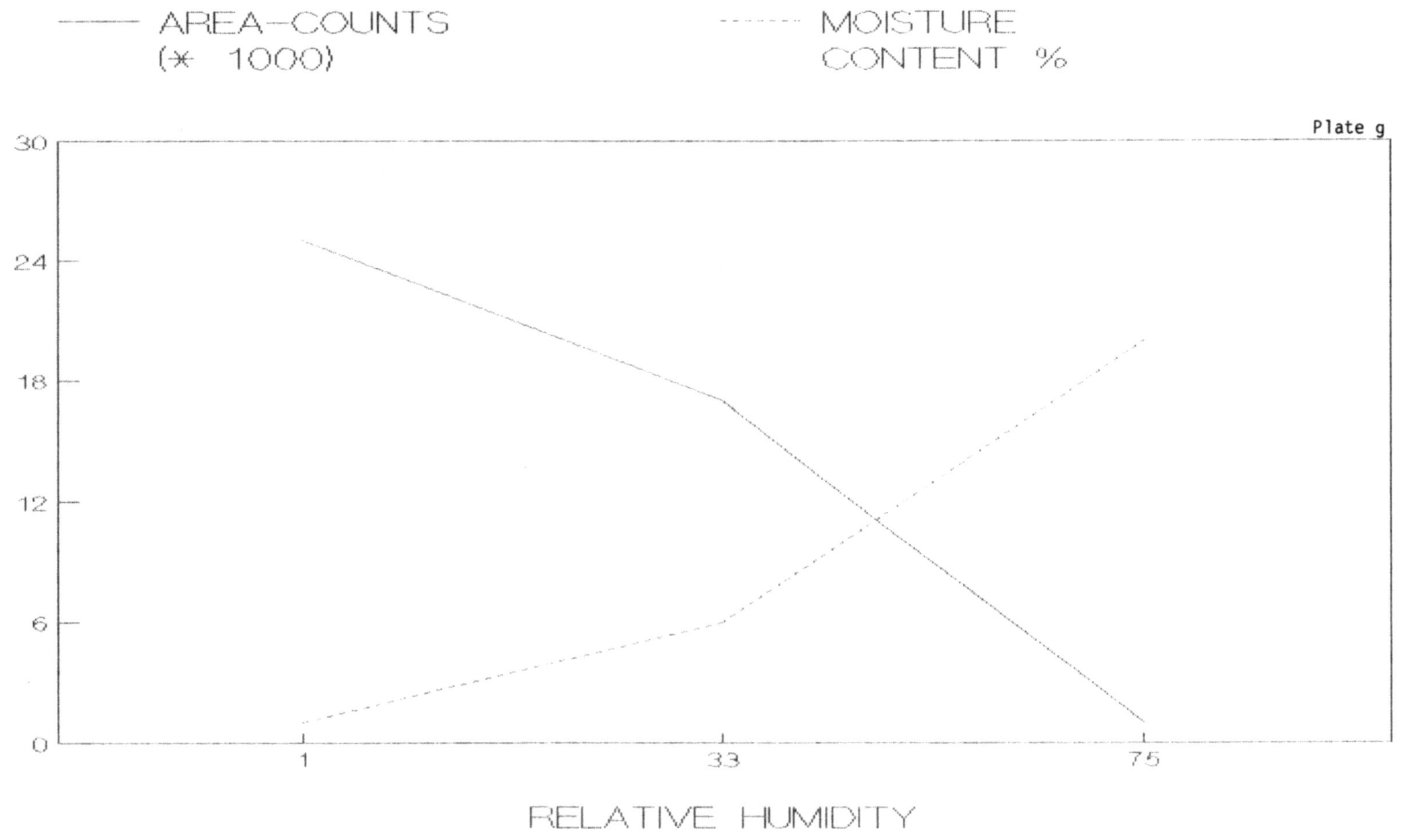

CONCLUSIONS

TYPE I ANTIOXIDANT IS EFFECTIVE DURING HEATING AND STORAGE.

TYPE II ANTIOXIDANT IS ONLY EFFECTIVE DURING COLD STORAGE.

TYPE III ANTIOXIDANTS ACT AS PRO-OXIDANTS AT LOW CONCENTRATIONS AND IN MANY CASES AS ANTI-OXIDANTS AT HIGH CONCENTRATIONS. DURING COOKING AND STORAGE THE PRODUCT SHOULD BE PREVENTED FROM DRYING OUT.

A STUDY OF AROMA RECOVERY DURING THE PRODUCTION OF CONCENTRATED JUICES

G. Konja*, A. Pozderović**, T. Lovrić*

* Prehrambeno-biotehnološki fakultet Zagreb
** Prehrambeno-tehnološki fakultet Osijek

Introduction

There is a growing demand for concentrated juices on the world market. Due to the wide range of possibilities of their use in the food industry and as consumer products, increasing attention is being paid to their production. Concentrated juice and concentrated aroma are produced for the market separately, since this has numerous advantages.

The most widespread method of concentrating fruit juices today is still evaporation at reduced pressure (in a vacuum), regardless of the poor selectivity of this process (1). Selectivity is one of the basic criteria for evaluating a concentration plant, especially in the case of fruit juices. The insufficient selectivity of concentration by evaporation, especially with regard to the aromatic components of juice, was one of the main reasons for the introduction of alternative concentration methods such as freeze concentration, reverse osmosis, etc. However, concentration by evaporation is still the prevailing method used for juices because it is economical, flexible and can be applied in plants with large capacities. Attempts have been made to overcome the disadvantage of this method with regard to selectivity by using various systems for aroma recovery and concentration. The different kinds of evaporation plant used nowadays (e.g. high-temperature, low-temperature, single-, two- or multistage, with or without thermocompression and, recently, plants with mechanical vapour compression (2), have been variously designed either to effect energy savings or to improve product quality. These plants are usually combined with aroma recovery and concentration units. Most of the latter use the principciple of rectification, gradual condensation of highly volatile components and absorption.

However, since the aromas of different kinds of fruit differ greatly, no universal rectification column has yet been devised which could separate and concentrate highly volatile and poorly volatile aroma components with equal effectiveness. Different aroma recovery plants have therefore been constructed which have been adapted to the more or less specific aroma composition of particular kinds of fruit. It is well known that, depending on the degree of juice concentration, some of the aroma components which are responsible for the entire aromatic complex in the fruit remain in the concentrated juice. Only when the presence of the aroma components which are the characteristic aroma carriers of a certain kind of fruit has been proved can the quality of an aroma concentrate and the effective operation of an aroma recovery and concentration plant be demonstrated.

When separating aromatic compounds from juices, the greatest problem arises with aroma components which have low boiling points. According to Pilnik (3), there is a difference between highly volatile aromas (e.g. apple aroma) and poorly volatile, azeotropic aromas (e.g. strawberry aroma, raspberry aroma). Highly volatile aromas can be totally separated from the juice by a greater or lesser degree of evaporation, while azeotropic aromas are partially retained even at high degrees of evaporation, i.e. in concentrate with a high dry solids content.

In his exhaustive survey of the dynamics of aroma separation from berries, drupe fruit and pome fruit, Šulc (4, 5) reports that in order to separate aroma components from fruit juice successfully, it is necessary to define the degree of juice evaporation required, which in turn depends on the quantitative and qualitative composition of the total aroma, i.e. on the mixture of highly and poorly volatile components of the fruit juice. It follows that data on the degree of aroma concentration, e.g. 1:100, is not a sufficient indicator of the quality of an aroma concentrate. In other words, aroma concentrates of the same degree of concentration can differ considerably from each other in quality, which means that the standardization of aroma concentrates is extremely difficult and that the quality of an aroma concentrate is not determined by the degree of mechanical concentration but by indicators such as the intensity, yield and purity of the flavour (6).

Numerous authors have been interested in the changes in aroma compounds which occur in different phases of the technological process of concentrated juice production.

In order to determine whether a plant for the separation and recovery of aroma compounds is operating effectively, a suitable analytic method must be used.

The method most frequently used for this purpose so far has been gas chromatography combined with methods of identifying aroma components.

The method of gas chromatography usually presupposes a certain preparation of the samples to be analysed, such as distillation, extraction and concentration. There are certain disadvantages inherent in the abovementioned operations of sample preparation, such as

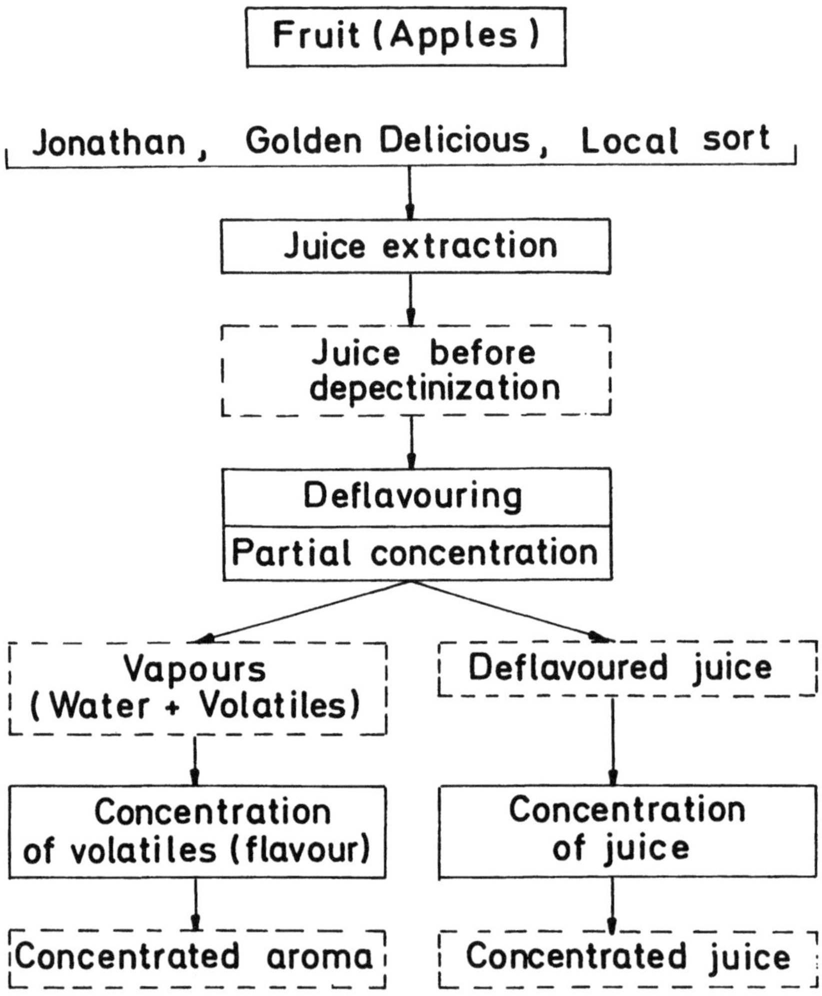

Fig. 1: Schematic process flow in concentrated apple juice production with flavour recovery

the unequal extraction of individual components due to their different properties and the loss of a greater or lesser amount of individual components during the removal of the solvent.

Recent investigations show that good results may be obtained by separating the volatile aroma components by extraction with a solvent followed by simultaneous distillation / extraction with a solvent (SDE) (7) or by extraction with supercritical CO_2.

The abovementioned disadvantages do not occur in the headspace method of gas chromatography, which is also simple and fast. The application of this method is limited to determination of the highly volatile components which are found in the vapour phase above the sample, but which are also mostly responsible for the specific olfactory sensations (8).

However, some authors consider that in spite of these advantages this method is not always suitable for the quantitative analysis of aroma components (9,10, 11) due to the significance of the influence of the matrix composition on the composition of the vapour phase.

The aim of this paper (research) was to investigate the qualitative and quantitative changes which occur in the aromatic components of apples during the production of concentrated juice with aroma recovery using four different systems of aroma concentration. The analyses of the aromatic components were made by gas chromatography, by headspace analysis and by the method of sample preparation by distillation, extraction and concentration.

Material and methods

Samples taken from four industrial plants for the production of concentrated apple juice with aroma recovery were investigated. The plants are designated by the letters A, B, C and D. They differ from each other in the design of their evaporation and aroma recovery systems and were constructed by four different equipment manufacturers. In plants A, B and C, the rectification of aroma compounds is carried out in a column with plates, while in plant D it is carried out in a column with Raschig rings.

Three varieties of apples (Jonathan, Golden Delicious and a local sort) were used as the raw material.

The basic flow diagram representing the process of concentrated juice and concentrated aroma production is shown in Figure 1.

The characteristics and operating conditions of each of the plants are shown in Table 1.

Samples were taken at the following stages in the production line:
- initial juice before aroma-stipping
- aroma-stripped juice (before depectinization)
- reflux
- concentrated juice (after ejection cooling)
- stillage (condensate from the rectification column)
- aroma concentrate (after cooling)

All the samples were kept in a freezer until the analysis.

Determination of the aroma components was carried out by gas chromatography. As already mentioned, headspace analysis (a) was used for aroma recovery systems A, B, C and D and, for purposes of comparison, the method with previous sample preparation (b) was used for system A.

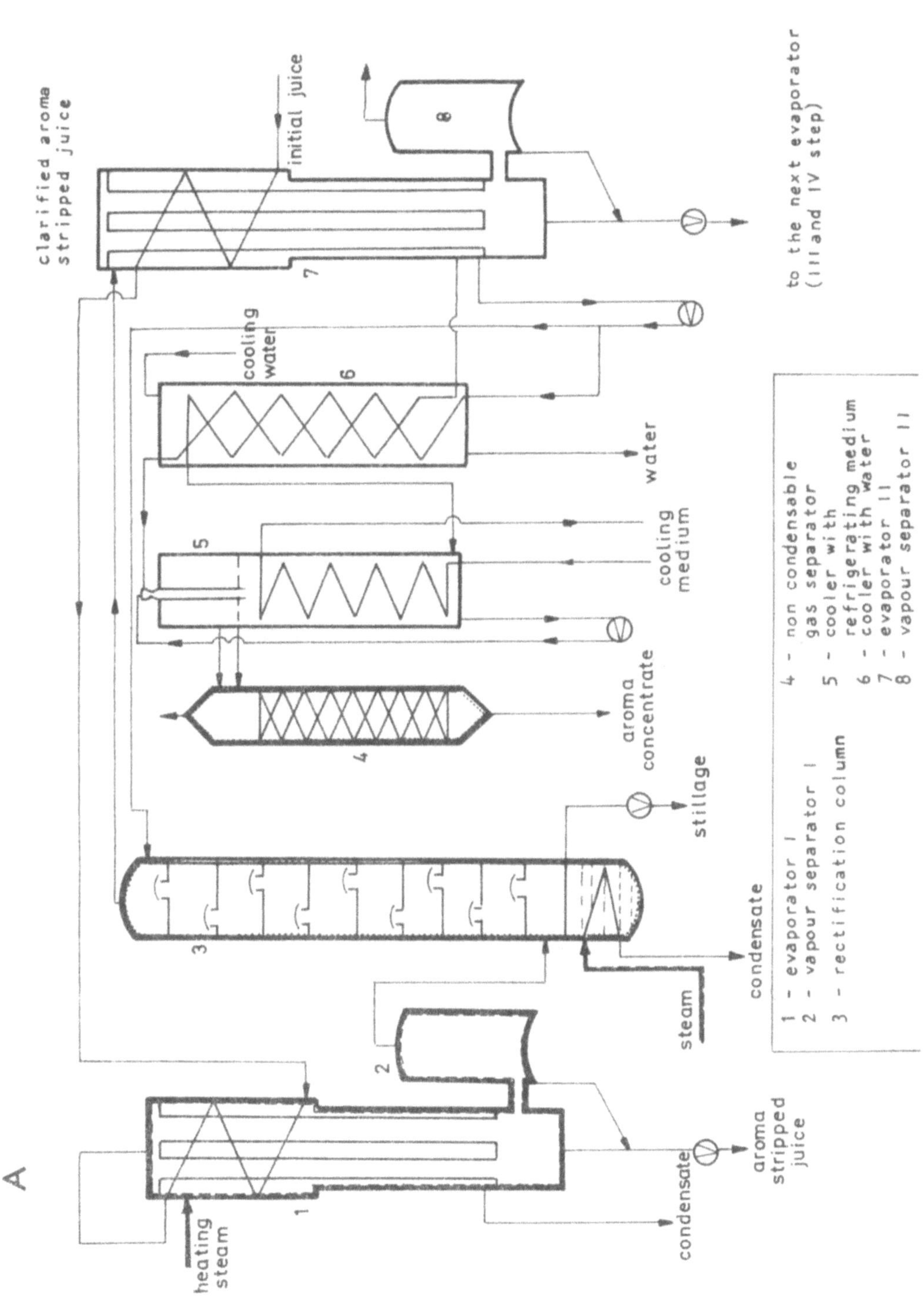

Fig. 2: Plant A for production of concentrated apple juice

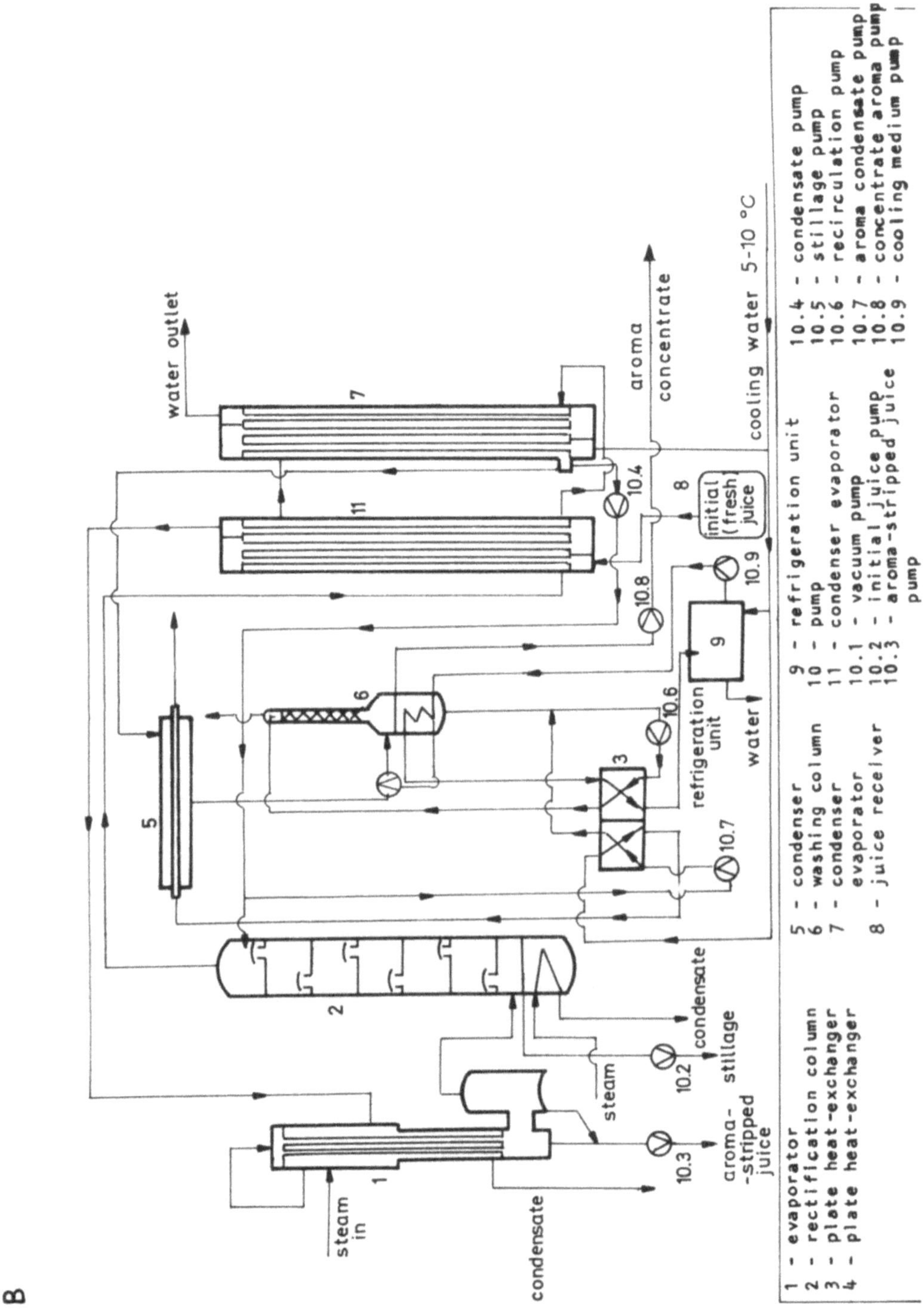

Fig. 3: Plant B for production of concentrated fruit juices

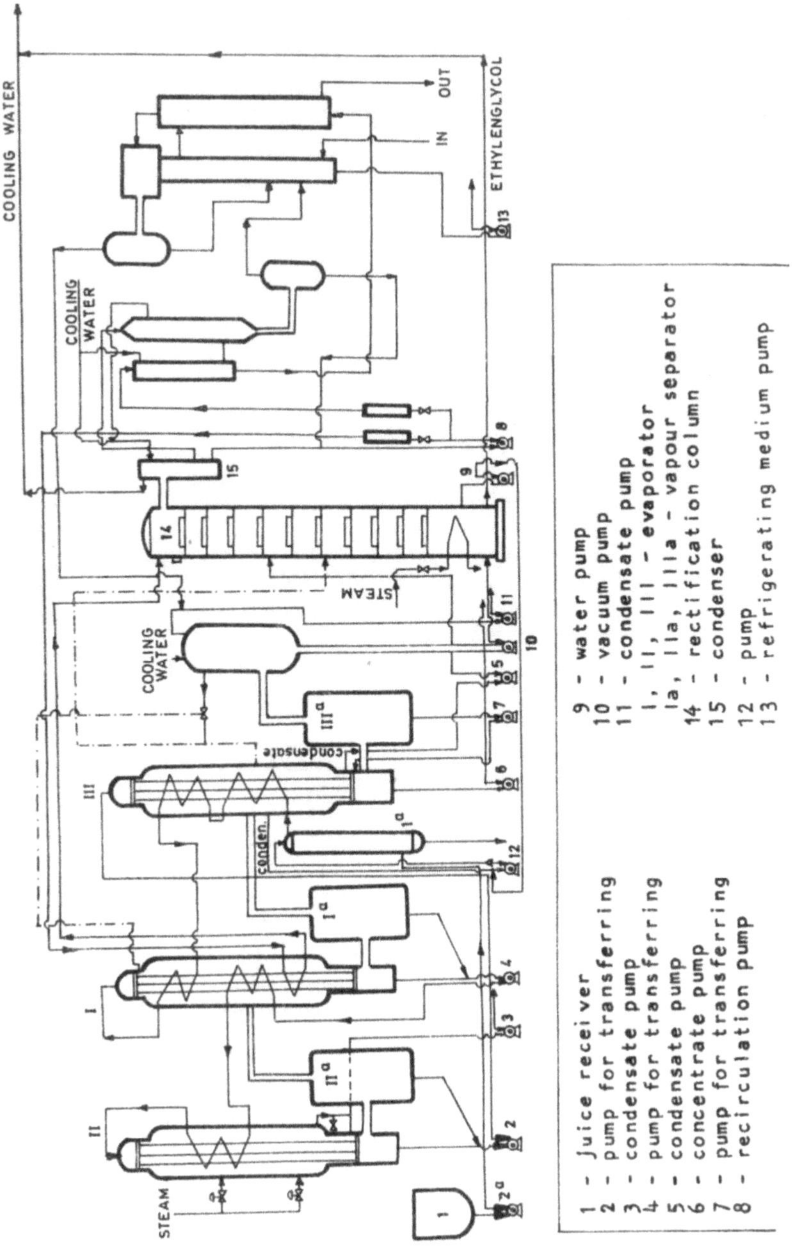

C

Fig. 4: Plant C for production of concentrated fruit juices

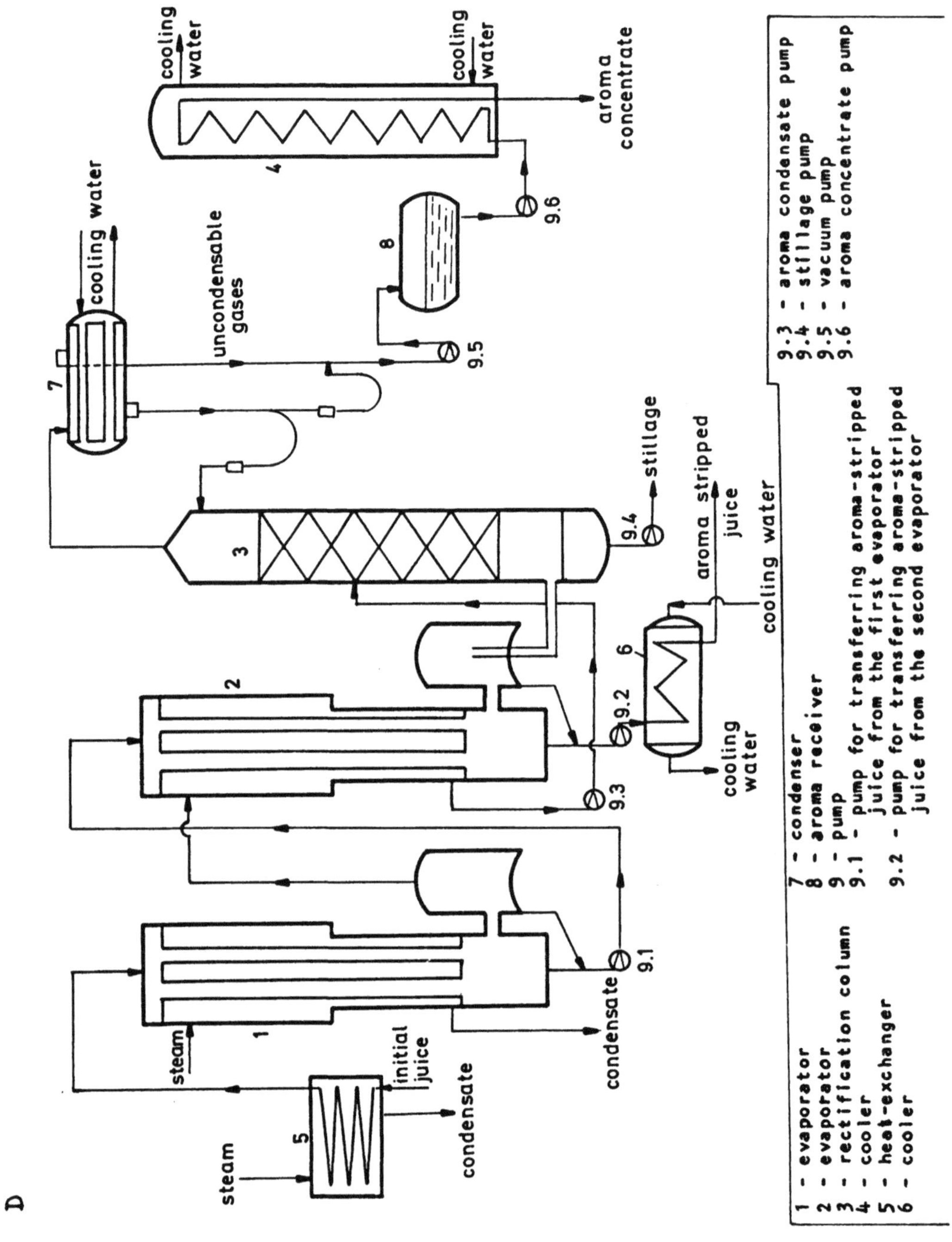

Fig. 5: Plant D for production of concentrated fruit juices

Table 1. Characteristics and operating conditions of the fruit juice concentration plants with aroma recovery

Plant designation	Characteristics and operating conditions of plant	
1	2	
A	- the aroma recovery plant is part of a 4-stage evaporation plant;	
	- the aroma recovery evaporator is the first stage of the 4-stage evaporation plant;	
	- total solid in the initial (fresh) juice before aroma-stripping	14 %
	- total solid in the aroma-stripped juice	15,5 %
	- water evaporated during aroma-stripping in relation to the total amount of evaporated water	12 %
	- degree of concentration	1 : 160
B	- the aroma recovery plant is a separate unit and is a one-stage evaporation plant with a rectification column etc.;	
	- the aroma-stripped juice is concentrated in a 4-stage evaporation plant;	
	- total solid in the initial (fresh) juice before aroma-stripping	11,4 %
	- total solid in the aroma-stripped juice	14,5 %
	- water evaporated during aroma-stripping in relation to the total amount of evaporated water	25 %
	- degree of concentration	1 : 130

1	2	
C	- the aroma recovery plant is part of a 3-stage evaporation plant without cooling of aroma condensate;	
	- the first stage of the evaporation plant is the aroma recovery evaporator	
	- total solid in the initial (fresh) juice before aroma-stripping	15,9 %
	- total solid in the aroma stripped juice	18,5 %
	- water evaporated during aroma--stripping in relation to the total amount of evaporated water	18,0 %
	- degree of concentration	1 : 170
D	- the aroma recovery plant is a separate unit and is a two-stage evaporation plant with a rectification column etc., without cooling of aroma condensate	
	- the aroma-stripped juice is concentrated in 4-stage evaporation plant	
	- total solid in the initial (fresh) juice before aroma-stripping	12,2 %
	- total solid in the aroma-stripped juice	13,5 %
	- water evaporated during aroma-stripping in relation to the total amount of evaporated water	12,0 %
	- degree of concentration	1 : 100

The headspace analysis was carried out on a Perkin Elmer gas chromatograph type Sigma 3 with an HS 6 block. Separation of aroma components was carried out with a packed glass column (3 % DEGS on a Chromosorb P 80-100 mesh) 2 meters long.

Before analysis the samples were diluted with distilled water to the dry solids content of the initial juice or in such a way that the analysis of all the samples was carried out at the same attenuation.

The samples were thermostatted at 90 °C for 4 min, and the injection time was 5 seconds.

The highly volatile aroma components were determined separately, and the poorly volatile aroma components separately.

The gas chromatograph operating conditions under which the highly volatile aroma components were analyzed:

- HS block temperature	90 °C
- detector and injector temperature	150 °C
- initial column temperature	40 °C isothermally 4 min
- set column temperature	40-60 °C/ 5 °C/min
- final column temperature	60 °C/34 min
- attenuation	4x1
- chart speed	0.5 cm/min
- carrier gas flow (N_2)	38 ml/min

The gas chromatograph operating conditions under which the poorly volatile aroma components were analyzed:

- column temperature	70 °C isothermally

and all the other conditions were the same as in the analysis of highly volatile components.

Identification of individual aroma components (14) was carried out by adding pure components to the samples.

The internal standard method was used for the quantitative calculation of the chromatogram (methyl ester of undecanoic acid).

When aroma components were determined by gas chromatography with previous preparation of samples (aroma--stripping and concentration), the analysis was carried out on a Carlo Erba chromatograph model Vega 6000. The conditions adopted were as follows: capillary column DB 5 "ICT", 60 m; split ratio 1:20, injector temperature 200 °C, detector temperature 250 °C; oven: 40 °C/5 min, 40-50 °C/2 °C/min, 50-180 °C/5 °C/min, 180-250 °C/7 °C/min, 250 °C/10 min; carrier gas pressure (H_2) = = 1,25 ml/min; attenuation 1 x 16; chart speed 1 cm/min; quantity injected 1 ul.

Four components were added to the samples tested for the quantitative and qualitative determination of individual components as follows: 1-hexanal, benzaldehyde, 2-methylbutyriol acid ethyl ester and nonanol dissolved in an ether:pentane mixture. The total surface and the surfaces of individual peaks were determined by an integrator. The values referring to the peaks made by ether, ethanol, pentane and the several components which are separated before ether were not taken into consideration.

The internal standard method (methyl — ester of undecanoic acid) was used for the quantitative calculation of the chromatogram.

To investigate the influence of dissolved substances, i. e. the matrix composition, on the results obtained by headspace analysis, a comparative analysis of aroma concentrate was carried out with samples diluted with water and with samples diluted with initial juice.

Table 2. The degree of concentration or loss of aroma components during the production of concentrated apple juice with aroma recuperation

Sample	Concentration or loss index (%)											
	More volatile aroma components				Less volatile aroma component				Total aroma			
	A	B	C	D	A	B	C	D	A	B	C	D
Initial juice* before aroma-stripping	100	100	100	100	100	100	100	100	100	100	100	100
Aroma-stripped juice	13.2	9.8	5.6	11.4	12.2	5.8	3.3	14.9	13.1	9.1	5.2	11.7
Concentrated juice diluted to the initial solid content	3.2	1.0	2.0	1.7	1.2	-	0.5	5.0	3.0	0.8	1.8	1.9
Reflux	511.8	-	2820	-	519.7	-	2583	-	512.6	-	2778	-
Stillage	6.3	3.6	43.2	4.3	4.1	2.3	18.8	10.8	6.1	3.4	38.8	4.9
Aroma concentrate	18688	12750	6495	2884	15168	12642	4420	2978	18340	12732	6126	2892

A, B, C, D - Plant designation

Results and discussion

The results shown in Tables 3 and 6 show that in all the samples analyzed, the largest fraction of the total aroma consisted of highly volatile compunds.

An analysis of the changes in the amount of aroma occurring during the various technological stages in the production of concentrated juice and concentrated aroma reveals the following:

During aroma stripping of the initial juice, there was a considerably greater degree of stripping in aroma recovery plants B and C (25 % in B, 18 % in C) than in aroma recovery plants A and D (Table 1).

The data presented in Table 2 shows that when the amount of water evaporated was 12 % of the total required amount, the quantity of aromatic components of the initial juice which evaporated amounted to 90 %. Only a slightly greater evaporation of aromatic components (up to 4 %) was achieved by doubling the amount of water evaporated (25 % B). Therefore, when aroma stripping apple juice using the aroma recovery plants described above, it is sufficient to evaporate 12 % of the total required amount of water.

During the concentration of apple juice, the aroma is almost completely lost. The maximum amount of "total" aroma retained in the concentrated juice is 3 % of the "total" aroma compared to the aroma of the initial juice.

An analysis of the results referring to the amount of aromatic components in the stillage of various plants shows that very little aroma (max. 6 %) was retained in the stillage of plants A, B and D compared to the aroma of the initial juice, while 38 % of the total aroma (43 % of the "highly" volatile and 18 % of the "poorly" volatile components) was retained in aroma recovery

Table 3. The fraction of more and less volatile aroma components in the total aroma during apple juice concentration with aroma recovery

Sample	Fraction %							
	More volatile aroma components				Less volatile aroma components			
	A	B	C	D	A	B	C	D
Initial juice before aroma--stripping	90,1	82,5	82,2	91,1	9,9	17,5	17,8	8,9
Aroma-stripped juice	90,8	88,8	88,7	88,6	9,1	11,2	11,3	11,4
Concentrated juice diluted to the initial solids content	95,9	-	94,6	77,4	4,0	-	5,3	22,5
Reflux	89,9	-	83,4	-	10,0	-	16,5	-
Stillage	93,3	88,1	91,4	80,3	6,6	11,9	8,6	19,6
Aroma concentrate	91,8	82,6	87,1	90,8	8,2	17,3	12,8	9,2

A, B, C, D - Plant designation

** Calculated on the basis of initial juice before aroma-stripping = 100

plant C. These results indicate that there was insufficient heating in the bottom part of the rectification column of plant C and that there was insufficient aroma stripping in this part of the column.

The amount of aroma in the reflux of aroma recovery system A was 5 times greater than in the initial juice while it was as much as 27 times greater in the reflux of aroma recovery system C.

The greater aroma enrichment in the reflux of aroma recovery system C was due to the condensation of a greater part of the aromatic vapours than was the case in plant A, and not to more efficient aroma concentration in the rectification column, since a large amount of aroma was found in the stillage of this plant.

The washing of uncondensed gases and vapours in plant C did not lead to a further significant increase in the amount of aroma since the cooling system (with ethylene-glycol) was not turned on during the operation of the plant so that the condensate was not cooled before washing.

In this aroma recovery plant, the concentration of aromatic components in the aroma concentrate was 61 times greater than in the initial juice.

In aroma recovery plant D, the cooling system was also turned off so the concentration of aromatic components in the aroma concentrate was only 29 times greater than in the initial juice.

In aroma recovery plants A and B, the aroma condensate was cooled to a temperature of 5-10 °C prior to the washing of uncondensed vapours and gases, so the degree of concentration in plants A and B was 183 and 127, respectively, compared to the initial juice.

The considerably lower degree of concentration of aroma components in aroma recovery plant D was due to incomplete condensation of the aroma vapours during the washing of uncondensed gases as well as, probably, to a different type of rectification column.

The lower degree of aroma concentration achieved with plant C is due partly to the same reasons as in the case of plant D (insufficient cooling and condensation) and partly to insufficient aroma stripping in the rectification column.

In the case of plants A and B, the lower degree of aroma concentration in plant B was probably due to the fact that a smaller amount of initial juice was used per one 1 of aroma concentrate, although losses of aroma compounds could also have occurred with uncondensed gases.

The results shown in Table 3 indicate that the fractions of "highly" and "poorly" volatile aroma components in the total aroma did not differ significantly in the initial juice and in the concentrated aroma. The fractions of some components (Tables 4 and 5) increased while those of some other components decreased, but taken as a whole the fractions of "highly" and "poorly" volatile components in the concentrated aroma remained approximately the same as in the initial juice.

The results discussed were above obtained by headspace analysis (b).

The results referring to the dynamics of hexanal, ethyl ester -2-methyl butyric acid, benzaldehyde and nonanol (Table 7) also show different degrees of concentration during the aroma recovery process. The greatest degree of concentration was achieved with ethyl ester -2-methyl butyric acid, the presence of which, along with hexanal, was proved in the stillage and, in a smaller amount, in the concentrated

Table 4. The degree of concentration or loss of individual aroma components during the production of apple juice concentratewith aroma recovery plants A and D

Peak No.	Component	Concentration or loss index (%)								
		Plant A					Plant D			
		Sample								
		II	III	IV	$V \times 10^2$	VI	II	IV	$V \times 10^2$	VI
1	Acet-Alde-hyde	21,1	487,5	3,5	149,4	6,8	16,5	5,4	40,8	10,7
2	Propanal	15,0	31,5	3,3	145,8	11,1	16,5	2,4	44,5	9,2
3	Non ident.	72,0	92,4	20,0	440,1	21,3	13,3	24,7	58,1	35,4
4	Butanal	10,8	1,0	2,6	277,8	5,3	13,9	3,8	24,4	4,5
5	Ethyl acetate	12,1	324,3	3,2	172,2	0,2	3,2	1,5	58,3	0
6	Non ident.	-	-	-	-	-	-	-	-	-
7	Methanol	37,5	562,5	14,6	28,2	0	89,1	108,8	52,7	10,8
8	Propanol	28,5	1339,8	14,2	134,2	0	57,5	39,2	52,5	0
9	Hexanal	4,6	239,0	1,7	299,9	1,3	6,2	0,4	15,3	0,3
10	n-Butyl acetate	3,9	290,8	1,6	224,1	0	5,6	1,2	20,9	0
11	Non ident.	5,4	72,7	0	202,1	0	2,6	3,7	13,4	0
12	Non ident.	22,1	0	22,4	200,0	0	-	-	-	-
13	Butanol	14,1	985,5	7,8	107,5	0,3	24,9	10,3	21,7	0,4
14	Non ident.	13,7	236,7	25,1	173,4	13,5	1,6	0,4	66,6	0,6
15	Hexenal	13,2	795,2	3,1	355,5	0	11,6	0,3	47,4	0
16	3-Methyl--1-Butanol	10,3	630,8	4,5	200,9	0,4	23,3	21,4	62,6	0,5
17	2-Methyl--1-Butanol	67,8	415,4	11,5	961,5	0	68,4	66,7	32,7	0
18	Pentanol	10,6	817,3	9,6	201,9	0	28,8	49,7	78,4	26,7
19	Non ident.	0	109,8	0	381,2	0	5,8	9,3	79,1	7,3
20	Non ident.	44,9	130,7	0	151,5	4,8	28,3	33,3	28,3	26,2
21	Hexyl acetate	5,2	152,3	0	228,7	0	0,4	0,6	89,7	0,3
22	Non ident.	13,5	382,1	9,4	151,6	6,7	61,1	64,5	12,3	6,8
23	Non ident.	10,6	623,2	5,2	136,7	0,4	26,9	15,2	71,2	1,7
24	Non ident.	33,9	128,8	0	677,9	0	47,8	33,3	37,1	7,4
25	Non ident.	33,3	120,2	0	0	6,5	46,1	101,9	62,1	5,5

I - initial (fresh) juice (calculation basis : initial juice = 100)
II - aroma stripped juice
III - reflux
IV - stillage
V - aroma concentrate
VI - concentrated juice diluted to the initial solid content

Table 5. The degree of concentration or loss of individual aroma components during the production of apple juice concentrate with aroma recovery plants B and C

Peak No.	Component	Concentration or loss index (%)								
		Sample								
		II	III	IV	$V \times 10^2$	VI	II	IV	$V \times 10^2$	VI
1	Acet-Aldehyde	7.1	2313.3	26.8	54.8	11.8	15.1	1.9	200.8	2.8
2	Propanal	9.4	-	-	31.3	-	17.5	-	-	-
3	Non ident.	41.9	-	-	41.8	-	21.9	-	156.3	-
4	Butanal	17.8	5714.3	28.6	62.5	25.7	12.8	-	87.8	8.7
5	Ethyl acetate	5.1	391.4	6.5	54.3	-	1.4	0.9	226.9	-
6	Non ident.	-	-	-	-	-	-	-	-	-
7	Methanol	-	-	-	-	-	-	-	-	-
8	Propanol	-	4166.6	145.8	63.2	-	24.1	12.8	86.5	-
9	Hexanal	11.7	2109.3	9.4	63.3	2.3	8.5	2.9	99.7	2.2
10	n-Butyl acetate	-	5684.2	61.6	60.6	-	4.9	1.9	80.3	-
11	Non ident.	5.6	1644.7	-	72.5	-	-	6.3	104.2	-
12	Non ident.	-	-	-	-	-	38.3	23.9	228.3	23.5
13	Butanol	4.1	4757.9	95.8	91.4	0.3	16.6	7.2	132.2	0.4
14	Non ident.	-	1095.8	-	35.5	-	0.9	0.5	42.5	0.3
15	Hexenal	116.6	7222.2	65.5	[illegible]	-	4.9	2.8	161.7	0.8
16	3-Methyl-1-Butanol	1.26	4423.6	46.5	65.7	-	10.4	4.1	99.5	-
17	2-Methyl-1-Butanol	-	3354.0	52.2	52.3	-	15.6	31.3	257.1	-
18	Pentanol	-	3750.0	50.1	75.0	-	8.0	-	100.0	-
19	Non ident.	150.2	-	-	3.6	-	7.2	-	53.6	-
20	Non ident.	240.0	-	-	-	-	-	-	-	-
21	Hexil acetate	7.9	2500.3	-	92.9	-	0.4	-	301.5	-
22	Non ident.	-	-	-	-	-	12.0	-	-	-
23	Non ident.	0.1	2539.1	16.4	39.2	-	8.3	2.8	-	-
24	Non ident.	-	-	-	-	-	-	-	-	-
25	Non ident.	-	-	-	-	-	-	-	-	-

See Table 4.

Table 6. The degree of concentration or less of aroma during the production of concentrated apple juice with aroma recovery

Sample	Aroma		
	$t_{R_{6,54-18,98}}$	$t_{R_{19,05-46,35}}$	Total
Initial (fresh) juice	100*	100*	100*
Aroma--stripped juice	24.2	7.7	22.0
Concentrated juice	7.6	1.3	7.3
Stillage	21,4	8.4	20,2
Aroma concentrate	5550	1880	6500

* Calculated on the basis of initial juice before aroma-stripping = 100

Table 7. The mass fraction change of some aromatic components in the production of concentrated apple juice with aroma recovery

Sample	Aroma component (ppm)			
	Hexanal	2-methyl-butyric acid ethyl ester	Benzaldehyde	Nonanol
Initial juice before aroma-stripping	100*	100*	100*	100*
Aroma-stripped juice	32.1	30.4	0	23.5
Concentrated juice	4.3	2.6	0	0
Stillage	15.5	37.1	0	62.0
Aroma concentrate	2876	13350	678	833

*Calculated on the basis of initial juice before aroma-stripping = 100

juice. These two aroma components, as the leading components of apple aroma, should not be retained either in the stillage or in the concentrated juice.

The presence of hexanal in the concentrated juice is unusual, considering the fact that hexanal is a relatively easily volatile substance, so it is difficult to axplain how it could have been retained in concentrate juice after evaporation. The explanation may lie in a change in the activity coefficient of hexanal during evaporation or in interaction between some juice components and hexanal.

A comparison of the results showing the changes which occurred in the amount of aroma during the individual steps of the technological process of producing concentrated juice in the case of aroma recovery plant A, which were obtained by two methods of determination (a and b) shows considerable differences, especially with regard to the amount of aromatic components in the aroma-stripped juice, the stillage and the aroma concentrate. The relatively smaller amount of these components in the initial juice and in the aroma concentrate is certainly not a consequence of their disproportionately greater loss during the preparation of samples for analysis. While the degree of aroma concentration obtained by headspace analysis was 1:183, it was 1:65 when determined by the method with previous preparation of samples. Conversely, the result showing the degree of concentration to be 1:183 is probably too high, according to the results of investigation of the influence of the matrix water-apple juice (Table 8), by 14 % on the average.

The results obtained indicate that special attention should be given to defining the quality of an aroma concentrate and to the effective operation of the plant investigated.

TABLE 8. COMPARISON OF "HEADSPACE" ANALYSIS DATA FOR CONCENTRATED AROMA DILUTED WITH WATER AND FOR CONCENTRATED AROMA DILUTED WITH INITIAL JUICE

COMPONENT	PEAK AREA (MM^2) DILUTED WITH WATER	DILUTED WITH JUICE	DIFFERENCE JUICE-WATER (%)
ACET ALDEHYD	143.2	102.2	- 28
PROPANAL	4.0	4.0	0
NON IDENT.	4.0	1.0	- 75
BUTANAL	14.5	21.0	+ 31
ETHYL ACETATE	117.7	97.0	- 17.5
NON IDENT.	-	-	-
METHANOL	-	-	-
PROPANOL	17.5	18.0	+ 2.7
HEXANAL	102.0	73.9	- 27.6
BUTYL ACETATE	99.4	78.0	- 21
NON IDENT.	6.4	4.0	- 37
NON IDENT.	-	-	-
BUTANOL	119.6	124.2	+ 3.7
NON IDENT.	39.0	26.0	- 33
HEXANAL	84.0	88.25	+ 4.8
3-METHYL-1-BUTANOL	72.15	70.2	- 2.7
TOTAL	823.55	708.25	- 14

*THE PEAK AREAS OF THE CORRESPONDING COMPONENTS OF THE JUICE WITH WHICH THE AROMA CONCENTRATE WAS DILUTED WERE SUBTRACTED PREVIOUSLY.

Conclusion

In aroma stripping of apple juice, it is sufficient to evaporate 12 % of the total amount of water to be evaporated during concentration, since approximately 90 % of the aroma components of the juice are then separated.

When the process is conducted properly and on an appropriate plant, the recovery and concentration of aroma components is sufficiently effective with regard to the "total" aroma and to the groups of "highly" volatile and "poorly" volatile components. However, disruptions in the quantitative relationships between individual components inevitably occur in individual phases of concentrated juice production and during aroma concentration.

Headspace analysis, as a simple and fast method, has been shown to be suitable for observing changes in the composition and concentration of aroma components during the production of concentrated apple juice with aroma recovery, especially if the influence of the matrix is duly taken into consideration.

Summary

The efficiency of operation of four different plants for aroma recovery during the production of apple juice concentrate was investigated. The research involved evaporation plants with different aroma recovery systems (A, B, C, D). Samples were taken from these plants in several characteristic steps of the technological process. Determination of apple aroma was carried out by gas chromatohraphy, for which the samples were prepared by distillation and extraction, and by the headspace technique. The results obtained showed that the degree of aroma concentration varied depending on the type of plant and on the method of aroma determination used. In the case of aroma determination by head-space analysis, the degree of "total"aroma concentration 1:183 in plant A, 1:127 in plant B, 1:61 in plant C and 1:28 in plant D was obtained.

The results obtained by gas chromatography with samples previously prepared by distillation and extraction amounted to 1:65, for plant A, compared to 1:183 determined by headspace analysis. This can be attributed to the loss of a part of the aroma which occurs during the preparation of samples for analysis. A check of the influence of dissolved solids on the results concerning the concentration of individual aroma components revealed that the results obtained using headspace analysis for quantitative determination were, on the average, higher by 14 % than the actual results for the so-called "total" aroma.

References

1. H.A.C.THIJSSEN,Advances in Preconcentration and Dehydration of Foods,Fundamentals of Concentration Processes,Applied Science Publishers LTD,London (1974).

2. H.BAIER,J.TORAN,Confructa-Studien,29,No.I,32-34 (1985).

3. W.PILNIK,Flüssiges Obst,32,526,(1965).

4. D.SULC,Flüssiges Obst,4,1 (1970).

5. D.SULC,Confucta Studien,28,No.III,300-317 (1984).

6. H.J.BIELIG and H.GRÜNDING,Proceedings:Topics in Flavour Research,Würzburg,173-199 (1985).

7. H.SUGISAWA,C.CHEN and K.NABETA,Analysis of volatiles, in P.Schreier(ed),W.de Gruyter,Berlin-New York,357 (1984).

8. F.CHIALVA,G.GABRI,P.A.P.LIDDLE and F.ULIAN,Journal of HRC&CC,5,4,182-188 (1982).

9. F.DRAWERT,W.HEIMANN,R.EMBERGER and R.TRESSEL, Chromatographia,2,77-78 (1969).

10. G.BAUMANN and K.GIERSCHNER, Riechstoffe,Aromen, Körperpflegemittel,3,62-68 (1974).

11. A.POZDEROVIC and T.LOVRIC, Prehambeno-tehnoloska i biotehnoloska revija,244,(2-3),101-108 (1986).

NEW RESULTS ON THE MOLECULAR ENCAPSULATION OF FLAVOURS BY CYCLODEXTRINS

L. Szente and J. Szejtli [1/]

Among the most recent and sophisticated flavour formulation processes cyclodextrin complexation represents a special way of encapsulation :entrapment of flavours on the possible lowest scale,which is called molecular encapsulation. In this process every single flavour constituent is included into the molecular cavity of cyclodextrin host molecule. The most significant practical advantages of the application of the molecular encapsulation in the food industry are the following:

Protection of flavours and fragrances against:
-evaporation,heat induced decomposition
-atmospheric oxydation
-light induced transformations

Elimination or reduction of :
-hygroscopicity
-amount of fibres and other non-flavoury undesired components
-unpleasant tastes and odours
-microbiological contaminations

Technological advantages:
-stable,standardizable compositions,with long storability
-simple dosing and handling
-reduced packing and storage costs
-manpower savings,simple economic technology

Major consequences of the cyclodextrin complexation

1. Transformation of liquids to solid form

One of the most obvious change occuring after cyclodextrin complexation of oily,liquid flavours is their microcrystalline,solid appearance. X-ray powder diffractometry proved,that the cyclodextrin complexed essential oils and other liquid flavour concentrates appeared to be solid,free-flowing powders of rather high crystallinity.A remarkable feature of cyclodextrin complexed flavours over the microencapsulated ones is their negligable hygroscopicity even under high humidity. The moisture sorption and clumping tendency of complexes and microcapsules were compared after a two-day storage under Rel.Humidity of 96% at room temperature. The results of hygroscopicity test are listed in Table 1.

[1/] Cyclodextrin Research Laboratory of CHINOIN Works Ltd.
H-1026 Endrödi S.38-40. Budapest,Hungary

Table 1. Moisture sorption of complexed and microencapsulated essential oils upon storage under 96% R.h. at room temperature

Samples tested	Moisture uptake in %						
	0.h	4.h	8.h	12.h	24.h	36.h	48.h.
peppermint-BCD[x]	0	0	2	5	5	8	8
peppermint-MC[xx]	0	8	11	14	21	26	33
lemon oil-BCD	0	0,2	0,7	1,1	4	4,5	6
lemon oil-MC	0	2,3	7	9,5	13	17,5	19
bergamotte-BCD	0	0,4	0,8	0,8	1,5	2,6	4
bergamotte-MC	0	0,8	2,2	5	11	14,3	15
vanilla-BCD	0	0,3	0,3	0,7	1,6	3	5,5
vanilla-MC	0	0,7	0,9	3,5	6	8,8	12

[x]BCD:beta-Cyclodextrin complex
[xx]MC: microcapsulated sample

2. Improved heat stability of flavours complexed

The molecular encapsulation of flavours by cyclodextrins reduces the equilibrium vapour tension of their volatile constituents. The macroscopic signs of this inclusion phenomena can be seen in heat stability tests in which complexed and adsorbed flavour samples are subjected to heat stresses. The most illustrative heat stability test is the Thermal Evolution Analysis/TEA/ that provides quantitative data on the rate of thermal release of volatiles. As an illustration of the usefulness of TEA-analyses, Table 2. summarizes the comparative heat stability of flavours in complexed and adsorbed form.

Table 2. Comparative heat stability of complexed and adsorbed flavours by Thermal Evolution Analysis

Samples	Loss of volatiles expressed as percentage of the total flavour content of formulations				
	85-90°C	130-145°C	150-160°C	175-190°C	220-250°
garlic-BCD[x]	2	5	12	20	23-25
garlic-A [xx]	94	2	-	-	-
caraway-BCD	3	5	15	22-25	30
caraway-A	88	6	-	-	-
onion-BCD	5	7	13	28	36-39
onion-A	90	4	-	-	-
dill-BCD	2-4	9	17	23-26	30
dill-A	96	-	-	-	-
grape fruit-BCD	3	7	16	28	36-38
grape fruit-A	94	3	-	-	-

[x]BCD:beta-Cyclodextrin complex
[xx]A: on lactose adsorbed flavour

The above promising heat-stability data indicate that cyclodextrin complexed flavours in dry,solid state may be stored for a longer periode of time without loosing considerable amount of volatile constituents. To get quick informations on the storability of the flavour-cyclodextrin complexes,samples were kept at 40° and 60°C for two weeks following the changes of the total flavour content as a function of time.The results of these "accelerated storage" studies are listed in Table 3.

Table 3. Accelerated storage stability of complexed and adsorbed flavours

Samples	Flavour content of the solid formulations/%/					
	0. day	7. day 40°C	14. day	0. day	7. day 60°C	14. day
jasmin-BCD[x]	9,8	9,4	9,6	9,8	9,0	8,8
jasmin-A[xx]	10	2,2	0,6	10	0,9	-
anise-BCD	9,2	9,4	9,2	9,2	9,0	9,2
anise-A	10	4,4	1,2	10	1,7	0,5
thyme-BCD	9,8	9,6	9,6	9,8	9,0	8,6
thyme-A	10	4,6	3,3	10	2,2	1,8
marjoram-BCD	9,0	8,7	8,9	9,0	8,4	8,0
marjoram-A	10	5,1	1,9	10	3,4	-
grapefruit-BCD	8,5	8,8	8,3	8,5	8,2	7,7
grapefruit-A	8,2	3,7	2,2	8,2	1,5	-

[x]BCD: beta-Cyclodextrin complex

[xx]A : lactose adsorbate

Storage experiments were also carried out at 80 and 100°C,but according to our observations at such high temperatures solid-phase chemical transformations may take place,between certain flavour components and cyclodextrin-hydroxyls./Monoterpene alcohols and phenolic compounds appear upon high temperature-storages,with the simultaneous formation of cyclodextrin-acetates,as a result of solid-phase transacylations./

3. Improved stability of flavours against oxydation

The majority of formulations of flavours also aim at the protection of some of their constituents against atmospheric oxydation.The stabilizing potency of cyclodextrin complexation can be characterized by using the classical Warburg-technique. The flavour cyclodextrin inclusion complexes and the corresponding adsorbates were exposed to pure oxygene in Warburg-apparatus at 37°C,and the oxygene consumption of samples was measured from time to time.The results of the Warburg measurements proved that the oxygene uptake of flavours entrapped in cyclodextrin was almost negligable,while the adsorbates of the same flavour content showed a fast and considerable oxygene consumption.

After the oxygen treatment, flavours were reextracted from complexes and mixtures,and their identity was checked by TLC and GLC respectively.The chromatographic investigations gave further support to the Warburg manometrical measurements,indicating that cyclodextrin complexation enables the storage of flavours even in pure oxygene atmosphere without considerable changes in their compositions.

4. Improved stability of flavours to light

The molecular encapsulation of essential oils and sveral flavours was found to improve the resistence of their light-sensitive constituents against daylight and ultraviolet irradiation. The photodecomposition of adsorbed and complexed flavours was tested both in solid state and in eqeous solutions. The results of the UV_{365}-light stability test are demonstrated to the example of model flavours,being easy to determine by UV-photometry. Table 4. illustrates the residual chemically identic flavour content of adsorbed and complexed citral,cinnamaldehyde and beta-ionone.

Table 4. Results of the UV_{365nm}light exposure test of solid flavour complexes and adsorbates at 22°C.

Samples	Residual flavour content of formulations/%/						
	0.h	4.h	8.h	12.h	24.h	36.h	48. hours
citral-BCD [x]	9,4	9,0	9,1	8,8	9,0	8,6	8,5
citral-A [xx]	9,7	6,1	4,0	2,4	0,7	-	-
cinnamaldehyde-BCD	9,0	9,1	9,0	8,7	9,0	7,0	6,6
cinnamaldehyde-A	9,0	4,4	3,4	3,7	2,0	2,2	-
beta-ionone-BCD	13,3	13,1	11	11	10,1	9,7	10,2
beta-ionone-A	13,5	10	8,8	4,8	3,3	3,7	3,4

[x]BCD:beta-Cyclodextrin complex

[xx]A: lactose adsorbate

The above formulations were also exposed to UV-irradiation in aqueous solutions. As it has been observed,the protective effect of molecular encapsulation against light induced alterations was significantly smaller,than experienced in solid-state investigations. This might be due to the partial release of entrapped flavours upon the dissociation of inclusion complexes in aqueous solutions.

5. Improved shelf-life flavour formulations via cyclodextrin complexation

The long term storage stability tests of flavour-beta-cyclodextrin inclusion complexes have been run since 1977,under non-stress conditions e.g. room temperature,normal humidity and light conditions. The cyclodextrin flavour complexes have been tested for their total flavour content as well as for their chemical composition.

The degree of preserving potency of cyclodextrin complexation is expressed with the simple comparison of the total flavour content of samples determined in 1977, and 1987, respectively. /See Table 5./

Table 5. Change of the flavour content of beta-cyclodextrin complexed spice flavours after a ten year storage under normal conditions

Samples	Flavour content/1977/ in %	Flavour content/1987/ in%
garlic-oil-BCD	10,2-10,5	10-10,3
onion oil "	10,4-10,6	10,1-10,4
caraway oil "	10,5	9,8-10,2
thyme oil "	9,4-9,3	8,9-9,2
lemon oil "	8,9-9,1	8,0-8,4
carrot oil "	8,9-9,0	7,9-8,3
anise oil "	9,0-9,2	9,0-9,3
peppermint "	9,4-9,7	9,0-9,2
marjoram "	8,8-9,3	8,0-8,2
tarragon "	10,2-10,5	9,0-9,4
cinnamon "	9,3-9,5	8,0-8,3
dill oil "	9,3-9,7	9,0-9,3
mustard oil "	10,8-11,0	11,0-11,2
smoke flavour"	8,0-8,2	4,5-4,7

6. Limits of the use of cyclodextrin complexation in flavour formulations

The limits of the application of molecular encapsulation of flavours by cyclodextrins are related to the inclusion phenomena between the host and guest molecules, and can be listed as follows:

-Cyclodextrin complexation can provide a limited flavour content in the powdery formulation, due to the ratio between host and guest, based on their molecular mass. The usual flavour content of flavour-beta-cyclodextrin complexes is around 10 %/w/w/, in contrast to the 50-70% flavour content of microcapsules.

-The water solubility of inclusion complexes is generally lower, than that of the starting cyclodextrin itself. Therefore cyclodextrin complexed flavours can limitedly be used for instant drinks.

-Cyclodextrins can also act as artificial enzymes, enhancing the rates of hydrolyses of certain ester-type flavour components, resulting in undesired alteration of flavours. /This effect can be particularly dangerous in the complexation of fruit flavours./

7. Legislation status of beta-cyclodextrin as food additive:

After having been declared as an enzymatically modified starch derivative, beta-cyclodextrin has been approved as a food additive in the

following European countries: Hungary/1983/,France /1985/, the Netherlands, Belgium, Luxembourg, West-Germany and Spain /1987/ using the CHINOIN toxicological and metabolic studies with the own data of AVEBE. Many other European governments are at present considering beta-cyclodextrin for use in the food industry.
Beta-cyclodextrin in food grade is available in tons at CHINOIN, after the industrial production at reasonable price of the product has been solved in Hungary.

STUDY ON THE EFFECT OF FAT IN MEAT FLAVOUR FORMATION

R. Schrödter, J. Schliemann and G. Wölm

Introduction

The formation of meat flavour requires definite conditions both in natural meat and in simulated systems. Material and processing parameters are involved. This means that the necessary flavour precursors like amino acids, peptides, sugar, purin and pyrimidin compounds, organic acids and others must be available. Furthermore the thermic treatment conditions as far as pH-value, heating time, temperature and water content concerned are to be regulated in order to produce a specific spectrum of meat aroma compounds. Cooking and roasting notes can be formed in this way and therefore the model systems should be adapted to the natural conditions. However, every system should contain the following groups of flavour precursors:

- a N-component such as pure amino acids mixture or hydrolysed proteins
- a C=O-compound, as a rule reducing sugars
- a S-source, first of all S-containing amino acids like cysteine or methionine

Using these ingredients the controlled Maillard reaction can only produce a basic meat aroma enhanced by addition of such ingredients like glutamate and 5'-ribonucleotides. Further factors which play an important role are fat and fat deterioration products. The formation of a species specific flavour depends on these fat components. According to the meat aroma pyramide [1, 2] we can describe the resulting flavour as top-note of the basic meat aroma impression.

Sensory evaluation of meat extracts from different species corroborates the hypothetical assumption of having only a basic meat flavour but not a species specific aroma. There is one exception: the poultry extract. The content of fat in poultry extracts is higher than in other meat extracts and it leads to a specific flavour impression. According to our own experiments significant differences of defatted broths (beef and pork) couldn't be recognized. For example by addition of polyunsaturated fatty acids or only sunflower oil Rothe et al. [3] changed the flavour of broths and they demonstrated the fat influence using the sensory profiling method.

---

1) Zentralinstitut für Ernährung Potsdam-Rehbrücke, Akademie der Wissenschaften der DDR, Arthur-Scheunert-Allee 114/116, Bergholz-Rehbrücke, DDR-1505

Material and methods

During the preparation of meat aroma concentrates on the basis of protein hydrolysates we couldn't estimate an influence of tallow and lard on the applicated aqueous phase. Compared to it in model systems we got a poultry-like flavour note in the aqueous phase using oxidized sunflower oil [4, 5]. In order to investigate this phenomenon we prepared the following samples:

- a basic model without oil with a meaty aroma note (see Fig. 1)
- the next ones are prepared with addition of edible oils and animal fats in fresh and oxidized form (Fig. 2). The peroxide value of oxidized oil samples is adjusted to about 30. The first column in Fig. 2 shows the PV before and the second one after Maillard reaction. The fat component was separated after reaction and the aqueous phase was analyzed by sensory evaluation and by gas chromatography.

		%
– basic composition		
	glutamic acid	45
	glucose	45
	cysteine HCL	5
	salt	5
– reaction conditions		
	water content	50 %
	heating time	60'
	temperature	100°C
	pH - value	6,0
– addition		
	fat / oil component	15 %

Fig. 1. Composition of basic meat aroma model

Nº	fat / oil	PV before	PV after
		reaction	
0	without addition of fat	—	—
1	sunflower oil	1,1	~0
2	-"- oxidized	34,8	13,1
3	lard	1,9	0
4	-"- oxidized	30,4	1,1
5	beef suet tallow	3,4	~0
6	-"- oxidized	32,2	~0
7	rape - seed oil	1,0	1,0
8	-"- oxidized	30,2	14,8
9	chicken (poultry) fat	0,3	~0
10	-"- oxidized	40,0	10,2
11	soybean oil	1,5	~0
12	-"- oxidized	28,5	7,7
13	corn oil	4,3	1,3
14	-"- oxidized	26,5	3,9

Fig. 2. Number of applied oil/fat phases and data of peroxide value (PV)

Results and discussion

Provided that we have an influence of fat the sensory difference between each of the aqueous phases of the 14 samples and the basic model must be significant.

At first we estimated significant differences between the fresh and oxidized samples in all cases and the changes in aroma of the aqueous phase were obvious. But is there any difference between oil samples compared with the basic sample (No. 0)? The aqueous phase of sample No. 1 - prepared with fresh sunflower oil - shows no difference to the basic model. Both - aqueous phases from the sample No. 0 and 1 - were united and used as standard. Fig. 3 contains the results of the difference test:

In general the aqueous phase of the sample prepared with oxidized fat can be distinguished significantly from the standard. The samples prepared with fresh fat are not different with the exception of the corn oil sample. But according to the ranking test and furthermore by descriptive analysis we recognized an off-flavour in corn oil samples so that we left them out of further consideration.

- 2,5 solution of test samples in water and adjusted at 0,8% NaCl content
- standard: mixture of sample № 0 and 1 (aqueous phase)
- differences between standard and modelsample
  (XXX – 99,9% XX – 99% X – 95% significant level, n = 20)

Model sample (aqueous phase)		difference in smell	difference in taste
2	sunflower ox.	XXX	XXX
3	lard	—	—
4	-"- ox.	XXX	XXX
5	tallow	—	—
6	-"- ox.	X	X
7	rape	—	—
8	-"- ox.	X	X
9	chicken	—	(X)
10	-"- ox.	X	XX
11	soy	—	—
12	-"- ox.	X	XX
13	corn	XXX	XXX
14	-"- ox.	XXX	XXX

Fig. 3. Results of difference test between the aqueous phase of standard solution (mixture of sample No. 0 and 1) and oil preparations

From the results of duo-trio-test we can draw the following conclusions:

- we have no effect of fresh oil whereas the oxidized one has an influence
- the samples differ in meat flavour impression

This may be due to the presence of deterioration products of fatty acids as well as to an increase of products caused by Maillard reaction.

First of all we had to estimate the quality of changing of the basic aroma within the oxidized samples. Using the ranking test and the descriptive analysis we got the results demonstrated in Fig. 4.

The best meat aroma sample was prepared by application of oxidized sunflower oil followed by poultry fat, soybean oil, rapeseed oil, tallow and lard. Because of the off-flavour the corn oil sample took the last place. 95 % and 99 % of confidence interval is given in the picture.

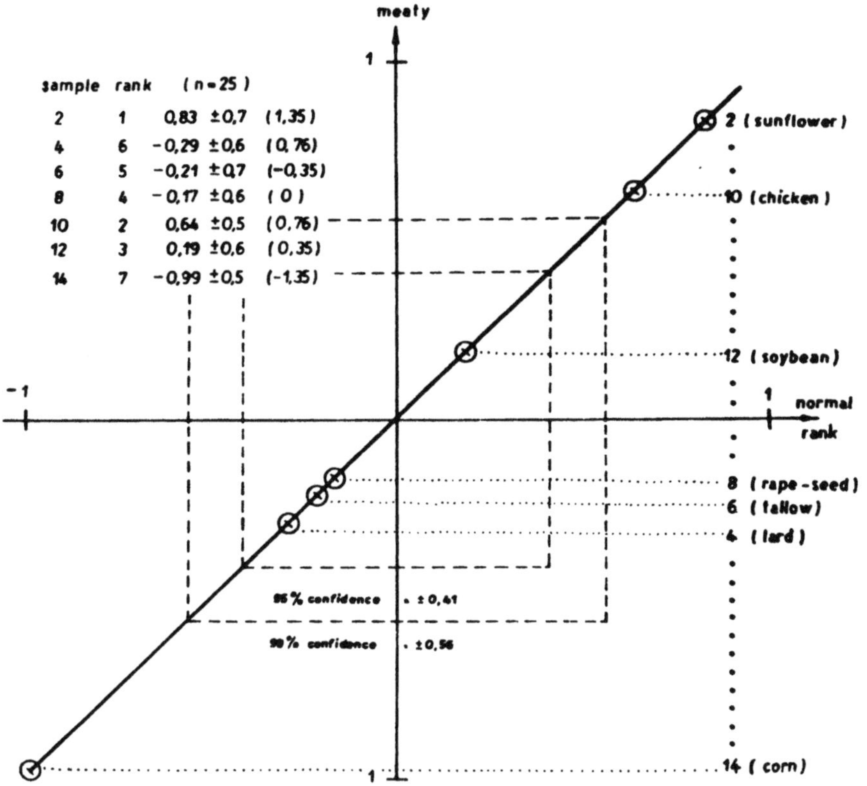

Fig. 4. Results of normal ranking test and transformation of the data according to Fisher and Yates, Table XX[1)]

From the rank order found we can draw the further conclusion that the positive influence exists on the aqueous phase of our meat model by such fats like sunflower oil, chicken fat and soybean oil which contain a high amount of polyunsaturated fatty acids especially linoleic acid (Fig. 5). In contrast to this the other fats applied also influence the basic meat aroma but in an unwanted manner as can be seen in Fig. 4.

Apart from the sensory tests we carried out the gas chromatographic analysis. We isolated the aroma compounds from the aqueous phase of the sample by combinated steam distillation and ether extraction [6]. After concentration the gas chromatogram was processed on a Varian 2800 using a Carbowax 20 M column (10 feet, 2 mm Ø). The comparison of a detected peak in the GC-profiles shows no change in the qualitative composition but there are differences in the quantitative response.

1) We should like to express our thanks to Dr. sc. K. Hoppe for supporting us in the statistical interpretation

fatty acid	lard	beef suet tallow	poultry fat	sunflower oil	corn oil	soybean oil	rape - seed oil
C_{12} Lauric	—	0,9	—	—	—	—	—
C_{14} Myristic	0,9	4,6	0,5	—	—	—	0,7
C_{16} Palmitic	29,0	25,4	19,0	5,2	10,0	9,4	2,4
C_{18} Stearic	14,0	19,8	7,5	4,3	2,4	3,3	2,1
C_{20} Arachidic	—	0,8	—	0,5	0,5	1,4	1,4
$C_{14:1}$	0,2	0,5	—	—	—	—	—
$C_{16:1}$ Palmitoleic	3,5	2,1	3,0	0,2	0,5	0,5	—
$C_{18:1}$ Oleic	46,0	41,0	47,0	21,8	31,1	21,7	24,0
$C_{18:2}$ Linoleic	8,5	2,8	21,5	55,0	39,0	52,8	15,2
$C_{18:3}$ Linolenic	0,5	0,5	1,5	0,5	1,0	7,5	5,2
$C_{20:4}$ Arachidonic	1,7	0,2	—	—	—	—	—
$C_{22:1}$ Erucic	—	—	—	—	—	—	49,7

Fig. 5. Fatty acid composition of applied oils and fat (g/100 g)

In the next step we selected peaks with high differences; the numbers are given in Fig. 6. To underline the conclusions drawn by sensory test we compared the changes in selected peaks of fresh fat with the oxidized one against the standard mixture (see Fig. 7). The variations and the increase of peak intensity from the sample without oil addition to the oil sample are obvious.

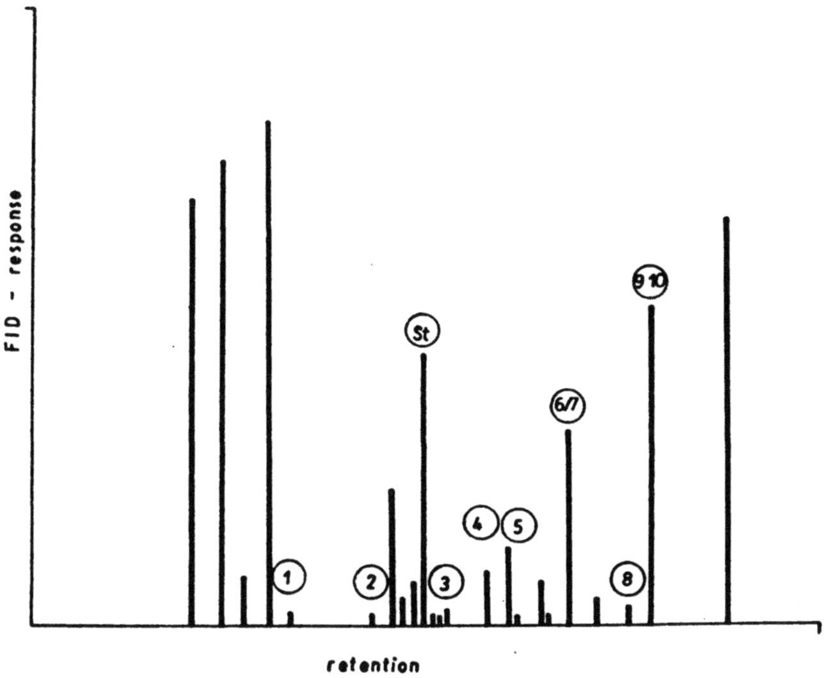

Fig. 6. Gas chromatogram of the aqueous phase (sample mixture) and number of selected peaks

The other samples show similar behaviour and properties so that we draw the conclusion that the influence of oxidized fat to the aqueous phase of our 'meat flavour' model system attributes to the increase of relevant precursors which are formed by degradation of fat and deterioration of polyunsaturated fatty acids.

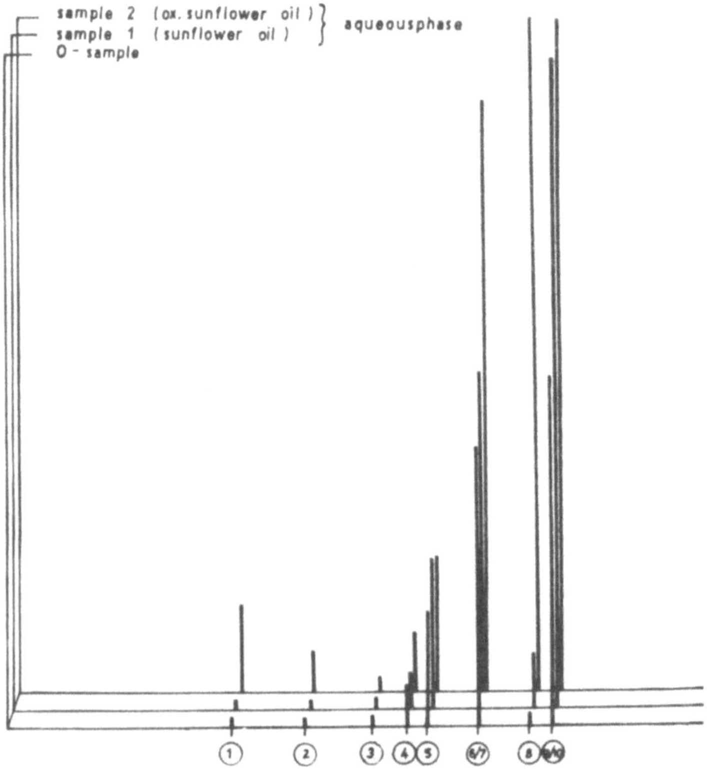

Fig. 7. Comparison of GC-profiles of standard mixture (O) and aqueous phases of oil-treated samples (sunflower oil fresh and oxidized)

Summarizing the results we have no direct influence of fat to the aqueous phase of our meat aroma model but an indirect one by its deterioration products. Further studies in the identification should unveil the nature of the recognized meat flavour changes.

References

[1] Solms, J.: Fleischwirtschaft 48 (1968), 287

[2] Sulser, H.: Extraktstoffe des Fleisches, Wiss. Verlagsges. GmbH, Stuttgart, 1978

[3] Rothe, M., E. Kirova and G. Schischkoff: Nahrung 25 (1981),543

[4] Schrödter, R., G. Wölm and J. Schliemann: Nahrung 26 (1982), 625

[5] Schrödter, R., J. Schliemann and G. Wölm: Nahrung 30 (1986), 799

[6] Römer, G. und E. Renner: Z. Lebensmittel-Unters. u, -Forsch. 156 (1974), 329

SULPHUR CONTAINING COMPOUNDS IN MEAT AROMA MODEL SYSTEMS

T. A. Misharina and R. V. Golovnya[1]

Meat products are of great importance in nutrition of mankind. During many years the structures of volatile compounds responsible for specific meat flavour are investigated to create artificial compositions with such aroma [1-3]. It is known that meat aroma volatiles are generated by Maillard reaction during heating of water solutions containing amino acids, carbohydrates and sulphur precursors especially cysteine [4-10]. In numerious studies of such type of model systems sulphur containing compounds were found as the most important in meat aroma. For example the reaction of 4-hydroxy-5-methyl-3(2H)furanone with hydrogen sulfide produces furanthiols and their disulfides which have meat odour properties [11]. These compounds were not identified in natural cooked meat aroma, but were found in volatile products of thermal degradation of thiamine [12], yeast extract [13] and reaction cysteine and thiamine with hydrolyzed vegetable protein without carbohydrates [14]. Earlier we have identified bis(2-methylfuryl)disulfide in volatiles of the model system containing bakery yeast autolysate and xylose [15].

Industrial protein hydrolysates are convenient and perspective sources for factory production of simulated meat flavour. In present paper volatiles of four model systems obtained from hydrolysates of yeast, bovine plasma or vegetable proteins were analysed and compared to find common components in meat aroma.

Experimental

Sample preparation.

Model systems were obtained from the water solutions containing amino acids mixtures of different sources, cysteine and xylose. There were four sources of amino acids mixtures: autolysate of bakery yeast(1), bovine plasma (2), acidic (3) and enzymic (4) hydrolysates of soya protein. The ingredients were transferred in the glass ampules,sealed heated at 100°C for 2 hr. The reaction mixtures had intensive aroma of boiled or stewed meat.

Volatiles from the reaction mixtures were extracted for 2 hr in a Likens-Nickerson apparatus with 50 ml diethyl ether [16]. Organic

1) Institute of Organo-Element Compounds, Academy of Sciences of USSR, Vavilova 28, 117 813 Moscow/USSR

extracts were treated with 1,0% water solution HCl (10ml), then 1,0% sodium hydrocarbonate solution (10ml) and dried over Na_2SO_4. The solvent was removed by distillation through a 30 sm Vigreux column until 100 μl residue. We obtained concentrates of volatile compounds of four model reaction mixtures : sample 1 - from yeast autolysate, sample 2 - from bovine plasma, sample 3 - from acidic soya hydrolysate, sample 4 - from enzymic soya hydrolysate. Each sample of concentrates (sample 1-4) was placed in glass capillary (10 sm x 1 mm), sealed and stored at -20°C for further analysis.

Gas chromatography (GC).
The samples 1-4 were analysed using 5717 A Hewlett-Packard gas chromatograph equipped with a flame ionization detector(FID) and a flame photometric detector (FPD) operated in S-mode with 394 nm narrow band filter. 1 μl of each sample was injected in two glass capillary columns (50 m x 0,32 mm i.d.) coated with 0,5 μm film of OV-101 or OV-17 stationary phases. Helium as carrier gas (1,5 ml/min) at inlet split ratio of 1:30 was used. The separation of samples 1-4 started at 50°C then the temperature was raised up to 250°C with a rate of 4°/min. The temperature of injector, FID and FPD was 200°C. 1 μl of n-alkanes C_5-C_{17} solution were analysed under similar conditions, their retention times were used for the calculation of retention indices[17].

Gas chromatography - Mass spectrometry (GC-MS).
GC-MS analyses of volatile compounds of samples 1-4 were performed using AEI-DBMS-1073 instrument. The separation of components was carried out on the fused silica capillary column (25 m x 0,32 mm i.d. , film thickness 0,25 μm) coated with SE-54 stationary phase connected with ion source. The flow-rate of helium - carrier gas was 1,5 ml/min at inlet split ratio 1:30. The separation was started at 50°C (4 min) then the temperature was programmed at a rate 2°/min up to 250°C (10 min). The injection port, the ion source and GC/MS interface were kept at 200°C. The energy of electron impact was 28 eV, scan range 30-300 amu, scan speed 3 s per decade mass. At the same conditions in the cases of GC peaks overlaping GC-MS analysis was carried out with using a computer-controlled multipeak monitoring unit.
Sensoric determination of meat-like aroma compounds in GC-eluate was carried out using OV-101 column at the same GC conditions as described above.

Results and discussion.

We have studied the volatiles concentrates (samples 1-4) of model systems which had been prepared from different sources: bakery yeast autolysate, bovine plasma, acidic and enzymic hydrolysates of soya proteins. The volatiles concentrates had an intensive meat aroma. This aroma remained if the samples 1-4 were diluted with water in 10^8-10^9 times. Using a sensory evaluation of GC eluates of samples 1-4 were found the common fractions with meat-like aroma. These fractions had retention indices in intervals 840-866 i.u.,885-900 i.u., 945-966 i.u., 1030-1040 i.u.,1130-1150 i.u. and 1500-1620 i.u. on GC capillary column with OV-101 stationary phase. The majority of GC peaks corresponded to sulphur compounds as it have been established by GC analysis with FPD. As an example, two chromatograms obtained on OV-101 column with FID and FPD for the sample 4 are given in Fig.1.
The identification of volatiles of the samples 1-4 was carried out by comparing retention indices and mass spectra of components with the data available on reference compounds [18]. There were many isomeric furanes and thiophenes containing alkyl, alkoxy, alkylthio, hydroxy, mercapto and carbonyl groups as have been determined using GC - MS. Some of these compounds with molecular masses 114, 126, 128 and 140 are given in Table 1. Such isomers have similar values of retention indices and the same mass fragments in mass spectra (Table 1). In order to identify such volatiles chemical modification of flavour concentrates was performed. Preparation of TMS derivatives from compounds containing hydroxy and mercapto groups was carried out using treatment of the samples with hexamethyldisilazane [19]. Trimethylsililation caused to the increase of molecular masses of compounds by 72 amu and of retention indices by 116-120 i.u. for hydroxy and 290 i.u. for mercapto containing compounds (Table 2). This procedure allowed us to identify inert to HMDS furanes and thiophenes with alkoxy and sulfide groups which overlap with hydroxy and mercapto containing compounds. The treatment of the samples with sodium borohydride followed by trimethylsililation of alkohols made it possible to identify carbonyl compounds (Fig.2). On the Fig.2 one can see chromatograms of the fractions which are changed by chemical modifications. Structures of isomeric furanes and thiophenes were determined using GC and GC-MS analysis of initial and chemically modified samples with account of regularities of retention indices and mass spectra changes after chemical modification.

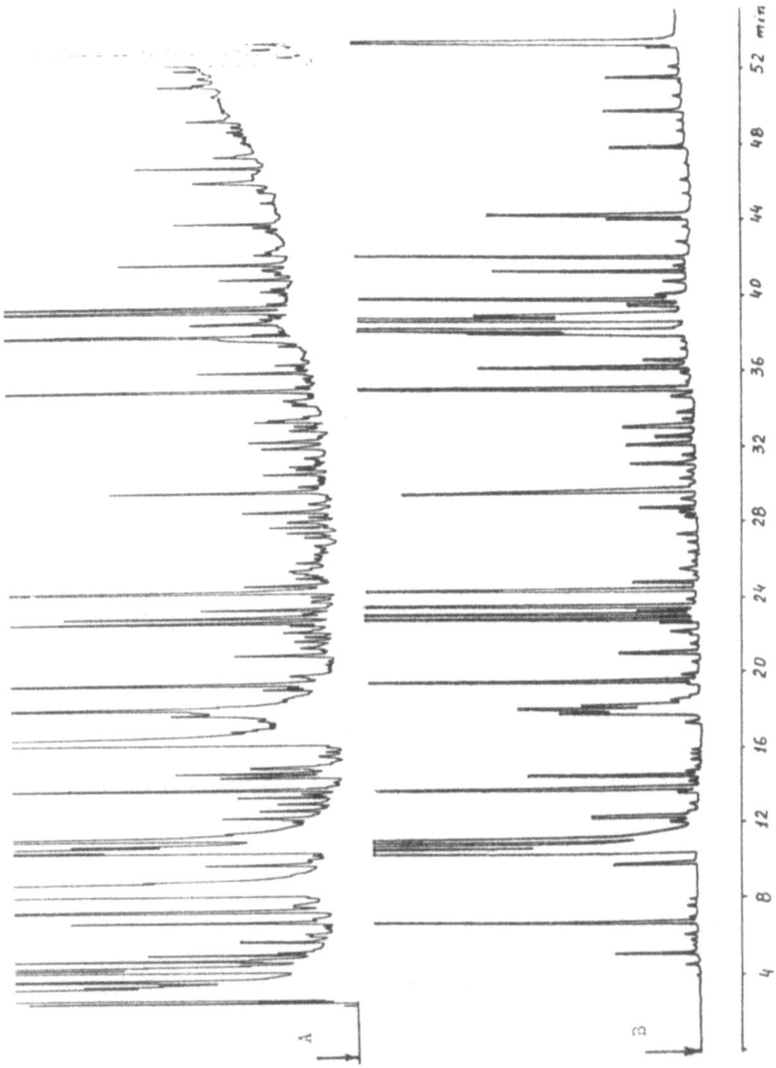

Fig.1. Chromatograms of meat aroma concentrate of model system from enzymic hydrolysate of soya proteins (Sample 4) registrated with : A - flame ionization detector, B - flame photometric detector. Glass capillary column with OV-101 (50 m x 0,32 mm, film thickness 0,5 µm); 50° - 4°/min - 250°.

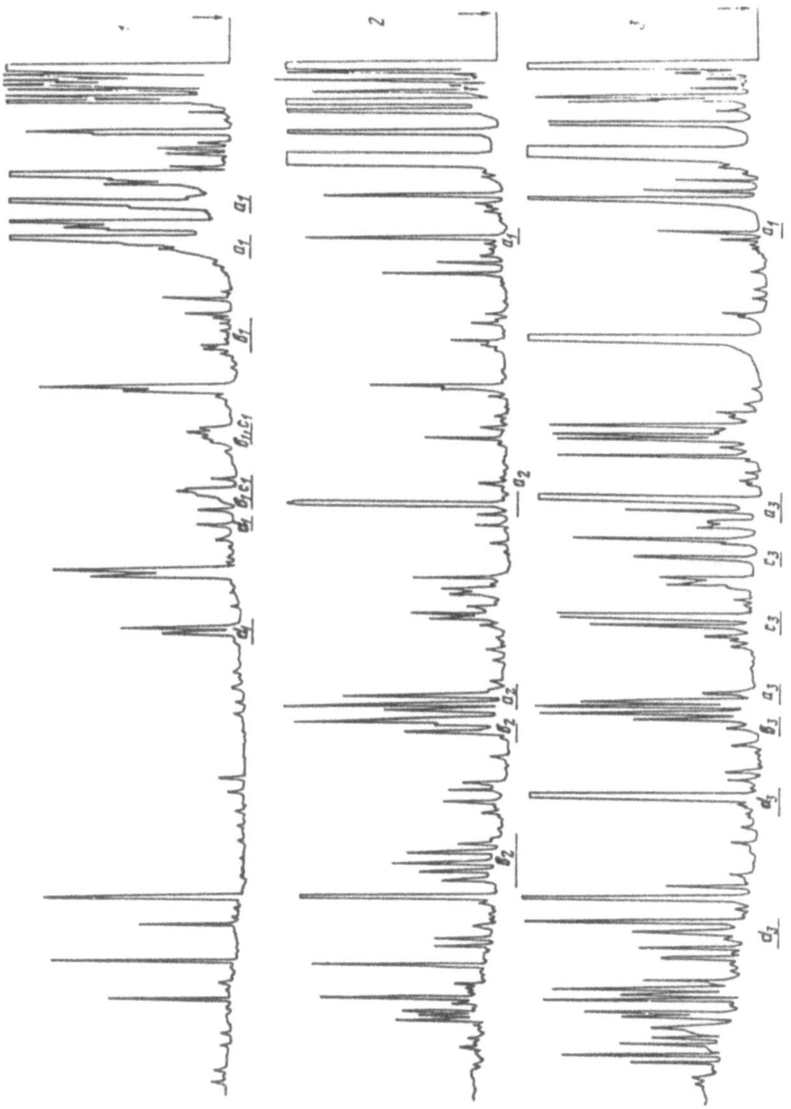

Fig.2. Chromatograms of meat aroma concentrate of sample 1.
1 - sample 1; 2 - sample 1 treated hexamethyldisilazane;
3 - sample 1 treated sodium borohydride and hexamethyldisilazane. Fractions containing compounds with the molecular masses:
1 : a_1-114; b_1-128; c_1-126; d_1-140;
2: a_2-186; b_2 -200;
3: a_3-186; b_3-200; c_3 -200; d_3 -214
are marked on chromatograms. Fused silica capillary column with SE-54 (25 m x 0,32 mm, film thickness 0,25 μm);
40°(4 min)-2° /min - 250°.

Table 1. Some series of isomeric thiophenes and furans with similar mass spectra and retention parametres.

Isomers with molecular mass 114 MS fragments(m/z): 114,113,97,85,81,53	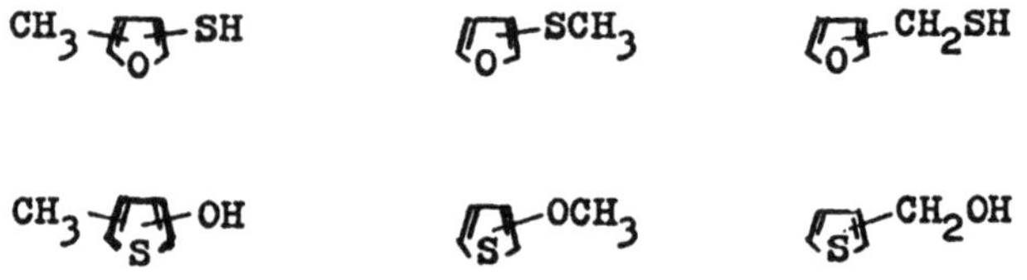
Isomers with molecular mass 126 MS fragments(m/z): 126,125,111,97,45	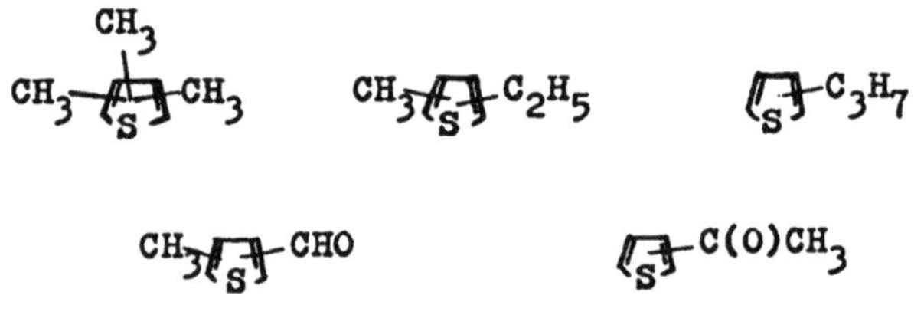
Isomers with molecular mass 128 MS fragments(m/z): 128,113,97,95,85, 81,53	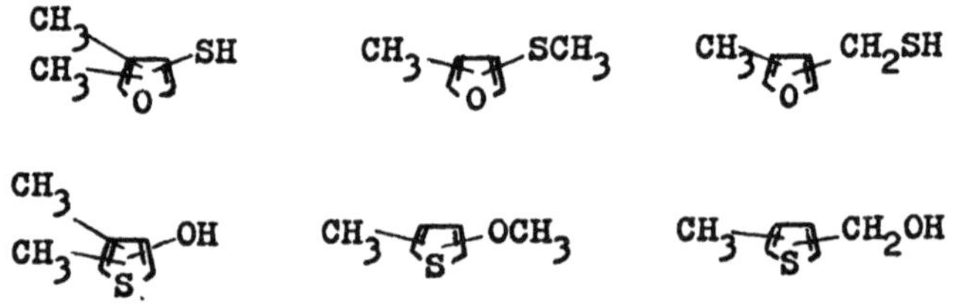
Isomers with molecular mass 140 MS fragments(m/z): 140,139,125,111, 97,45	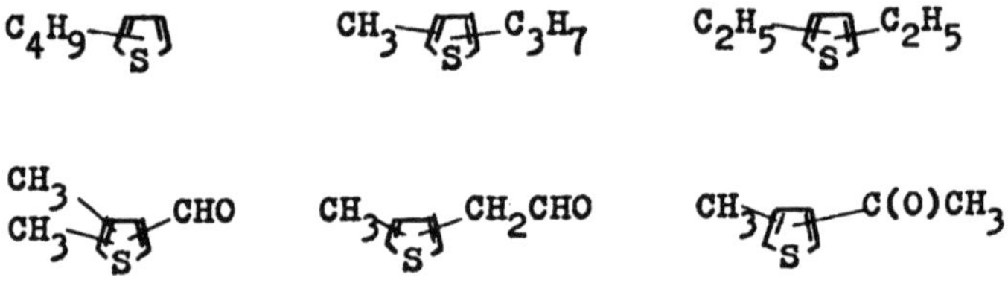

Table 2. The changes of retention index values and mass spectra of standard compounds after chemical modification.

Compound	Retention index in program. temperature	Index difference $\delta I_{i,j}$	Mass spectrum m/z
1. $-CH_2SH$	907		81,53,114
2. $-CH_2SSi(CH_3)_3$	1194	$\delta I_{2,1}=287$	81,186,171, 53
3. C_6H_5CHO	950		105,106,77
4. $C_6H_5CH_2OH$	1040	$\delta I_{4,3}=90$	79,108,107, 77,91
5. $C_6H_5CH_2OSi(CH_3)_3$	1156	$\delta I_{5,4}=116$ $\delta I_{5,3}=206$	165,180,91
6. $-CHO$	995	$\delta I_{7,6}=35$	111,112,83, 45,58
7. $-CH_2OH$	1030	$\delta I_{8,7}=120$	85,114,113, 97,81,53
8. $-CH_2OSi(CH_3)_3$	1150	$\delta I_{8,6}=155$	97,171,186, 141,73
9. CH_3-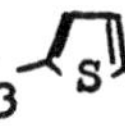$-CHO$	1110	$\delta I_{10,9}=8$	125,126,97, 53,45
10. CH_3-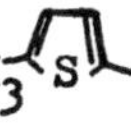$-CH_2OH$	1118	$\delta I_{11,10}=121$	128,111,113, 97,95,45
11. CH_3-$-CH_2OSi(CH_3)_3$	1239	$\delta I_{11,9}=129$	111,200,185, 97,45
12. $-C(O)CH_3$	1085	$\delta I_{13,12}=-31$	111,126,83, 43,39
13. $-CH(OH)CH_3$	1054	$\delta I_{14,13}=116$	85,128,113, 110,84,45
14. $-CH(CH_3)OSi(CH_3)_3$	1160	$\delta I_{14,12}=75$	185,111,200, 73,45

The study of volatiles compositions of the samples 1-4 showed that all four model systems have produced a number of common compounds. In Fig.3 two chromatograms of samples 1 and 2 are given. The chromatograms were obtained in the same GC conditions using capillary column coated with OV-101 as stationary phase. These chromatograms allow to see that both of the model systems contain a great number of the same compounds that also was proved by the GC and GC-MS analysis as described above. In addition we have investigated the volatile substances produced by model system without amino acids only from xylose and cysteine. The GC-MS results obtained allowed to estimate the substances have been formed in this two-components system. These compounds are asterisked in Table 3. They are present in samples 1-4.

The comparison of the volatiles compositions of samples 1-4 gives a possibility to clear up the role of the substances common for all samples in the formation of meat flavour of model systems under study. Some furanes and thiophenes are particularly important for meat flavour because most of these compounds possess cooked meat aroma [19]. Some of found compounds possessing green, butter-like, vegetable-like, coffee and another food-like odours contribute to the characteristic cooked meat odour of model systems. All aroma compounds can be separated on three groups: aliphatic compounds, thiophenes and furanes. 2,3-Butanedione and 2,3-pentanedione were found in the samples 1-4 and in model system of xylose/cysteine. These substances possess a butter-like odour and are important both as potential contributors in meat aroma and as possible precursors of mercaptoketones (Table 3). The mercaptoketones are generated by the reaction of α-diketones and hydrogen sulfide [5] and can intensify the meat flavour [20]. According to the data [6] diketones can be formed from the xylose as precursor. Besides the above mentioned identified compounds, methional was found in the all investigated model systems. It has the odour of cooked vegetables and potato and is an important constituent of food aroma. Methional is a product of the Strecker degradation of methionine and it was found in natural meat [21]. During the Strecker degradation of cysteine the formation of hydrogen sulfide and 2-mercaptoacetaldehyde occurs [22]. It was shown that mercaptothiophenes are formed during condensation of two mercaptoacetaldehyde molecules [23]. The thiophenes identified in model systems are listed in Table 3. As was shown earlier, the thiophene derivatives are Maillard reaction products of cysteine/cystine with ribose [5] or xylose [4,10]. Thiophene aldehydes were found in natural meat [16,17], in reaction pro-

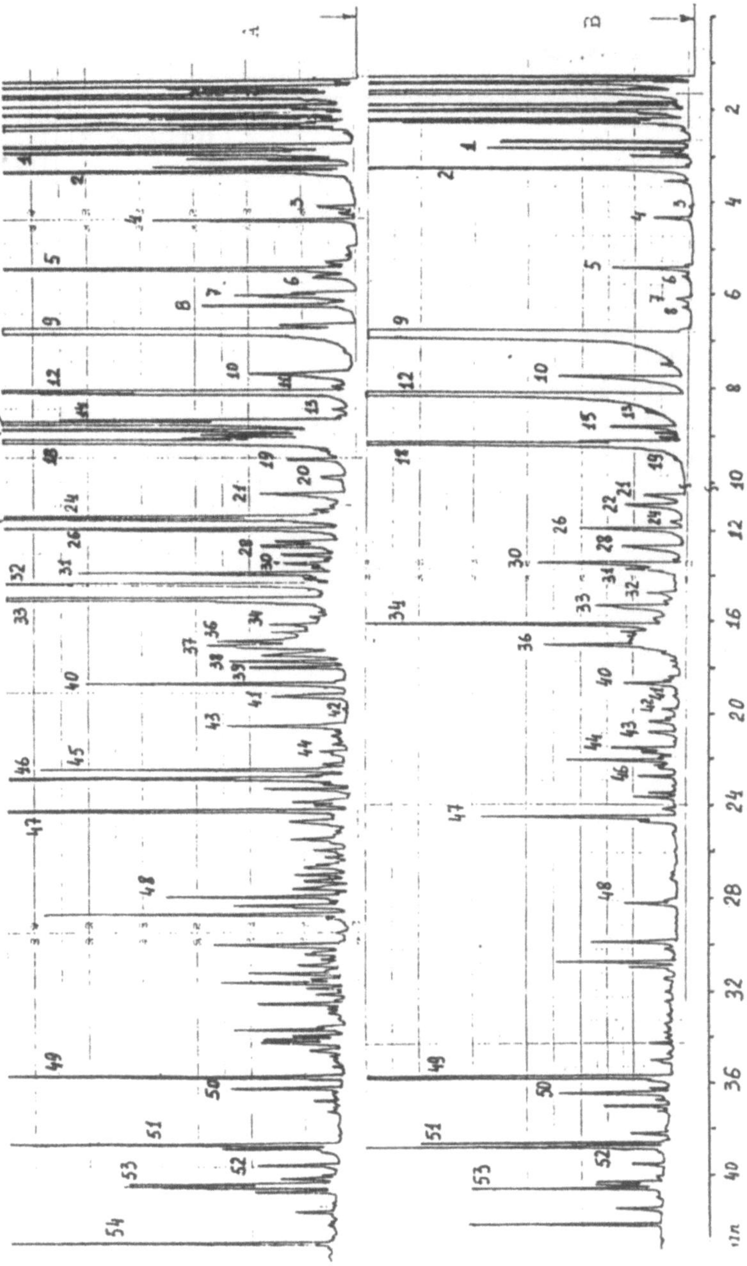

Fig.3. Chromatograms of concentrates of model systems with meat aroma (samples 1 and 2) obtained from bakery yeast autolysate (A) and bovine plasma (B). Glass capillary column with OV-101 (50 m x 0,32 mm, film thickness 0,5 µm);50°-4°/min-250°. Peak numbers correspond with those in Table 3.

Table 3. Volatile compounds identified in four model systems with meat flavour.

Compound	Peak (Figure 3)	Retention index in prog. temperature
Aliphatic compounds		
2,3-Butanedione *	1	630
Mercaptoethanal *	2	650
2,3-Pentanedione *	5	727
2-Mercaptobutan-3-one	6	745
3-Mercaptopentan-2-one *	15	875
2-Mercaptopentan-3-one *	16	880
1-Mercapto-1-methylthioethane	19	899
3-Methylthiopropanal	20	903
Thiophene derivatives		
2-Methylthiophene *	7	782
Tetrahydrothiophen-3-one	21	930
3-Mercaptothiophene	25	960
2-Methyltetrahydrothiophen-3-one *	26	968
2-Formylthiophene*	30	998
2-Methyl-3-mercaptothiophene *	34	1038
2-Acetylthiophene	36	1070
2-Methyl-4-mercaptotetrahydrothiophen-3-one*	38	1084
(two isomers)	39	1096
2-Methyl-5-formylthiophene	40	1100
3-Methyl-2-formylthiophene*	42	1109
2-Methyl-5-acetylthiophene	43	1118
Furane derivatives		
2-Methyltetrahydrofuran-3-one *	3	710
Furfural *	9	806
Furfuryl alcohol	11	838
2-Methyl-3-mercaptofuran *	12	849
2-Methyl-4-mercaptofuran	13	853
2-Methyl-3-mercapto-4,5-dihydrofuran	17	885
Furfuryl mercaptan *	18	890
2-Methyl-3-furylsulfide	24	947
Furfurylmethylsulfide	27	976
2-Methyl-5-acetylfuran	33	1021

Table 3 cont'd

(2-Methyl-3-furyl)methyldisulfide	44	1133
2-Methyl-3-oxa-8-thiabicyclo/3,3,0/-1,4-octadiene	45	1148
Furfuryldisulfide	47	1170
Bis(2-methyl-3-furyl)disulfide*	49	1509
(2-Methyl-3-furyl)-(2-methyl-4,5-dihydro--3-furyl)disulfide	50	1544
(2-Methyl-3-furyl)furfuryldisulfide	51	1558
Bis(2-methyl-4,5-dihydro-3-furyl)disulfide	52	1598
Bis-furfuryldisulfide *	53	1607

ducts of cystine with 2,3-butanedione [7] and in products of thermally degraded cysteine and cystine [23,24]. The formation pathways of thiophene derivatives in the Maillard reaction were discussed in the number of papers [5,12,23,25].

For the first time the 2-methyl-3-mercaptothiophene and two isomers of 2-methyl-4-mercaptotetrahydrothiophen-3-one were found by us in model systems (samples 1-4) and in reaction products of xylose and cysteine. These three substances have a meat-like odour and may be important constituents of meat flavour.

Twenty derivatives of furan were identified in samples 1-4 (Table 3). Thio-derivatives of furan are of great importance as the components of meat flavour. Mercaptofurans and their disulfides are the key compounds of meat aroma [11,14,19]. Furthermore, these compounds have extremely low odour threshold values. For instance, the odour threshold of bis(2-methyl-3-furyl)disulfide was found to be 2 parts in 10^{14} parts of water [12]. 2-Methylfuranethiol-3 and its disulfide are major components of model systems volatiles (see Fig.3,peaks 12 and 49). Probably, these substances may be responsible for meat flavour of model systems under study. Earlier 2-methylfuranethiols and some difuryldisulfides were found in reaction products of 4-hydroxy-5-methyl-3(2H)furanone with hydrogen sulfide [11], in products of degraded thiamine [12], in volatile components of a yeast extract [13], in a model system obtained by refluxing a water mixture of cysteine/thiamine/hydrolyzed vegetable protein (carbohydrate-free) [14]. Bis(2-methyl-3-furyl)disulfide was identified also by us in volatile components of Maillard reaction of a bakery yeast autolysate with xylose [15]. Probably, 2-methylfuranethiols can arise from reaction of 2-methyldihydro and 2-methyltetrahydrofuran-3-ones with hydrogen sulfide. Similar reaction of furfural or 4-hydroxy-5-methyl-3(2H)furanon with hydrogen sulfide is known [11,26]. Furfural and substituted furanons are formed during the xylose decomposition [4-6,10]. Oxidation of mercapto containing compounds leads to the disulfides mixture. Seven disulfides with methyl, furyl, furfuryl and thienyl groups were found in our model systems. These compounds possess a meat-like and coffee-like aroma;they are significant contributors in meat flavour.

Thus, the investigation of model systems with meat aroma obtained by heating of hydrolysed yeast,animal or vegetable proteins with cysteine and xylose showed that independently on protein sources,a number of similar compounds including the key-compounds with meat aroma have been formed.

Literature

[1] Schutte,L.: Crit.Rev.Food Technol.4(1974)457-505.

[2] Golovnya,R.V. and M.Rothe: Nahrung 24(1980)2,141-154.

[3] Schrödter,R.,and G.Wölm: Nahrung 24(1980) 2,175-183.

[4] Ledl,F. and Th.Severin: Chem.Microbiol. Technol. Lebensm. 2(1973) 155-160.

[5] Mulders, E.J.: Z.Lebensm. Unters.-Forsch. 152(1973)193-201.

[6] Scanlan, R.A., S.G. Kayser, L.M. Libbey, and M.E. Morgan: J.Agric. and Food Chem. 21(1973) 673-675.

[7] Hartman, G.J. and C.-T. Ho: Lebensm.-Wiss.u.-Technol. 17(1984) 171-174.

[8] Hartman, G.J.,J.D.Scheide, and C.-T. Ho: Lebensm.-Wiss.u.-Technol. 17(1984) 222-225.

[9] Hartman, G.J., J.T. Carlin, J.D. Scheide, and C.-T. Ho: J.Agric. and Food Chem. 32(1984) 1015-1018.

[10] Mussinan, C.J. and I. Katz: J.Agric. and Food Chem. 21(1973) 1, 43-45.

[11] Van den Ouweland, G.R.M., and H.G. Peer: J.Agric. and Food Chem. 23(1975) 3,501-503.

[12] Buttery, R.G., W.F. Haddon, R.M. Seifert, and J.G. Turnbaugh: J.Agric. and Food Chem. 32(1984) 3, 674-676.

[13] Ames, J.M., and G. Mac Leod: J.Food Sci. 50(1985) 125-131.

[14] Evers, W.J., H.H. Heisohn, B.J. Mayers, and A. Sanderson. In: Phenolic,Sulfur and Nitrogen Compounds in Food Flavors. ACS Symp.Ser. 26,(Ed.) Charalambous, G. and I.Katz. ACS Washington. 1976,p.184.

[15] Golovnya, R.V., T.A. Misharina, V.G. Garbusov, and F.A.Medvedev: Nahrung 27(1983) 3, 237-249.

[16] Likens, S.T., and G.B. Nickerson: Am.Soc.Brew.Chem.Proc. (1964) 5-13.

[17] Van den Dool, H., and P.D. Kratz: J.Chromatogr. 11(1963)463-467.

[18] Garbuzov, V.G., T.A. Misharina, A.F. Aerov, and R.V. Golovnya: Z.Analit.Khim.40(1985)4, 709-720.

[19] Misharina, T.A., S.V. Vitt, R.V. Golovnya, and V.M. Belikov: Z. Analit.Khim. 41(1986) 10, 1876-1881.

[20] Oser, B.L., and R.L. Hall: Food Technol. 25(1972) 1, 35-41.

[21] Van Straten, S., Fl. Vrijer, and J.C. Beauveser: Volatile Compounds in Food. Krips Repto, Zeist 1977.

[22] Lien, Y.C., and W.W. Nawar: J. Food Sci. 39(1974) 911-918.

[23] Shu, C.-K., M.L. Hagedorn, B.D. Mookherjee, and C.-T. Ho: J. Agric. and Food Chem. 33 (1985) 3, 438-446.

[24] Boelens, M., L.M. van der Linde,P.J. de Valois, H.M. van Dort, and H.J. Takken: J.Agric. and Food Chem. 22(1974) 6, 1071-1076.

[25] Shibamoto, T.: J.Agric. and Food Chem. 28(1980) 2, 237-243.

[26] Shibamoto, T.: J.Agric. and Food Chem. 25(1977) 1, 206-208.

COMPARATIVE STUDY ON SELECTED SENSORY AND GAS CHROMATOGRAPHIC DATA OF CHICKEN FLAVOUR

J. Schliemann, G. Wölm and R. Schrödter[1)]

Within the last few years we have dealt with possibilities and problems of the simulation of chicken flavour. On the basis of a thermal reaction between vegetable protein hydrolysates, carbohydrates, sulphur sources and slightly oxidized oils we tried to produce a chicken-like flavour concentrate.

Of course, the primacy of these investigations was to find relations between process and recepture parameters on one side and sensory properties of the simulates on the other side in order to optimize the process of concentrate production. But additionally we got such an amount of material that we became able to use it for the search for relations between sensory and analytical, especially gas chromatographic results. At present time these studies are in an intermediate state so that we can present only some temporary results.

Table 1: Statistical test plan for the optimization of concentrate production - process and recepture parameters

parameter \ level	$-\alpha$	-1	0	+1	$+\alpha$
x_1 = peroxide value	2,4	30	50	70	97,6
x_2 = amount of oil (% in dry mass)	0,5	6	10	14	19,5
x_3 = Cys-HCl (% in dry mass)	0,22	1,25	2,00	2,75	3,78
x_4 = Dry mass	16,2	30	40	50	63,8
x_5 = Reaction time (min)	6,2	20	30	40	53,8

1) Zentralinstitut für Ernährung Potsdam-Rehbrücke, Akademie der Wissenschaften der DDR, Arthur-Scheunert-Allee 114/116, Bergholz-Rehbrücke, DDR-1505

Our test program was a Box-Hunter plan with 5 variables each of them in 5 levels (Tab. 1). This program contains those influence factors that have been recognized as being very important for the concentrate production in different statistical preinvestigations. First: the peroxide value. It is well-known that mainly carbonyl compounds are intended to be of special importance for the chicken flavour. They come from degradation of the fat among the flavour precursors. Slightly oxidized oil therefore seems to be suitable to be used as precursor. Logically the amount of oil was likewise interesting and became our second influence factor. Furthermore important for meat flavours are sulphur containing substances. They are intended to be responsible for the basic meat flavour but not for the species specifity at the same extent. Hence the amount of Cys-HCL was our third factor. Last not least the dry mass and the time of thermal reaction were included in the plan.

The sensory evaluation was carried out by means of a multiple comparison. Both in odour and taste we evaluated the total quality and the partial impressions 'chicken-like', 'hydrolysate-like' and 'off-flavour'. Reference samples were the central samples of the plan. Each panelist had to decide whether the samples were better or worse than the references and to register the intensity of the differences.

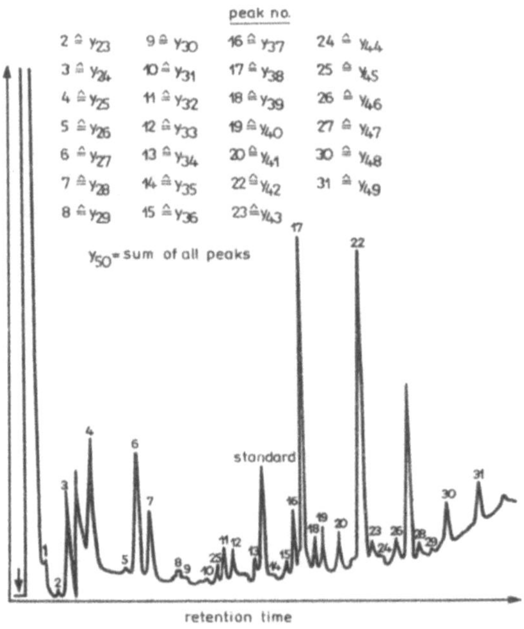

Figure 1: Gas chromatogram of the volatile compounds of a selected concentrate

According to our technical possibilities first of all we used packed Carbowax 20 M columns under temperature programmed conditions for the separation of volatile compounds. The result of the separation is shown in Fig. 1 and those peaks that were calculated statistically are mentioned. Additionally the total intensity of all peaks was considered. For isolating the volatile compounds we used a combined water vapour distillation and ether extraction following Renner and Römer.

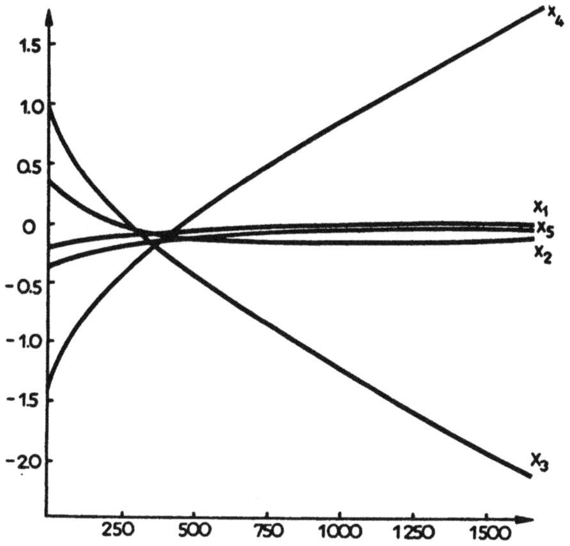

$$y_{38} = 860.5 - 11.61 x_1 + 0.035 x_1^2 - 27.47 x_2 + 0.773 x_2^2$$
$$+ 97.03 x_3^2 + 0.476 x_4^2 - 20.85 x_5 + 0.282 x_5^2$$
$$+ 0.163 x_1 x_4 + 0.369 x_2 x_4 - 17.35 x_3 x_4 + 3.299 x_3 x_5$$

Figure 2: Regression function of y_{38} (furfural, peak no. 17) and corresponding crest line analysis

Fig. 2 represents an example for the relationships between gas chromatography and the different influence factors. Having carried out a multiple regression analysis we got the following regression function for the y-value 38 which was identified as furfural. From this equation it becomes obvious that all parameters influence y_{38}. The crest line analysis according to Hoerl also to be seen in this picture proves this statement (both axis of this analysis are subdivided into the normatted values for the influence factors and the y-values). But in some cases the main effects (or single effects) of a variable are reduced by interaction effects as for example in the case of x_1

(peroxide value), x_2 (amount of oil) and x_5 (reaction time). In this example an increase of the Cys-HCl content leads to a decrease of y_{38}. That could mean that more Cys-HCl in the system causes either a stronger reaction with the aldehyde or an inhibition of the aldehyde formation. In both cases the result is a decrease in aldehyde concentration. x_4 is the dry mass and obviously an increase of it leads to an extended formation of furfural.

Table 2: Statistical data according to selected regression functions from y_{23} to y_{51} (GC)

y-value	value of determination(B)	average of y-values
y_{33}	0,9118	27,82
y_{34}	0,7528	23,22
y_{38}	0,9271	363,33
y_{39}	0,7557	35,52
y_{40}	0,8202	41,54
y_{41}	0,9729	52,97
y_{42}	0,9732	350,70
y_{44}	0,9040	15,09
y_{47}	0,8412	159,39
y_{48}	0,8950	48,40

Calculation of significances according to DUFEK (1971)

$p = 0{,}01 \rightarrow B = 0{,}3345$

Regarding the other gas chromatographic peaks in principle we got similar results. But because it is impossible to discuss all regression functions in detail we can only give a selection of y-values (Tab. 2). In this table only those values are listed, the values of determination of that are greater than 0.750. To give an impression of the order of magnitude of the corresponding peaks the average peak intensities are also listed.

In order to get relations between sensory and gas chromatographic results the gas chromatographic material was brought into a cluster analysis following Lance and Williams. The resulting dendrogram is to be seen in Fig. 3. Using a separation level of 2.5 we can observe 4 clusters with 1, 5, 10 and 36 samples. By statistical means the first cluster consisting of only one sample could be recognized as pull out. The separation of the clusters one from each other is marked by the dark lines.

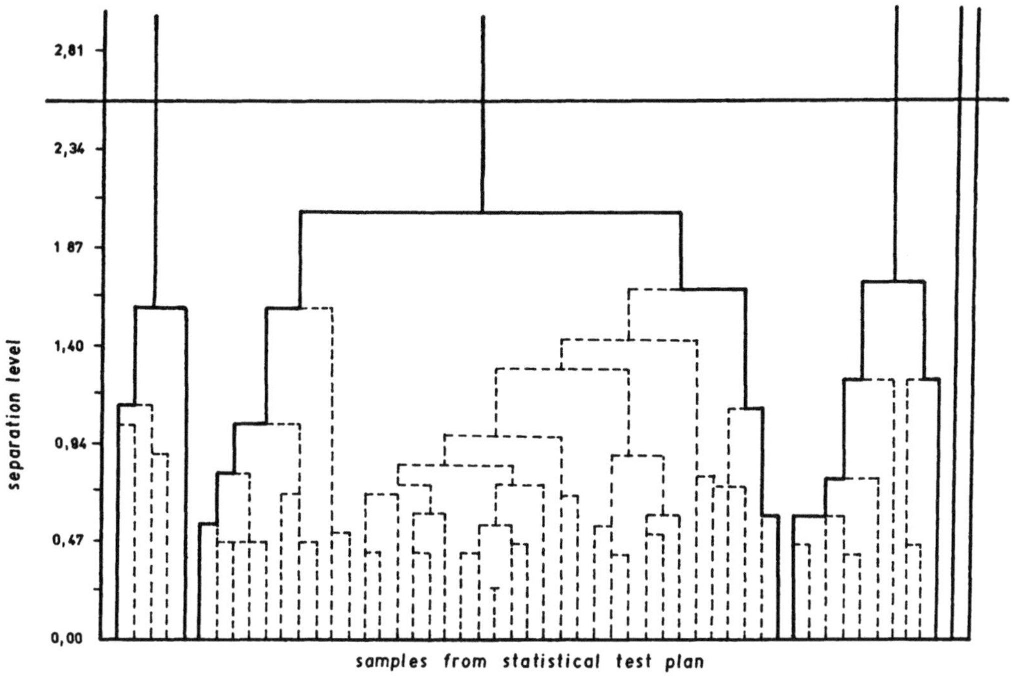

Figure 3: Dendrogram of the cluster analysis of the gas chromatographic data

The next step was to establish whether there is any coincidence between sensory and gas chromatography. We simply substituted the gas chromatographic data of each sample by the corresponding sensory ones. Because we were mostly interested in the behaviour of the partial impression 'chicken-like' we focussed our attention on what happened in this case (Fig. 4). In the odour a separation of the clusters one from each other is nearly impossible. The standard deviation marked by the circles are widely overlapping. That means that the gas chromatographic data of the clusters differ but the sensory ones do not. Hence, the GC data cannot be used for the characterisation of the odour impression 'chicken-like'. In the case of the taste impression a good separation of all clusters one from each other can be observed from the sensory point of view.

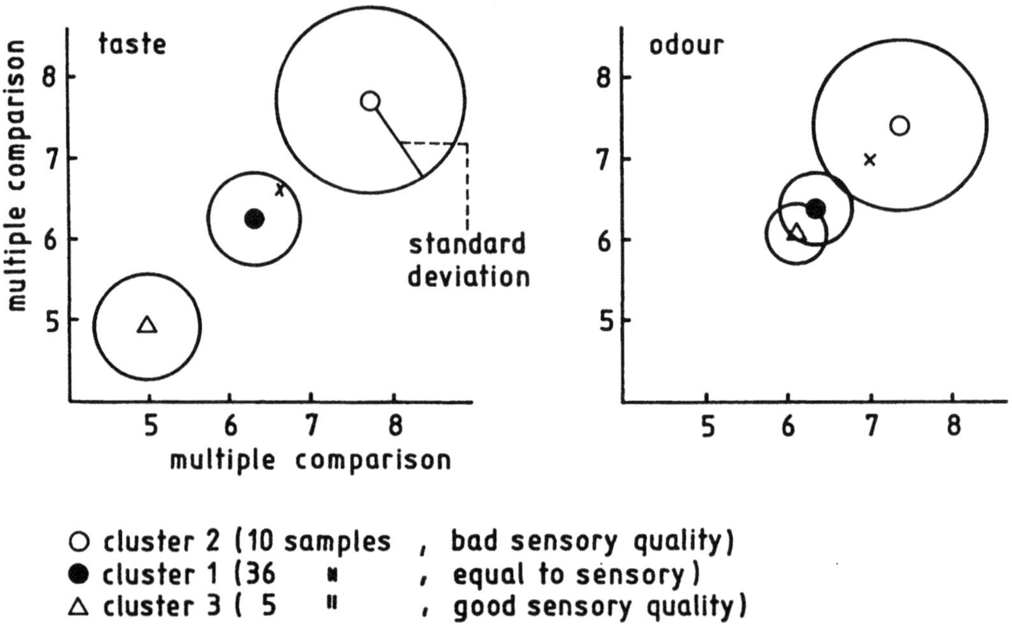

Figure 4: Cluster analysis of samples in the sensory impression 'chicken-like' (average and standard deviation)

Because the third cluster contains those samples with the best sensory evaluation we have concentrated on this cluster. Unfortunately we had to register that the peaks of the corresponding samples had only very low intensities. That means that these y-values (or volatile substances) are only as far useful for the characterisation of chicken-like taste as they aspire to a concentration minimum. They are not interesting for the chicken-like taste characterisation. But vice versa they are suitable for the description and characterisation of off-flavour notes. Similar consideration of the clusters for the 'hydrolysate-like' and 'off-flavour' impression proved this statement.

Of course, it is very problematic to discuss these or related results without knowing anything about the chemical identity of the separated volatile compounds. Consequently we recently investigated our concentrates by means of directly coupled gas chromatography and mass spectrometry using a fused silica column with Carbowax 20 M as stationary phase for the separation of isolated compounds.

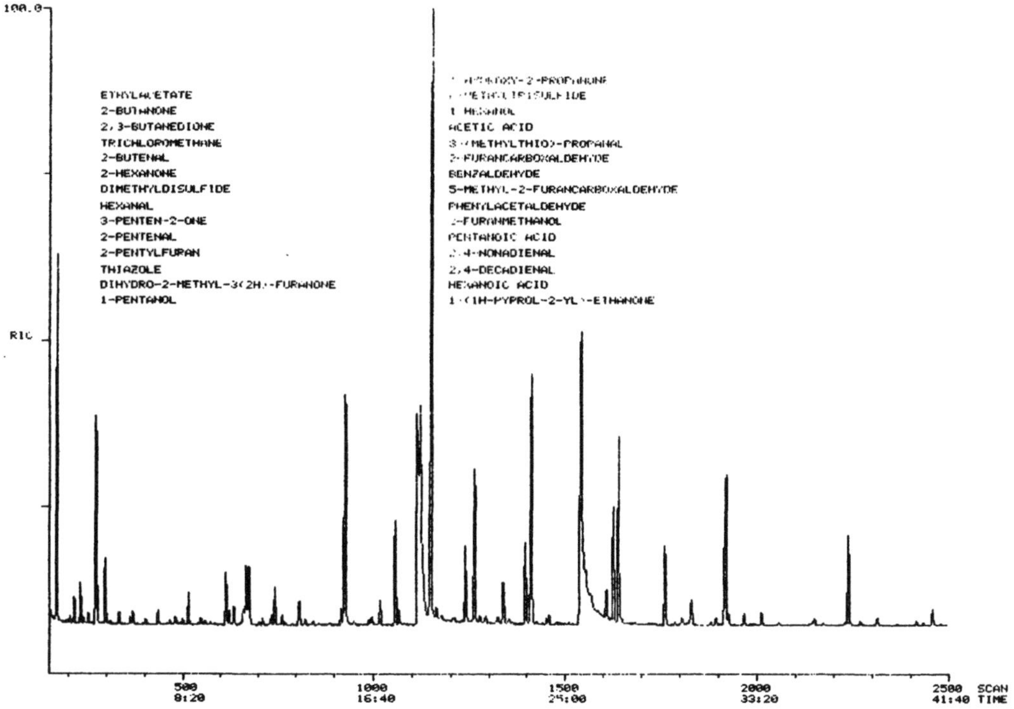

<u>Figure 5:</u> Volatile compounds of a chicken flavour concentrate identified by GC/MS

Fig. 5 illustrates the separation and it also contains a list of those compounds that have been identified surely. Additionally we found a number of further compounds the identification of which is doubtful at present time. This last group of substances obviously consists of numerous carbonyl and sulphur compounds. But returning to the listed substances in this figure we have to state that most of them are known from natural chicken flavour. Exceptions are 3-penten-2-one, the dihydro-2-methyl-furanone, thiazole and the acids. As far as we know from literature no one of these compounds has a chicken-like odour or taste impression. Gas chromatographic sniffing technique with our concentrates proved this expectation, thus supporting the results of the cluster analysis.

Last not least two additional remarks are necessary. First: the also identified 2,4-nona- and -decadienal which are known from literature to be mainly participated in chicken flavour formation have not been considered in our statistical evaluation because of their gas chromatographic retention behaviour and their low concentration. Second: for our GC/MS investigations the concentration of the volatile compounds was to small. The chances for better results in this field are shown in Fig. 6. It is a strongly enhanced chromatogram in comparison to Fig. 5 but it represents the same sample. This figure illustrates that there are numerous further volatile compounds which have to be considered in our studies. But in the actual case the signal/noise-ratio did not permit a strict identification by coupled gas chromatography and mass spectrometry. It will be a task of future to improve this situation.

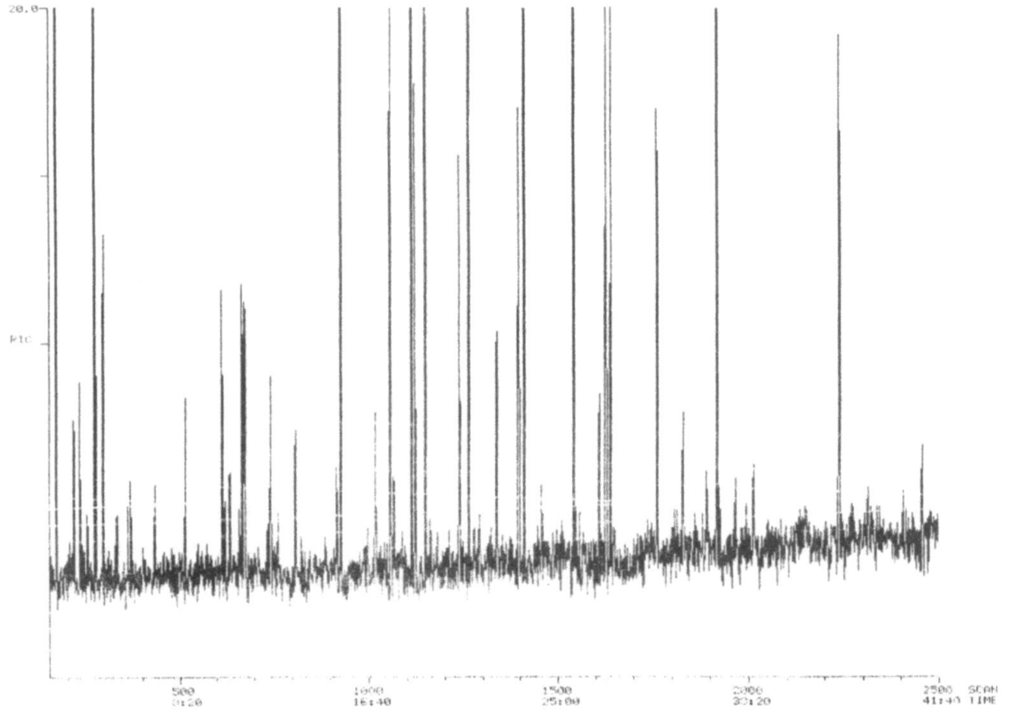

<u>Figure 6:</u> Repetition of Fig. 5 under strongly enhanced GC response conditions

Closing this short paper we can summarize in the following way:
We succeeded in the search for relations between selected parameters of chicken flavour concentrate production and gas chromatographic data of the concentrates. By means of a cluster analysis we additionally got some relations between sensory and gas chromatographic data. But we have to establish that all of the considered volatile compounds are not suitable for the characterisation of chicken-like aroma but very well suitable for characterising off-flavour notes. This can be import-

tant for the process control in the concentrate production. Nevertheless we are of the opinion that the selected way can also lead to the recognition of relations between characteristically chicken-like flavour notes and gas chromatographic data if some main correctures will be considered in future investigations.

BREAD FLAVOUR: QUALITATIVE AND QUANTITATIVE ANALYSIS

W. Grosch and P. Schieberle[1)]

Introduction

The flavour of bread depends on the formula, the conditions of the dough fermentation, the baking process and the storage [1]. To determine the influence of these parameters on the flavour, it would be of obvious advantage to have indicator substances that contribute significantly to the flavour of bread. Therefore, investigations were started to identify key aroma compounds of the bread flavour which could act as indicators and to develop methods for their quantification.

Odour unit concept

It is a general problem in flavour research that the progress in instrumental analysis has led to long lists of volatiles [2] but not to the evaluation of their relative importance to the flavours of foods.

The first approach to estimate the importance of a flavour compound in a particular food was the calculation of the ratio of its concentration to its flavour threshold [3] or to its odour threshold [4, 5]. The result was denoted "aroma value" [3], "odour value" [5] or "odour unit" [4]. The latter term will be used in the following discussion.

The odour unit concept was applied by Rothe and Thomas [3] and by Mulders [5] to the volatiles isolated from wheat bread. The data listed in Table 1 show, that from the substances tested by Mulders [5] in the headspace of wheat bread, acetaldehyde showed the highest odour value followed by dimethyl disulfide and some alcohols. Consequently the author concluded that acetaldehyde contributed the most to the aroma.

The critical points and the limitations of the odour unit concept have been discussed by Rothe [6], by Frijters [7] and by Meilgaard and Peppard [8]. However, these authors agree that it can be used as a

---

1) Deutsche Forschungsanstalt fuer Lebensmittelchemie, 8046 Garching/ F.R.G.

Table 1. Odour unit of some components present in the headspace of wheat bread [5]

Compound	Odour unit	Compound	Odour unit
Acetaldehyde	57	Dimethyl sulfide	5
Dimethyl disulfide	16	1,1-Diethoxyethane	4
Ethanol	11	3-Methylbutanol	3
2-Methylpropanal	10	2-Methylpropanol	2
3-Methylbutanol	10		

guide to aroma compounds that contribute significantly to the flavour of a food.

Determination of FD-factors

One drawback in the calculation of odour units is that it is very laborious, since quantitative and threshold data must be estimated for a great number of volatiles. Furthermore, in many cases, it is not certain whether the character impact compounds are among the volatiles identified. For these reasons, the odour unit concept was only applied to a relatively small number of foods. An impressive example is the work of Meilgaard [9] on the flavour of beer.

The drawbacks in the calculation of odour units are avoided by a combination of a chromatographic procedure with sensory analysis.

It is well-known that in an extract of the volatile compounds which have been isolated from a food, the odourants can be distinguished from the odourless volatiles by a combination of sensory analysis and gas chromatography. This method which is denoted as "GC-effluent sniffing" has been used increasingly in flavour research. However, GC-effluent sniffing detects all the volatile compounds whose concentrations in the GC-effluent are higher than their odour thresholds. It does not allow a differentiation between those odour compounds that contribute intensely to the flavour of a food and those which are only components of the background flavour.

The important odour compounds can be evaluated by the GC-effluent sniffing of a dilution series of the original aroma extract from the particular food. Two variations of this technique were developed by Acree et al. [10, 11] and by us.

Acree et al. [10, 11] used a video-terminal in addition to the gas chromatograph. They calculated CHARM-values on the basis of the duration of the sensory responses which were maintained during GC-effluent sniffing of dilutions of the original extract. CHARM-values are directly proportional to odour units.

We used a simpler method for the determination of the relative odour importance of the volatile compounds [12, 13]. The aroma extract is stepwise diluted with a solvent until no odour-active region is detected in the GC-effluent. The highest dilution at which a substance is still smelled is its flavour dilution factor (FD-factor)[1]. The undiluted sample has then, by definition, an FD-factor of one. For example, an FD-factor of 20 for a substance means that one volume of the initial concentrate was diluted by 19 volumes of the solvent, and that this was the highest dilution of which the substance was still smelled during the gas chromatographic procedure. In other words: the concentration of the odourant in the aroma extract was 20 times its odour threshold perceived by the GC-effluent sniffing. This reveals that the FD-factor of a flavour compound is proportional to its odour unit. Therefore, the flavour compounds occurring in an aroma extract are actually ranked according to their odour units. The new technique which is designated as "aroma extract dilution analysis" is less time consuming than the calculation of odour units from concentration and odour threshold data. Furthermore, and this is the great advantage in comparison to the calculation procedure, the determination of FD-factors indicates the potent odour compounds which should be identified.

Odourants of wheat and rye bread crusts

The aroma extract dilution analysis was applied to aroma concentrates which were prepared from the crusts of wheat and rye breads [14].

GC-effluent sniffing of extracts which had been previously freed from acidic compounds showed 31 odourants in the case of the wheat crust

---

1) The term "dilution value" (D-value) previously used [13, 14] was denominated "FD-factor" to distinguish it from the "decimal reduction value" (D-value) used in microbiology and enzymology.

and 43 in that of the rye bread crust (Table 2). There were also significant differences in the aromagrams[1)] of the two extracts (compare Fig. 1a to 1b), since the FD-factors of most of the wheat bread odourants were low: As seen in Fig. 1a only 6 odourants had FD-factors greater than 32.

The highest FD-factor was found for 2-acetyl-1-pyrroline (Acpy), followed by 2(E)-nonenal (green, tallowy) and 3-methylbutanal (malty). On the basis of its high FD-factor and its roasty, sweet odour which resembles that of the wheat bread crust, Acpy is the character impact compound of the wheat bread crust.

Table 2. Flavour compounds of wheat and rye bread crusts: results of the GC-effluent sniffing and the identification experiments [14]

No.[a)]	Compound	RI[a)]	Odour description[b)]	Odour[c)] Wheat	Odour[c)] Rye
1	3-Methylbutanal	910	Malty	+	+
2	Diacetyl	970	Buttery	+	+
3	Hexanal	1095	Green	+	+
4	2-Heptanone	1170	Soapy, fruity	o	+
5	3-Methylbutanol	1205	Malty	+	+
6	2(E)-Hexenal	1220	Fatty, green	o	+
7	4(Z)-Heptenal	1230	Biscuit-like	+	+
8	Unknown	1250	Fatty	+	+
9	2-Octanone	1285	Soapy	o	+
10	1-Octen-3-one	1290	Mushroom-like	+	+
11	2(E)-Heptenal	1310	Fatty	o	+
12	2-Acetyl-1-pyrroline	1320	Roasted, sweet	+	+
13	Hexanol	1360	Green, floral	o	+
14	Unknown	1365	Resinous	+	+
15	Nonanal	1385	Soapy	o	+
16a	Trimethylpyrazine }	1395	Potato-like }	+	+
16b	2-Methyl-3-ethylpyrazine }		Roasted, burnt }		
17	2(E)-Octenal	1420	Fatty, nutty	+	+
18	2,5-Dimethyl-3-ethylpyrazine	1435	Potato-like	+	+
19	Unknown	1445	Sweet, cereal-like	+	+

1) An "aromagram" is the plot of the logarithm of the FD-factors for all odour active substances in an aroma concentrate vs. their retention indices (RI).

Continuation of Table 2

No.[a)]	Compound	RI[a)]	Odour description[b)]	Odour[c)] Wheat	Rye
20a	Furan-2-al	1455	Sweet, cereal-like	o	+
20b	2,6-Dimethyl-3-ethylpyrazine		Potato-like	+	+
21	Unknown	1485	Potato-like	+	+
22a	Unknown	1490	Green	+	+
22b	2-Acetylfuran		Smoky	o	+
23	Benzaldehyde	1495	Almond-like	o	+
24	2(E)-Nonenal	1520	Green, tallowy	+	+
25	Unknown	1525	Green, pungent	+	+
26	Diethylmethylpyrazine	1535	Potato-like	+	+
27	5-Methyl-2-furaldehyde	1560	Almond-like	o	+
28	2(E),6(Z)-Nonadienal	1575	Cucumber-like	+	+
29	2-Acetylpyridine	1585	Roasted	+	+
30a	Acetylpyrazine	1605	Roasted	+	+
30b	5-Methyl-6,7-dihydro-(5H)-cyclopenta(b)pyrazine				
30c	2-Acetyl-1,4,5,6-tetrahydropyridine				
31	Phenylacetaldehyde	1625	Honey-like	+	+
32	Unknown	1665	Sweaty	o	+
33	Unknown	1680	Waxy, fatty	+	+
34	5-Ethyl-(3H)-furan-2-one	1690	Spicy	+	+
35	Unknown	1705	Waxy	o	+
36	5-Ethyl-(5H)-furan-2-one	1745	Spicy	o	+
37	Unknown	1780	Mushroom-like	+	+
38	2(E),4(E)-Decadienal	1795	Fatty, waxy	+	+
39	Unknown	1805	Cooked apple-like	+	+
40	Unknown	1825	Hydrolysed protein-like	o	+
41	Unknown	1850	Burnt,sweet	+	+
42	Unknown	1980	Sweet, fatty	+	+
43	Unknown	2000	Sweet, fatty	+	+

a) Number and retention index (RI) of the odour active region.

b) Description of the odours recognized during GC-effluent sniffing of the bread crust extracts.

c) Odour of the compound was (+)/was not (o) perceived at the sniffing port.

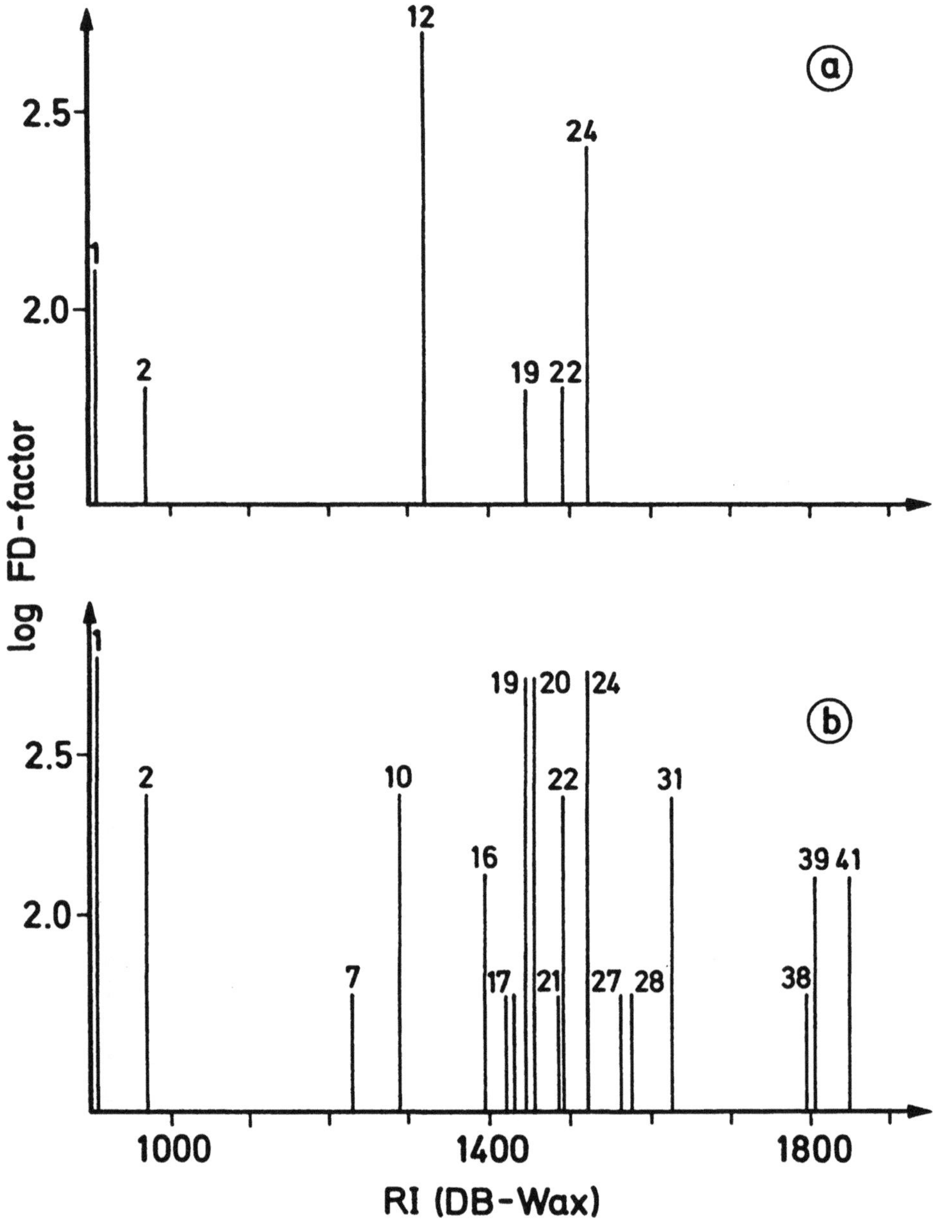

Fig. 1a, 1b. Aromagrams of the aroma extracts of (a) wheat bread crust and (b) rye bread crust. The assigned numbers correspond to those in Table 2

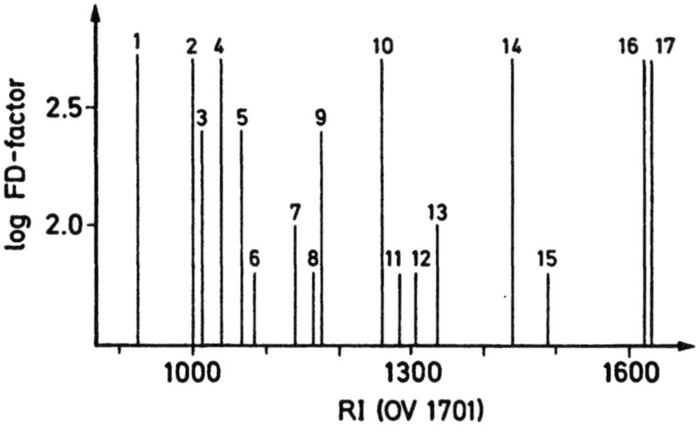

Fig. 2. Aromagram of volatile compounds obtained by simultaneous destillation/extraction from cooked beef [16]

(1) 2-methyl-3-furanthiol, (2) unknown, (3) 2-acetyl-1-pyrroline, (4) methional, (5) 1-octen-3-one, (6) 2-octanone, (7) 2-acetylthiazole, (8) 2(E)-octenal, (9) phenylacetaldehyde, (10) 2(E)-nonenal, (11) 2-decanone, (12) unknown, (13) 2(E), 4(E)-Nonadienal, (14) 2(E),4(E)-decadienal, (15) 2-dodecanone, (16) ß-ionone, (17) bis(2-methyl-3-furyl) disulfide

The aromagram of the rye bread crust is more complex: 18 odourants appeared with FD-factors greater than 32 (Fig. 1b). Four of them showed the highest FD-factors: 3-methylbutanal (malty), an unknown compound (sweet, cereal-like), 2,6-dimethyl-3-ethylpyrazine (potato-like) and 2(E)-nonenal (green, tallowy). Acpy was also found in the rye bread crust but its FD-factor was 32 times smaller than in the wheat bread crust.

Acpy was at first found by Buttery and Ling [15] in the aroma of a rice variety which produced a popcorn-like flavour during cooking. Recently this compound was also found in cooked beef [16], as seen in the aromagram shown in Fig. 2. Sulfur compounds and lipid oxidation products predominate in this aromagram, but Acpy (No. 3 in Fig. 2) together with 2-acetylthiazole (No. 7) contribute significantly to the roasty note which can be perceived in the odour profile of cooked beef.

It should be mentioned that the data given in the aromagrams are only an approximation to the real values, since the FD-factors were not corrected for differences in the yields obtained after the isolation procedure. Furthermore, the aroma of a food is partially affected by the food composition, since the volatility of the odourants is influenced by their binding to lipids, proteins and carbohydrates, while in the aroma extract dilution analysis all extracted odourants are completely volatilized.

Quantitative analysis

If more exact data are desired, the results obtained by aroma extract dilution analysis have to be complemented by quantitative measurements. In addition, quantitative methods for potent aroma compounds which can be used as indicators for distinct aroma notes of a food are also of interest, e.g. the Acpy as indicator of the crust-like odour note of wheat bread.

However, quantification of odourants is a difficult task, since they are volatile and often very reactive. Furthermore, the concentration especially of the odourants showing the highest FD-factors can be so low that they show no peak in the gas chromatogram of the original aroma extract. Examples are the Acpy in extracts obtained from wheat bread [17] and cooked beef [16] and the 1,5(Z)-octadien-3-one in an extract obtained from a soya oil with a reversion off-flavour [18]. For this reason the odourants have to be enriched by extraction and separation steps before its gas chromatographic determination is possible.

However, the volatility and in numerous cases also the reactivity of the odourant can cause great losses during its isolation and enrichment. In particular, an accurate quantification of unsaturated carbonyl compounds and thiols is difficult because they degrade easily during isolation and enrichment. In consequence, the yields of these odourants are often relative low and not reproducible.

The use of an isotope dilution assay is the best method to quantify such labile odourants. We applied it for the determination of the Acpy and three other odourants which exhibit roasty odour notes. The details of the method in which the corresponding odourants labeled with deuterium were used as internal standards are described elsewhere [19]. Since these standards have the same chemical and physical properties as the odourants to be quantified, the yields after enrichment do not influence the results. Model experiments showed that the differences between theoretical values and the values measured using this isotope dilution assay amounted to not more than 10 % [19].

The four odourants were quantified in the crusts of different types of wheat and rye breads [19]. The results are summarized in Table 3. The high level of Acpy in the crusts of the wheat breads was the most striking difference between the wheat and the rye breads. This result agreed with the difference found for both bread types in the aromagrams (Fig. 1a, 1b) which were discussed before. The quantitative data in combination with the very low threshold of 0.1 ppb (water) [20] confirm that Acpy is the character impact odour compound of the wheat bread crust.
A comparison of the wheat breads showed that the pyrazines but not the Acpy increased when the bread was prepared by addition of a baking improver (Table 3).

Table 3. Concentrations (µg/kg) of the 2-acetyl-1-pyrroline and three pyrazines with roasty odours in the crusts of wheat and rye breads [19]

Compound	Wheat bread[a)]		Rye bread[b)]	
	W 1	W 2	R 1	R 2
2-Acetyl-1-pyrroline	78	72	1	4
2-Methyl-3-ethylpyrazine	66	75	41	94
5-Methyl-5H-cyclopenta(b)pyrazine	15	29	17	37
Acetylpyrazine	19	34	51	80

a) Wheat bread prepared without (W 1) and with (W 2) a baking helper.
b) One-stage (R 1) and three-stage (R 2) sourdough process.

The two types of rye breads investigated differed in the odour strength. As expected the odour of the bread prepared by a one-stage sourdough process was weaker than that obtained by a three-stage process. The data shown in Table 3 reflect this difference. In particular the 2-methyl-3-ethylpyrazine, which showed the highest FD-factor within the odourants of rye bread having a roasty note (Fig. 1b), was more than doubled. This is obviously due to the longer fermentation time during the three-stage sourdough process, which leads to an increase in the precursors and subsequently in the flavour compounds during baking.

Origin of 2-acetyl-1-pyrroline [21]

The isotope dilution assay was also used for the determination of Acpy in the studies of the origin of this important flavour compound of wheat bread.

Tressl et al. [22] reported that small amounts of Acpy are formed when proline/monosaccharide mixtures are heated. Therefore the investigations were started with experiments which should indicate the role of free proline.

Table 4. Amount of 2-acetyl-1-pyrroline in wheat bread in relation to the amount of proline added to the dough[a)]

No.	Proline added[b)] (mg/kg flour)	2-Acetyl-1-pyrroline (µg/kg flour)	2-Acetyl-1-pyrroline (µg/100 mg proline added)
1	without	34	-
2	120	41	5.8
3	200	46	6.0
4	500	66	6.4
5	2 000	132	4.9
6	10 000	453	4.2

a) The dough was prepared from 500 g wheat flour (type 550), 30 g yeast, 15 g NaCl, 10 g sucrose, 15 ml shortening and 230 ml water.

b) L-proline was dissolved in the water used for the preparation of the dough.

A wheat bread was prepared from a dough in which the free proline content increased during fermentation from 12 mg/kg to flour 24 mg/kg. Analysis of the bread showed the production of 34 µg/kg Acpy (Table 4). Addition of proline to the dough enhanced somewhat the formation of Acpy but the increase amounted only 4.2 to 6.4 µg per 100 mg proline added (Table 4). The relatively small increase in Acpy, which was dependent on the amount of proline added, suggests that the reaction conditions for the formation of Acpy were not significantly altered by the increase in the proline concentration. Obviously a relatively small but nearly constant fraction of the proline added was always converted into Acpy during the baking process.

Since there were no indications that the reaction route was changed by the addition of proline, the small increase in Acpy which was observed after addition of proline, led to the conclusion that still another source supplied a precursor of the Acpy.
Further experiments revealed the contribution of yeast. As shown in Table 5, the production of Acpy was significantly lowered when the yeast was replaced by a leavening agent. Compared with bread from yeast fermented dough, only 28 % of the Acpy was found in bread which was produced with a leavening agent. This difference agreed with the difference in the crust flavour impression whose intensity was significantly lower in the case of the bread prepared with the leavening agent.

Table 5. 1-Acetyl-1-pyrroline of wheat bread made using yeast or a leavening agent

Dough[a)] containing per kg flour	2-Acetyl-1-pyrroline (µg/kg flour)
60 g yeast	34.2
32 g leavening agent	9.6

a) See footnote "a" of Table 4.

Yeast (Saccharomyces cerevisiae) as source of the Acpy was studied in the model experiments summarized in Table 6. Higher amounts of Acpy were formed in the presence of sucrose when the yeast cells were destroyed by grinding with silica gel (experiments No. 3 and 4 in Table 6). Only 5 % of the Acpy was found when the grinding of the yeast/sucrose/silica gel slurry was omitted (No. 5). An increase in the sucrose level led to an increase in the Acpy formed (No. 3, 4 and 6), but the increase was not proportional to the amount of sucrose added. Replacement of the sucrose by fructose enhanced the Acpy production (No. 8), while glucose had the opposite effect (No. 7).

The results of these experiments allow the conclusion that the yeast/sucrose system is an additional important source for the Acpy which is formed during the baking process. Studies aimed at identifying the

Table 6. Formation of 2-acetyl-1-pyrroline from yeast (*Saccharomyces cerevisiae*)

No.	Additions to 30 g yeast		2-Acetyl-1-pyrroline[c] (µg)
1	None		1.3
2	23 g siliga gel[a]		1.5
3	1 g sucrose[a]	+ 23 g silica gel	6.3
4	10 g sucrose[a]	+ 23 g silica gel	10.3
5	10 g sucrose[b]	+ 23 g silica gel	0.6
6	30 g sucrose[a]	+ 23 g silica gel	20.3
7	10 g glucose[a]	+ 23 g silica gel	5.1
8	10 g fructose[a]	+ 23 g silica gel	26.9

a) The mixture was ground in a mortar for 10 min.

b) The grinding procedure was omitted.

c) The compound was determined by an isotope dilution assay [19] in an extract which was obtained by simultaneous distillation/extraction of the reaction mixture for 2 h.

component of the yeast which acts as precursor of the Acpy are in progress.

Summary

A screening procedure has been developed for the evaluation of the potent odour compounds occurring in an aroma extract. The procedure which is denoted "aroma extract dilution analysis" yields the flavour dilution factors (FD-factor) of the odour active regions in the gas chromatogram. The FD-factor of an odour compound is proportional to its odour unit (ratio of its concentration to its odour threshold). The method has been used for the study of the wheat and rye bread crust aroma. 2-Acetyl-1-pyrroline (roasty, sweet odour) was identified as character impact compound of wheat bread crust. The aroma extract obtained from the rye bread crust was more complex; the following compounds having high FD-factors were identified: 3-methylbutanal, 2(E)-nonenal, 2,6-dimethyl-3-ethylpyrazine, diacetyl, 1-octen-3-one and phenylacetaldehyde. The formation of 2-acetyl-1-pyrroline was quantified in different bread types and in model experiments using a stable isotope dilution assay. The model experiments revealed that the

system yeast/sucrose is an important source for the 2-acetyl-1-pyrro-
line formed during the baking process.

Literature

[1] Rothe, M.: Aroma vom Brot. Akademie-Verlag, Berlin 1974.
[2] Van Straten, S. and H. Maarse: Volatile Compounds in Food. Qualitative Data. Division for nutrition and food research TNO, Zeist, 1983.
[3] Rothe, M. and B. Thomas: Z.Lebensm.Unters.Forsch. 119 (1963) 302-310.
[4] Guadagni, D.G., R.G. Buttery and J. Harris: J.Sci.Food Agric. 17 (1966) 142-144.
[5] Mulders, J.: Z.Lebensm.Unters.Forsch. 151 (1973) 310-317.
[6] Rothe, M.: Nahrung 20 (1976) 259-266.
[7] Frijters, J.E.R.: Chem.Senses Flavour 3 (1978) 227-233.
[8] Meilgaard, M.C. and T.L. Peppard: The Flavour of Beer. In: Food Flavours. Part B. The Flavour of Beverages. Morton, I.D. and A.J. MacLeod (eds.). Elsevier, Amsterdam, 1986.
[9] Meilgaard, M.C.: J.Agric.Food Chem. 30 (1982) 1009-1017.
[10] Acree, T.E., J. Barnard and D.G. Cunningham: Food Chem. 14 (1984) 273-286.
[11] Cunningham, D.G., T.E. Acree, J. Barnard, R.M. Butts and P.A. Braell: Food Chem. 19 (1986) 137-147.
[12] Schmid, W. and W. Grosch: Z.Lebensm.Unters.Forsch. 182 (1986) 407-412.
[13] Ullrich, F. and W. Grosch: Z.Lebensm.Unters.Forsch. 184 (1987) 277-282.
[14] Schieberle, P. and W. Grosch: Z.Lebensm.Unters.Forsch. 185 (1987) 111-113.
[15] Buttery, R.G. and L.C. Ling,: Chem.Ind (London) 1982 (1982) 958.
[16] Gasser, U. and W. Grosch: Z.Lebensm.Unters.Forsch., in press.
[17] Schieberle, P. and W. Grosch: Z.Lebensm.Unters.Forsch. 180 (1985) 474-478.
[18] Ullrich, F. and W. Grosch: In preparation.
[19] Schieberle, P. and W. Grosch: J.Agric.Food Chem. 35 (1987) 252-257.
[20] Buttery, R.G., L.C. Ling, O. Bienvenido, O. Juliano and J.G. Turnbaugh: J.Agric.Food Chem. 31 (1983) 823-826.
[21] Schieberle, P.: In preparation.
[22] Tressl, T., B. Helak, N. Martin and D. Rewicki: Formation of flavour compounds from L-proline. In: Topics in Flavour Research. Berger, R., S. Nitz and P. Schreier (eds.). H. Eichhorn, Marzling-Hangenham, 1985.

THE EFFECT OF THERMALLY TREATED MALT ADDITIVES TO BREAD FLAVOUR

E. Kamiński, E. Wąsowicz and R. Przybylski[1)]

1. I n t r o d u c t i o n

Bread flavour is relatively toolow in precursors to develop flavour during the baking process. To improve the quality of bakery products in some countries malt or malt extracts are used as a source of sugars and free amino acids. As shown in our previous paper typical bread flavour could be obtained as a result of the Maillard reaction, by heating sugars with alanine, leucine, arginine, lysine, glycine, aspartic acid and aspargine at proper values of pH and temperatures (1).

Malt or malt extracts from wheat, barley and rye contain all the amino acids and sugars required for developing bread flavour in Maillard type reaction. The contents of free amino acids in different malts was from five to ten times higher than in non-germinated grain (fig. 1) (2).

The sugars found in malts were: arabinose, xylose, fructose, glucose, maltose and maltotriose (3).

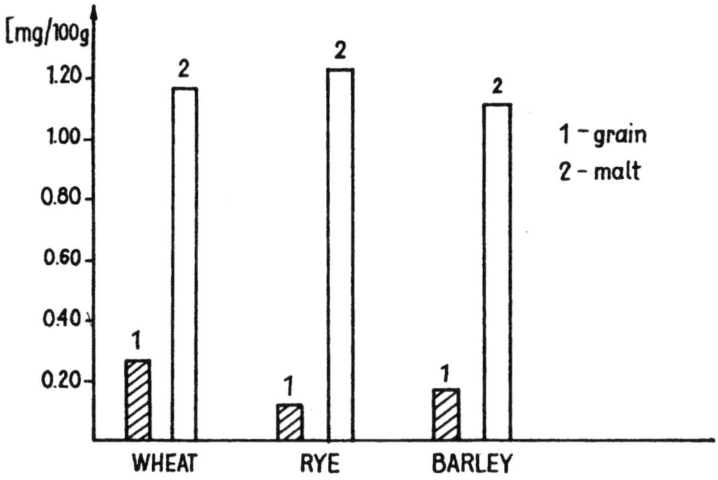

Fig. 1. Contents of free amino acids in grains and malts

1) University of Agriculture, Institute of Food Technology of Plant Origin . 60-624 Poznań, Poland

2. Changes in the contents of free amino acids and sugars in malt extract subjected to heat treatment

Malt extracts subjected to heat treatment change colour, odour, contents of amino acids and sugars. The data presented in table 1 illustrate the changes in the contents of amino acids in a wheat malt extract heated at 120°C for different periods of time. The most rapid changes were stated with glutamine, glutamic acid, lysine, cystine, arginine and tyrosine.

The most stable amino acids were: alanine, leucine, isoleucine and methionine. The changes in the contents of sugars in a wheat malt extract heated at 120°C are presented in table 2. With the time of heating the contents of glucose, maltose and maltotriose decreased while the contents of fructose increased, this being due to the conversion of glucose to fructose.

Table 1

Contents of free amino acids in wheat malt extracts of 40% dry weight and changes in their contents on heat treatment at 120°C

Amino acids	Wheat malt extract (mg/ml)	Amino acids remaining after heat treatment (%)		
		Time of heat treatment		
		15 min.	30 min.	60 min.
Alanine	800	88.6	72.3	60.3
Valine	1075	73.4	55.0	45.6
Glycine	505	64.3	50.9	46.4
Isoleucine	775	77.8	60.0	51.4
Leucine	1850	83.4	55.3	47.2
Threonine	545	66.6	61.2	44.9
Proline	2950	50.3	41.4	35.6
Serine	850	63.7	63.6	37.7
Methionine	390	75.0	61.4	55.2
Phenylalanine	1565	56.3	46.4	39.0
Asparagine+aspartic acid	1475	52.7	46.2	40.8
Glutamine +glutamic acid	4715	16.3	11.4	8.3
Tyrosine	265	52.9	29.4	25.8
Lysine	465	35.4	27.4	23.1
Arginine	440	69.4	46.0	26.1
Cystine	330	-	48.6	16.6
	% of the total amino acids	50.2	40.1	33.6

Table 2

Contents of sugars in wheat malt extracts of 40% dry weight and changes in their contents on heat treatment at 120°C

Sugar	Wheat malt extract (mg/ml)	Sugars remaining after heat treatment (%)		
		Time of heat treatment		
		15 min.	30 min.	60 min.
Fructose	4.26	114.0	143.0	183.0
Glucose	24.10	94.1	86.9	63.9
Maltose	89.40	95.9	95.1	92.9
Maltotriose	3.02	89.3	87.3	83.3

3. The effect of temperature on the formation of bread flavour in heat processed malt extracts

The effect of heating on the formation of bread flavour in wheat, barley and malt extracts has been studied already (4). The most desirable sensory properties of malt extracts were obtained at pH 4.0-5.5 and temperatures 120-140°C. As an example the results of sensory evaluation of a rye malt extract heated at different temperatures and pH 5.5 are presented in fig. 2. Heating at 80°C for 3 hours had no detectable effect on flavour. The malt extract heated at 120°C and 140°C showed a very strong flavour, closely resembling bread flavour. The volatiles isolated from thermally treated malt extracts were separated by GLC technique (5). The separation of volatiles isolated from rye extracts on a packed column (Carbowax 20 M) gave about 30 fractions (table 3). The main fractions yielded the following odours: like buttery, roasted grain, peanuts, coffee, sour, almond, biscuits, caramel, dried fruits, bread crust and like vanillin.
The amounts of volatiles were found to depend on the parameters of heat treatment applied. With the increase of temperature from 80°C to 160°C the amounts of some volatiles increased considerable. For example, the fraction with the odour like vanillin (retention index 1767) increased 66 times.
Some fractions were detected only in malt extracts heated at higher temperatures. The fraction yielding a like acetic acid odour (retention index 954) did not occur in malt extracts heated at temperatures below 120°C.

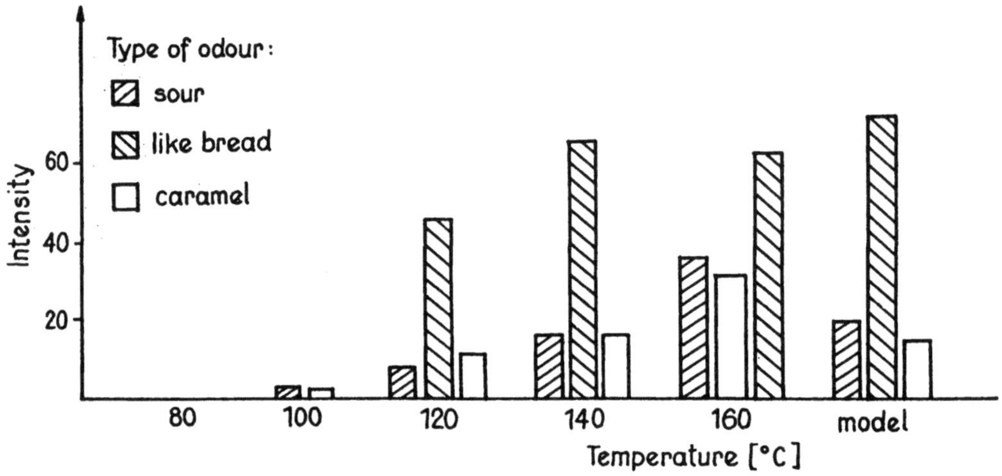

Fig. 2. Sensory profiles of thermally treated rye malt extract at different temperature for 180 min at pH 5.5

Table 3

Gas chromatography characteristics of volatiles from rye malt extracts heated for 3 hours at 140°C and pH 5.5

Retention Index[a])	Contents of individual fraction (%)	Description of odour
671	0.37	Oily
716	2.50	Buttery
760	0.08	Bread crust
901	0.79	Roasted grain
915	1.43	Putrid
920	0.07	Coffee
954	15.21	Sour (Acetic acid)
1000	10.03	Almond
1080	1.70	Sour, heated grain
1126	1.36	Fried potatoes
1135	1.31	Caramel
1162	0.85	Biscuits
1169	31.42	Roasted grain
1230	0.24	Herbal
1262	0.73	Roasted, sour
1269	0.04	Heated grain
1275	1.03	Straw-Like
1303	0.92	Pharmaceutic
1349	0.36	Dried fruits

continued table 3

Retention	Contents of individual fraction (%)	Description of odour
1411	13.76	Dried apples
1461	1.31	Herbal, dried fruits
1513	0.06	Bread crust
1522	0.45	Heated grain
1535	0.29	Floral
1552	0.16	Roasted grain
1567	1.98	Herbal
1640	0.25	Roasted grain
1767	9.44	Vanillin
1947	1.45	Undetermined

a) Gas chromatographic column: length - 2 m, inner diameter 2 mm, packed with carbowax 20 M on chromosorb P, 100/120 mesh (Ultrabond).

4. The effect of rye malt additives on the quality of bread

Rye malt has been found to be a rich source of precursors of bread flavour compounds. The study concerned the evaluation of the effect of different malt additives on the quality, particularly on the flavour of wheat and mixed wheat - rye bread.

The following additives were tried:

a) Malt flour

Malt flour was obtained by grinding dried rye malt with a milling device MIAG. Before grinding the dry malt was tempered to the moisture contents 14.5%. The extraction of the malt flour amounted to 70%.

b) Thermally treated flour

Dry wheat or rye malt was moistened to about 50% moisture contents and dried at 130°C for 1 hour. The malt was colled and then ground with a laboratory mill. The extraction of the flour amounted to 65%.

c) Thermally treated malt extract

The malt extracts were prepared as follows: milled rye malt was mixed with water (1:5) and extracted under gradually rising temperature - from 30°C to 70°C - until full saccharification of starch. During extraction time intervals were applied securing optimal action of proteolytic and amylolytic enzymes.

The resulting extract was concentrated in a rotary vacuum evaporator to 40% d.w. The concentrated extract was heated at 130°C for 2 hours.

d) The essence from thermally treated malt extract

The concentrates of volatiles from thermally treated rye malt extract were prepared by a 6-hour extraction with a solvent system- ether : pentane (1:1). The resulting extract was dried with anhydrous sodium sulphate, concentrated under nitrogen and diluted with 50% ethanol.

4.1. The effect of malt additives on the quality of wheat bread

The baking of the wheat bread was carried out under laboratory conditions using wheat flour type 850 by one-phase method. The recipe was as follows:

flour - 500 g
water - 300 ml
yeasts - 15 g
salt - 7,5 g
the addition of malt preparations; in relation to the weight of flour - 1-10%.

The baking of bread was carried out at 240°C and it took 40 minutes. The evaluation of bread was done 18-24 hours after baking. The odour and taste of the bread were evaluated by the multi-comparative method (6).

The evaluation made it possible to differentiate taste and odour of the samples enriched with malt preparations and to establish their optimal concentration. The data presented in table 4 illustrate the effect of optimal doses on the quality of bread.

A 2,5% addition of malt flour caused no significant changes on the qualities of bread. A 5% addition of this preparation brought about a decrease in the volume of bread, and affected the porosity and elasticity of its crumb. The addition of the malt flour amounting to 10% deteriorated the taste and aroma of bread. The above was confirmed by the desirability test: the percentage of the non--accepted samples was above the disqualifying minimum for both qualities.

The addition thermally treated malt flour, amounting to 2.5%, 5% or 10% had no significant effect on the elasticity and porosity of the crumb, only the crust and crumb of the bread were darker as compared to the control samples. The addition of malt flour thermally treated amounting to 2.5% considerably improved the taste and aroma of bread. The addition of malt extract thermally treated in the amounts 1%, 2.5% and 5% caused no significant changes in physical qualities of bread, excepting the colour of crumb. The bread enriched with 2.5% and 5% additives had a darker crumb than the control bread. 1% and 2.5% additives improved both the taste and aroma of bread. The taste of the crumb was evaluated as desirable flavours as compared to the control bread.

Additives of essence from thermally treated malt extract in amounts 1% to 5% had a favourable effect on the taste and aroma of bread. Besides the addition of the essence caused no changes in the colour of the crumb.

Table 4

The effect of rye-malt additives on the quality of wheat bread

Quality factor	Bread control sample	Bread with the addition of malt preparations:			
		Rye-malt flour 2.5%	Rye-malt flour thermally treated 2.5%	Rye-malt extract thermally treated 2.5%	Essence of rye-malt extract thermally treated 2.5% x)
Bread crust colour	brown	brown	brown	brown	brown
Crumb:					
colour	cream	grey-cream	brown	brown	cream
flexibility	good	sufficient	good	very good	good
porosity	uniform	uniform	uniform	uniform	uniform
Bread volume (cm^3/100 g flour)	450	440	458	470	450
Odour	typical of bread	typical of bread	typical of bread	typical of bread, intensive, aromatic	typical of bread, intensive aromatic
Taste	typical of bread	typical of bread	typical of bread	typical of bread, intensive	typical of bread, intensive

x) The amount of added essence cor(r)esponded to amount of added extract to bread.

4.2. The effect of malt preparations on the quality of mixed wheat-rye bread

The wheat-rye bread was produced in an industrial bakery in Poznań. The dough was prepared according to the following recipe:

sour from a rye flour, type 800	1.6 kg
wheat flour, type 850	2.19 kg
water	1.05 kg
salt	0.05 kg
yeasts	0.01 kg
malt additives	2.5% (in relation to the weight of flour).

A preparation of the extract from a thermally treated rye malt and an alcohol essence of this preparation were used as malt additives. The fermentation time of the dough amounted to 70 min. The baking was effected in an industrial oven. The initial baking temperature amounted to 260°C and the final 215°C.

The baking took 42 min. The weight of the loaves amounted to 0.8 kg. The analysis of the bread was done 18 and 48 hours after baking.

The results of quality evaluation of wheat-rye bread (descriptive method) enriched with malt additives are presented in table 5.

The bread with the addition of malt extract had darker crust and crumb as compared to the control bread.

The flavour of the bread was better than that of the control sample however. The colour of the crumb of the bread obtained with the addition of an essence of malt preparation resembled that of the control, but its taste and aroma were more intensive.

The addition of malt preparations was found to improve storability of wheat-rye bread. After 48-hour storage the quality of enriched bread was higher than that of the control samples. As shown by the desirability evaluation the percentage of the non-accepted samples was below the disqualifying minimum. The control bread stored for 48 hours received no acceptance of the evaluating panel.

Table 5

The effect of the rye-malt additives on the quality of wheat - rye bread

Quality factor	Bread (Control)	Bread samples with the addition of thermall treated malt preparations:	
		Rye - malt extract 2.5%	x) Essence of rye-malt extract 2.5%
Bread crust			
colour	brown	deep brown	brown
surface	smooth, glossy glittering	smooth, glossy	smooth, glossy
Crumb:			
colour	cream	brown	cream
flexibitlity	good	good	good
porosity	uniform, fine	uniform, fine	uniform, fine
Odour	typical of bread	typical of bread, aromatic intensive	typical of bread, aromatic intensive
Taste	typical of bread	typical of bread	typical of bread

x) The amount of added essence cor(r)esponded to amount of added extract to bread.

5. The effect of malt additives on volatile odour substances in bread

The volatiles occuring in wheat-rye bread and in bread enriched with malt additives were isolated by the distillation-extraction method in Likens-Nickerson apparatus. The analysis of the volatiles was carried by gas chromatography technique using a packed column. The concentration of the volatile determined by this method in the bread enriched with an alcohol essence from a preparation of rye malt extract (2.5%) was twice higher than in that prepared with no additives. The bread with the addition of the essence was found to be higher in the fractions yielding the following odours: buttery, roasted grain, roasted peanuts, grassy, cucumbers, dried apples (table 6).
The bread was lower however in the fractions yielding the odours: acetic acid, yeast, burnt rubber, fried potatoes and herbal.

Table 6

Gas chromatography characteristics of mixed wheat-rye bread and bread with the addition of an essence from malt extract

Retention index a)	Contents of individual fractions (%)	Per cent increase or decrease of volatiles in bread with the addition of essence	Description of odour
1131	0.03	-	
1163	0.16	- 86.84	Yeast-like
1183	84.37	125.29	Buttery
1237	0.08	152.63	Roasted grain
1251	0.14	- 43.75	Oily
1274	0.22	-	Oily-buttery
1318	0.55	94.69	Grassy
1400	2.92	6.66	Burnt-rubber
1447	0.34	- 66.66	Oily-nut
1495	0.03	12.50	Oily
1561	0.06	107.14	Dried fruits
1571	0.02	-	Oily nut
1600	0.25	57.62	Roasted peanuts
1626	0.42	- 19.19	Roasted coffee
1640	0.22	- 43.13	Peanuts
1693	0.11	- 46.15	Bread crust
1706	3.52	- 49.88	Acetic acid
1733	0.15	-	Straw-like
1760	1.77	- 13.30	Burut straw
1773	0.21	- 56.00	Fried potatoes
1800	0.26	57.14	Potato-oily
1820	0.31	5.40	Fried potatoes
1880	0.32	84.21	Fresh cucumbers
1920	0.37	244.31	Roasted grain
1940	0.55	343.93	Roasted grain
1960	0.27	16.92	Floral
2000	0.04	-	Dried fruits
2050	0.13	160.00	Dried apples
2100	0.64	- 8.49	Dried apples
2175	0.18	40.47	Unpleasant
2200	0.10	25.00	Woody
2216	0.86	- 49.01	Herbal
2316	0.41	-	Dried fruits

a) Gas chromatography column: lenghth 3 m, inner diameter 2 mm, packed with 15 Carbowax 20 M, TPA on Chromosorb WAW DMCS

L i t e r a t u r a

1 Kamiński, E., L. Kozłowska and R. Przybylski: Roczniki Akademii Rolniczej w Poznaniu 141 (1982), 57-73.

2 Kamiński, E., L. Gruchała and R. Przybylski: Hodowla Roślin Aklimatyzacja Nasiennictwo 6 (1978), 401-410.

3 Kamiński, E., R. Przybylski and L. Gruchała: Nahrung 25 (1981), 507-518.

4 Przybylski, R. and E. Kamiński: Nahrung 27 (1983), 487-496.

5 Kamiński, E., E. Wąsowicz and R. Przybylski: Acta Alimentaria Polonica 7 (1981), 59-79.

6 Baryłko-Pikielna, N.: Zarys analizy sensorycznej żywności, Wydawnictwo Naukowo-Techniczne, Warszawa 1975.

INFLUENCE OF BULGARIAN LACTIC ACID LEAVEN ON THE FORMATION OF AROMA COMPOUNDS IN PRE-FERMENT AND IN BREAD PREPARED BY INTENSIVE KNEADING

P.Bratovanova[1)]

M.Tultchinskiy found greater quantities of acetylmethylcarbinol in wheat bread prepared with "liquid" ferment and lactic acid bacteria, as compared to the same bread prepared only with pressed ferment or with "liquid" ferment only [6] .Today V.Kusminskiy, V.Patt, R.Rolandova and others suggest the use of lactic acid leaven in express technologies for wheat bread preparation. [2,3]

This article presents the results obtained at the use of some Bulgarian lactic acid leaven in the production of wheat bread on intensive kneading aggregates and the influence of these leaven on the formation of aroma compounds (carbonyl compounds) in pre-ferment and in bread.

1.Material and Methods

The pre-ferment was prepared with 30% wheat flour type 500 and 50% yeast of the total amount at 2,0 kg yeast used for 100 kg flour.Lactic acid leaven "37-18" or "B_6" were added in the pre-fermentation stage.H.Salovaata and G.Spĭcher found in wheat leaven presence of lactic acid microorganisms L.plantarum and L.brevis.

In our conditions after 5 h fermentation period we found predomination of Str.lactis and L.plantarum in the pre-ferment without leaven.The pre-ferment prepared with leaven "37-18" was enriched in L.bulgaricus and Str.thermopilus and the pre-ferment prepared with leaven "B_6" was enriched in Str.cremosis, Str.citrovorus and Str. diacetilactis.The leaven had been lyophilized and activated in skimmed milk.The optimal quantities obtained by us were: 6,6 g for "37-18" and 13,4 g for "B_6" per 100 kg flour.

We established the technological indices for pre-ferment fermentation periods of 2,3,4,5 and 6 h.However in this article are presented only the results of 5 h fermentation period.

By application of correlative and regressional analysis methods put in computer JF-800 BASIC program we studied the interrelation between technological and qualitative indices of bread prepared with lactic acid leaven. The deformation characteristics of crumb were established by penetrometer VFD.

The aroma compounds (carbonyl compounds) were established to be 2,4-dinitrophenylhydrazones by ascending paper rechromatography (zonal chromatography) at immobile phase; N,N-dimethylformamid and mobile phase n-hexan [1] .

The separated 2,4-dinitrophenylhydrazones of carbonyl compounds were establ-

1)Higher Institute for National Economy "D.Blagoev", №77 Lenin bul, 9002 Varna, Bulgaria

ished thorugh their absorption spectra within the range of 220 to 400 nm. For this purpose the 2,4-dinitrophenylhydrazones of the substances were chromatographically purified, dissolved in methanol free of carbonyl compounds and spectrophotometred on detecting spectrophotometer SPECORD-VV-VTS. The separated carbonyl compounds were identified through 47 references also spectrophotometred.
Fig.1 shows the scheme of zonal paper chromatographic separation of carbonyl compounds into 2,4-dinitrophenylhydrazone, established in pre-ferment and in crumb.
Fig.1

Scheme of paper rechromatographic separation of 2,4-dinitrophenylhydrazones into carbonyl compounds contained in pre-ferment and crumb of white wheat bread.

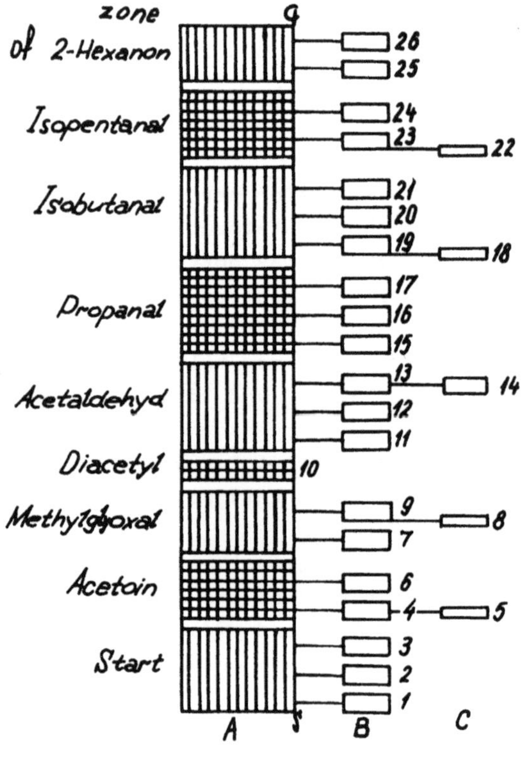

Illustration:

A. Zones of basic chromatogram /Ist fraction/

B. Zonal /rechromatographic/ separation /IInd fraction/

C. Rechromatographic separation /IIIrd fraction/

Carbonyl compounds from Figure 1

Nº of Carbonyl compounds	max at 220-400 nm
1.Unidentified	348,0
2.Unidentified	347,0
3.Unidentified	346,0
4.Acetoin (acetylmethylcarbinol)	357,0
5.Unidentified	355,0
6.Ethandial	351,0
7.Unidentified	354,0
8.Unidentified	350,0
9.Methylglyoxal	351,0
10.Diacetyl	353,0
11.Unidentified	352,0
12.Unidentified	361,0
13.Acetaldehyd	358,0
14.Unidentified	362,0
15.2-Butenal	376,0
16.Aceton	364,0
17.Propanal	359,0
18.Unidentified	359,0
19.Isobutanal	361,0
20.2-Butanon	364,0
21.2-Methylbutanal	361,0
22.2-Nonenal	373,0
23.Isopentanal	362,0
24.Hexanal	360,0
25.2-Hexanon	361,0
26.2-Heptanon	359,5

The quantities of carbonyl compounds were established with spectrophotometre Sph-26 at λ 350 and 360 nm, measurement error 9,5 x 10^{-3} to 18,9 x 10^{-3} in absolute values and relative quantities from 0,48 to 0,95 µg%.

2.Results and comment

2.1.Technological indices of pre-ferment and qualitative indices of bread prepared with lactic acid leaven

Table 2.Technological indices of pre-ferment prepared with lactic acid leaven

Indices	Pre-ferment without leaven	Pre-ferment with leaven "37-18"	"B_6"
1	2	3	4
1.Humidity in %	63,80	62,80	62,90

1	2	3	4
2.Dynamic viscosity in mPa.s	1424,68	1756,74	1651,29
3.Acidity in ^{o}N	3,84	4,62	4,46
4.Active acidity /pH/	5,08	4,49	4,52
5.Oxireductional potential /eH/ in mV	1,06	1,38	1,36
6.Lifting capacity in min	15'00"	12'10"	14'25"
7.Volatile carbonyl compounds in μg per 100 g fresh trial	518,24	598,86	652,61
8.Fermentation losses in %	1,276	1,137	1,098

The pre-ferment prepared with lactic acid leaven is of higher acidity. The pH level was considerably diminished and its lifting capacity was increased.

Table 3.Qualitative indices of white wheat bread prepared with lactic acid leaven

Indices	Bread without leaven	Pre-ferment with leaven "37-18"	Pre-ferment with leaven "B_6"
1.Mass in g	832,00	832,00	830,00
2.Dry substance in %	62,02	62,05	62,10
3.Humidity of crumb in %	43,40	43,60	44,00
4.Acidity in ^{o}N	1,28	1,96	1,80
5.Active acidity /pH/	5,57	5,00	5,06
6.Oxi(dation-)reduction potential /eH/ in mV	0,76	1,08	1,06
7.Porosity in %	75,00	76,00	76,80
8.Volume in cm^3	2510,0	2600,0	2680,0
9.Relation of bread height to length $\frac{H}{L}$	0,260	0,295	0,290
10.Volatile carbonyl compounds in μg at 100 g fresh trial	470,20	513,46	561,22

The white wheat bread prepared with intensive kneading aggregates with added lactic acid leaven is characterized by better qualitative indices: acidity, porosity and volume.

Regressional equations[1] of interrelation between:

A.Acidity in ^{o}N /x/ and pH /y/ of pre-ferment:

$$y = a + bx + ct \qquad /1/$$

$$y = 8,57036 - 0,81988x - 0,01826t \qquad /2/$$

(error from -5,30% to +4,98%)

Note: Regressional equations refer to indices of pre-ferment, dough and bread prepared with wheat flour type 500.

B.Active acidity /pH/ /x/ and oxireductional potential /eH/ /y/ of pre-ferment:

$$y = a + bx \qquad /3/$$

$$y = 3{,}73304 - 0{,}52321x \qquad /4/$$

(error from -2,05 to +4,24%)

C.Humidity of dough /x/ and dry substance /y/ of bread:

$$y = a + bx \qquad /5/$$

$$y = 102{,}147 - 0{,}919839x \qquad /6/$$

(error from -0,43 to +0,66%)

D.Porosity of bread /x/ and volume of bread /y/:

$$y = 4318{,}04 + 91{,}3234x \qquad /8/$$

$$y = a + bx \qquad /7/$$

Table 4.Deformation characteristics of white bread crumb prepared with lactic acid leaven

Deformation characteristics	Fresh bread	Storage /h/		Bread with "37-18" leaven fresh.	Storage h		Bread with "B_6" leaven Fresh.	Storage h	
		24	48		24	48		24	48
1.Plastic deformation	185	150	110	220	195	160	235	210	180
2.Residual deformation	96	98	70	91	88	80	96	84	80
3.Elastic deformation	89	52	40	129	107	80	139	126	100

2.2.Aroma compounds (carbonyl compounds) of pre-ferment and of bread prepared with lactic acid leaven

Carbonyl compounds are typical and important ingredient of the bread aroma complex [4] .In pre-ferment 26 volatile carbonyl compounds were found The greater part of these compounds are to be found in crumb as well.The 2-nonenal substance is not to be found in crumb while substance Nº 14 is to be found only in pre-ferment prepared with "B_6" leaven. Lactic acid leaven contribute to the total increase of the amount of carbonyl compounds in pre-ferment and in crumb (tables 2 and 3). Tables 5 and 6 show the contents of the basic carbonyl compounds in pre-ferment and in crumb of bread prepared with lactic acid leaven.

Table 5.Volatile carbonyl compounds in pre-ferment prepared with lactic acid leaven.

Carbonyl compounds in μg per 100 g fresh trial	Pre-ferment without leaven	Pre-ferment with leaven "37-18"	"B_6"
1.Acetoin	35,30	56,48	67,00
2.Methylglyoxal	72,80	89,60	112,00
3.Diacetyl	42,60	64,40	78,40
4.Acetaldehyd	153,55	160,76	167,95
5.Propanal	16,20	31,62	15,75
6.Isobutanal	76,50	68,00	80,50
7.Isopentanal	102,00	110,50	114,75

Table 6.Volatile carbonyl compounds in crumb prepared with lactic acid leaven

Carbonyl compounds in μg per 100 g fresh sample	Bread without leaven	Bread with leaven "37-18"	"B_6"
1.Acetoin	42,50	60,00	58,00
2.Methylglyoxal	58,80	72,80	84,00
3.Diacetyl	48,72	61,60	67,20
4.Acetaldehyd	139,15	143,96	155,95
5.Propanal	27,00	18,00	13,50
6.Isobutanal	63,50	63,50	68,00
7.Isopentanal	85,00	89,00	93,50

The quantities of acetoin, methylglyoxal and diacetyl are higher in pre-ferment and in bread prepared with lactic acid leaven.The quantity of acetaldehyd also rises to certain extent.The quantities of isobutenal and isopentanal are not much affected.The greatest amount of aroma compounds is to be found in pre-ferment and in bread prepared with lactic acid leaven "B_6".

3.Conclusion

1.Lactic acid leaven added in pre-ferment preparation stage contribute to the improvement of its technological indices: higher acidity, lower pH, better oxireductional potential, improved lifting capacity, etc.

2.Lactic acid leaven improve the quality of bread: higher crumb acidity, larger porosity and volume of bread, more nutritive compounds and higher biological value of bread, better appearance, fragrance and savour, higher plasticity and elasticity of bread.

3.In pre-ferment and in bread prepared with lactic acid leaven most prominent is the quantitative increase of carbonyl compounds: acetoin, methylglyoxal and diacetyl.The quantity of aroma compounds is highest in pre-ferment and in bread prepared with "B_6" leaven

4.Literature

1.Bratovanova P. "Izsledvane na aromatichnite veshtestva-carbonilni suedinenia na hliaba pri proisvodstvo i suhranenie", Dissertatsia, Plovdiv, VIHVP, 1977.

2.Kusminskiy P.V, V.A.Patt "Expresnaya technologia prigotovlenia bulochnih isdelii", Hlebopekarnaya i konditerskaya promishlenost, 1978,II,25-27.

3.Polandova R.D, G.F.Dremucheva,"Biochimicheskiy metod uluchshenia kachestva pshenichnovo hleba",Hlebopekarnaya i konditerskaya promishlenost, 1984, 5, 20-22.

4.Rothe, M.Aroma von Brot.Akademie-Verlag, Berlin, 1974.

5.Salowata H, G.Spĭcher, "Use of wheat sourse to improve bread quality"

JCC 12, Kongr.1986, Inst Ses.Getreidewiss und-technol", S.I. s.a.26-35.
6.Tulchinskiy M.N, "O soderjanii acetilmetilcarbinolia v hlebe i o metodike evo opredelenia" Hlebopekarnaya i konditerskaya promishlenost, 1939, 7, 5-7.

THERMODYNAMIC APPROACH TO FLAVOUR RESEARCH

R.V.Golovnya[1)]

The studies of food flavour components cover in fact the total problem of analysis of trace amounts of organic substances in complex mixtures of unknown composition, which are in vapours over a sample. The flavour concentrate is a mixture of a few hundreds of components in amounts varying from a few micrograms to parts of picograms. The thermodynamic approach allows to specify two variants of "head-space" analysis. Head-space (h.s.) analysis is known as the analysis of volatile substances evolved from a sample into its environment.

The first variant of head-space analysis is realized when the sample is a system in thermodynamic equilibrium, as for example, juices, drinks, wine, mineral waters, etc. In this case we are dealing with vapour-phase analysis since the vapour of each component is in direct contact with its condensed (liquid or solid) phase present in the sample. The possibilities of vapour-phase analysis were described in monographs [1,2]. In systems in thermodynamic equilibrium the composition of the vapour phase over a sample depends on distribution coefficients, component concentration and structure, as well as temperature and pressure, but does not depend on the sample mass. The analysis of systems in thermodynamic equilibrium may result not only in qualitative, but also quantitative composition of volatile substances both in vapour and in the sample.

The studies of flavour components of the texturedfoodstuffs are related to the investigation of the thermodynamically non-equilibrium systems having some essential specific features affecting the results of h.s. analyses (second variant). In this case the flavour components result from the combination of chemical and enzymatic reactions and are transported from the food sample into the environment. There is no direct proportionality between the amount of volatile substances in vapour over the sample and their content in it. Thus the results of h.s. analysis of flavour components can't give exhaustive information about the content of volatile substances in the textured sample or the amount of precursors they are formed from. (See scheme 1).

The h.s. analysis of the thermodynamically non-equilibrium systems

1) Institute of Organo-Element Compounds, Academy of Sciences of USSR, Vavilova 28, 117 813 Moscow/USSR

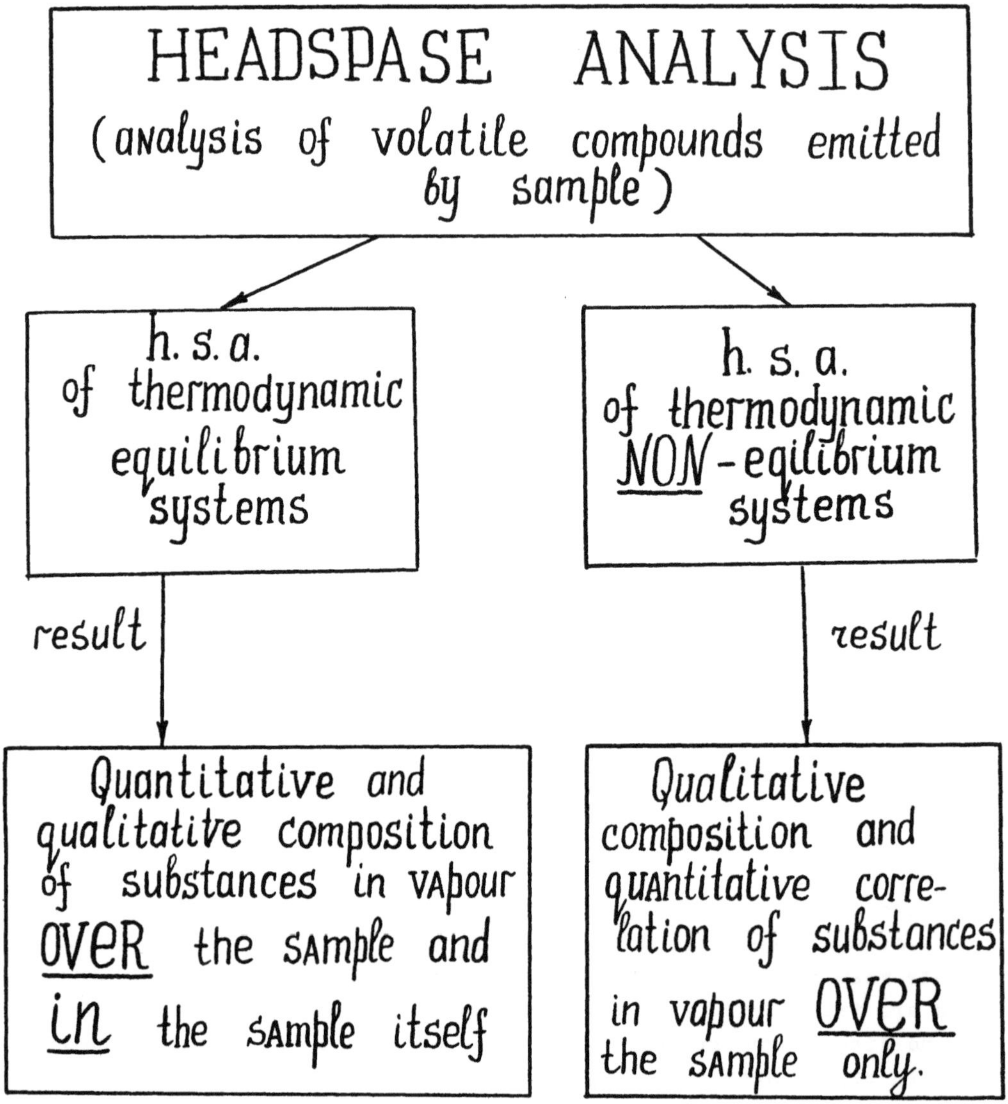

TALANTA 1987,v.34(1),p.51-60

"DETERMINATION OF VOLATILE COMPONENTS OF SYSTEMS NOT IN THERMODYNAMIC EQUILIBRIUM"

will result only in qualitative composition and in relationship of components over the sample under study. Besides, it should be taken into account, that in texturedfoodstuffs the results of flavour component determination will depend on structure integrality. Partial or complete destruction of sample integrality (cutting, peeling, crushing, etc.) essentially affects the qualitative composition and quantitative relationship of volatile components. In this sense there is a relationship between the results and sample mass. The results of the analysis are substantially affected by the time selected for subject investigation, i.e. whether it is freshly prepared or taken after storage, whether it has a ripening stage, etc. It is clear that the h.s. analysis of the non-equilibrium systems requires a special methodology.

The thermodynamic approach to h.s. analysis, methodological aspects of flavour component identification and GC/computer algorithm of identification are described in reviews [3,4].

Unified methodology for analysis of complex mixtures of trace organic substances of biological origin reveals great possibilities of effective application of the results of flavour studies to practice. Flavour research enables:

- differentiation of three categories of the existing food aroma composition: natural flavour, nature-identical and simulated natural flavour;
- standardization and quality control of spices, essences and natural extracts;
- determination of storage time and optimal time of marketing fresh, canned and frozen foodstuffs; (See Fig.1 and Fig.2)
- standardization of organoleptic quality of natural, traditional and new forms of foodstuffs;
- determination of aroma components of natural foodstuffs and employment of the data in search of model reactions producing food flavour;
- optimization of processes of food aroma composition synthesis on the basis of model reactions. (See Fig. 3 and Fig. 4)

What is the reason? Why such a wide range of problems can be solved on the basis of flavour research results? How we can explain the high flavour information with respect to food making and its quality? Is it possible to widen the range of practical application?

It should be mentioned that food ingredients, i.e. proteins, fats and carbohydrates, are nonvolatile compounds devoid of odor or flavour. Accumulated knowledge in field of flavour research allows, in

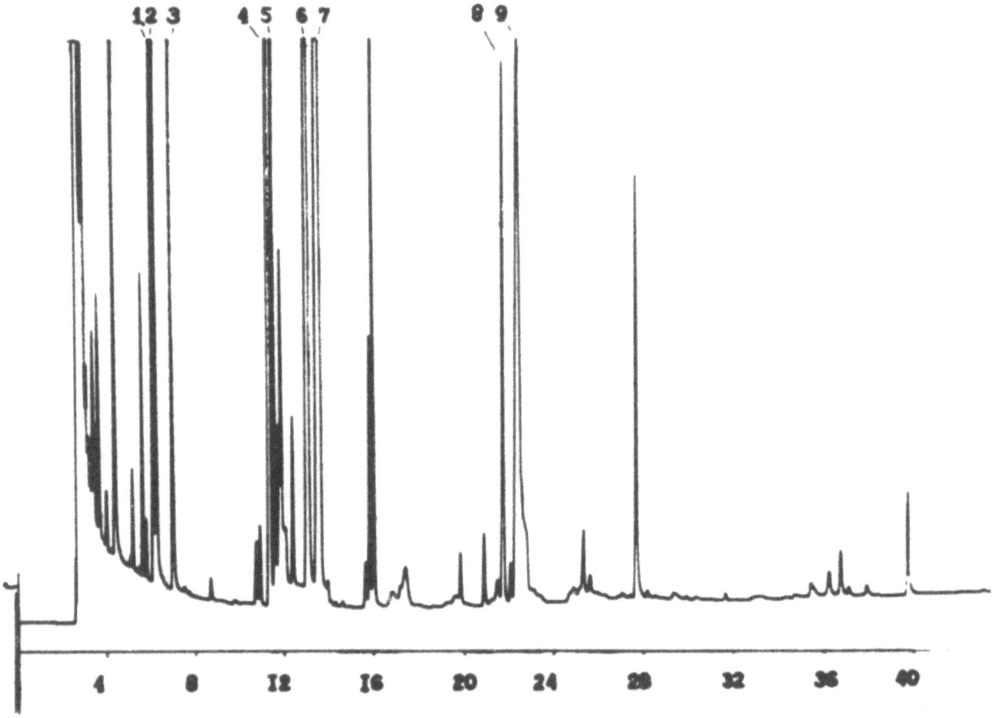

Fig.1. Chromatogram of fresch onion flavour compounds [5]. Glass capillary column (50mx0,32mm) with OV 101, 50°C - 4°/min - 250°C, sample - 1µl. The squares of peaks 1-10 were used for determination of storage time.

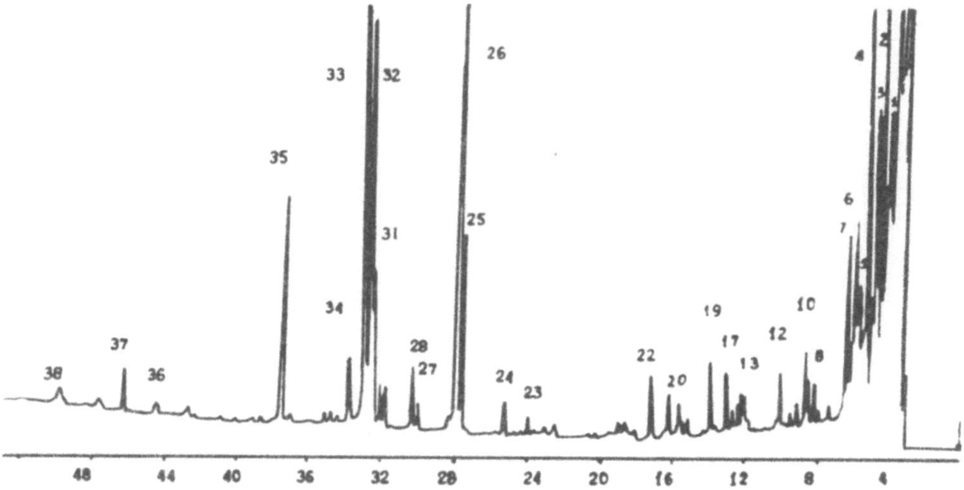

Fig.2. Chromatogram of salmon flesh [6]. Glass capillary column (50mx0,25mm) with OV 101, 40°C - 4°/min - 250°C, sample - 1μl, GC-MS. The square of peaks 4 and 7 were used for determination of time of selling.

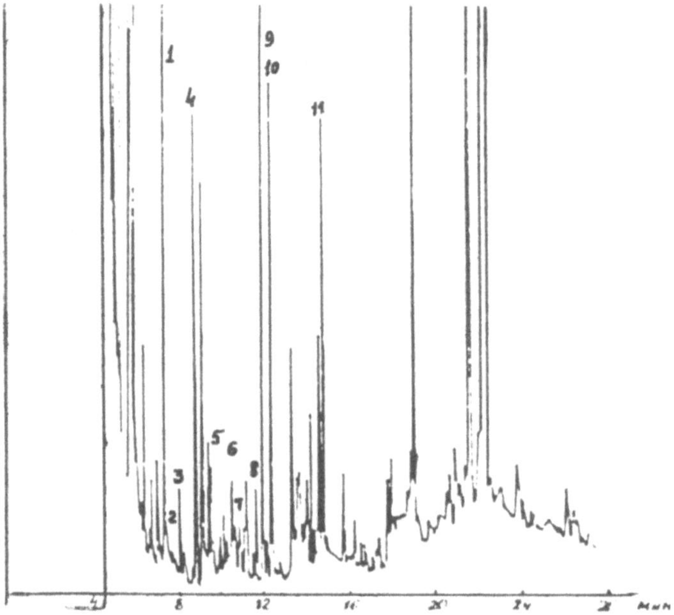

Fig.3. Chromatogram of volatile nitrogen bases from Antarctic krill E. Superba Dana [7]. Glass capillary column (40mx0,28mm) with PEG-40M/KF, 80°C - 4°/min - 160°C, sample - 2μl.

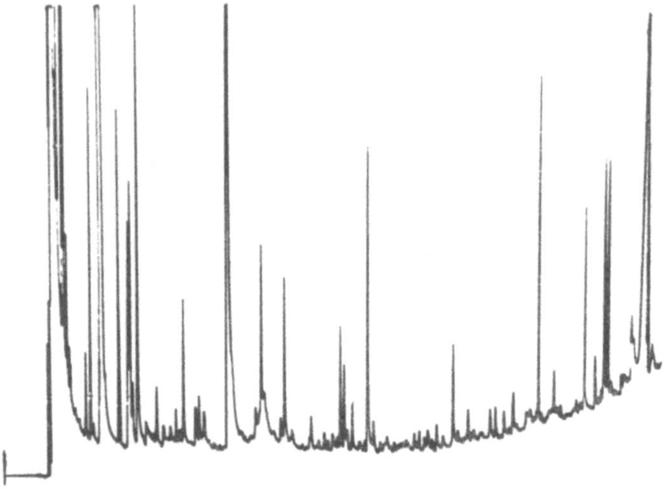

Fig.4. Chromatogram of meat flavour of model system [8]. Column (25mx0,30mm) with SE-54 (Oribond), sample - 1μl, 50°C - 4°/min-250°C.

our opinion, to consider the volatile components as a result of numerous reactions with low degree of transformation. The precursors of many aroma compounds are the products of decomposition of polymer and high-molecular substances ofthe basic food components. The final products of some reactions can serve as starting substances for other transformations. The formation of intermediate compounds, including catalysts and inhibitors of a reaction series, is possible. A few hundreds of various transformations can occur in foodstuffs producing volatile substances. This transformations can't be completly independent, a specific self-coordination between them must be. The mixture of volatile substances responsible for flavour formation is produced by a system of self-coordinated reactions. The composition and quantitative relationship of volatiles can change but within rather restricted limits, thus producing no substantial effect on flavour quality. In a slightly disturbed system the composition of volatile substances is capable of self-recovery within certain limits, since precursors are present in the initial sample, and the system of coordinated reactions did not suffer irreversible changes. The flavour reflects a certain "stability" of the chemical, biochemical and enzymatic transformations of the main constituents of product. Hence, the composition of volatile components may serve as a criterion of "harmony" of the processes occurring in high-quality foodstuffs. Food deterioration caused by environmental effects and microbiological contaminations affect the system of self-coordinated trans formations and disturb its stability. Withdrawal of the system beyond the limits of equilibrium recovery, results in irreversible consequences, accumulation of individual components, emergence of new ones. It will have an influence on the typical flavour and give information about the changes in food quality and its nutritional value.

This approach allows to consider a composition of volatiles as a valuable source of information about the processes occuring not only in their preparation stages. The information can be used for control of completeness of different technological stages of traditional food production to ensure high quality of foodstuffs.

In this paper we shall concentrate on the studies that open new possibilities of practical uses of flavour research.

It was suggested earlier that volatile organic bases in odour, whose precursors are the products of protein transformations, including the proteolytic processes, should depend on intensity of such transformations [1].

Lactobacilli cultures

The volatile nitrogen-containing bases produced by food microorganisms, their composition changes according to growthing phase were first studied on the bacteria Str. lactis 670_6 [9,10]. As it was established by gas chromatography, a great number of volatile amines are metabolic products of lactobacilli cultures. The concentration of them reaches its maximum in logarithmic phase. (See Fig. 5).

This work was a matter of principle, since in 1969 it was first shown that the composition of volatile amines changed during the culture growth, i.e. in the stages of the living of biological organisms. Some working hypotheses were given.

1. The amines, including the secondary and tertiary ones,probably were synthesized by microorganism during growth.
2. The composition of volatile amines may be specific for each microorganism species and serve as a taxonomic indication.
3. The chromatographic profile of amines, as metabolic products, canbe used for determination of microorganism growth stage and productivity of them.

The ideas were found to be fruitful. The hypotheses were confirmed by many findings of recent years. A number of studies on pathogenic bacteria were conducted. It has been shown that the volatile amines are bacterial metabolites and can serve as their chemotaxonomic markers [11-17].

A number of studies on volatiles of pure cultures of lactobacilli used in food industry were carried out at the same period [18-20].

The lactobacilli Str. lactis 127, Str. acetoinicus 28943 [19], Lbc. acidophilus [20], used in food industry, and Str. lactis-MGU, nisin food antibiotic producer [18] were investigated. The volatile amines determined in the starting period, logarithmic and stationary phases and death period. From 31 to 44 different organic nitrogen containing bases were identified according to cultures. Amine composition and analysis technique are presented in a paper by M.B. Terenina and I.L. Zhuravleva [21]. The chromatograms of volatile amines in various growth stages of Str. lactis-MGU strain are given as an example in Fig. 6 [18]. An increase of quantity of amines metabolic products of strains (peaks No 7, 10, 16, 22 and 25) at the stationary phase can be distinctly traced up.

It turned out that the amine chromatographic profile also changes in the death period (Fig. 7). The type of changes is specific for each culture [18-20].

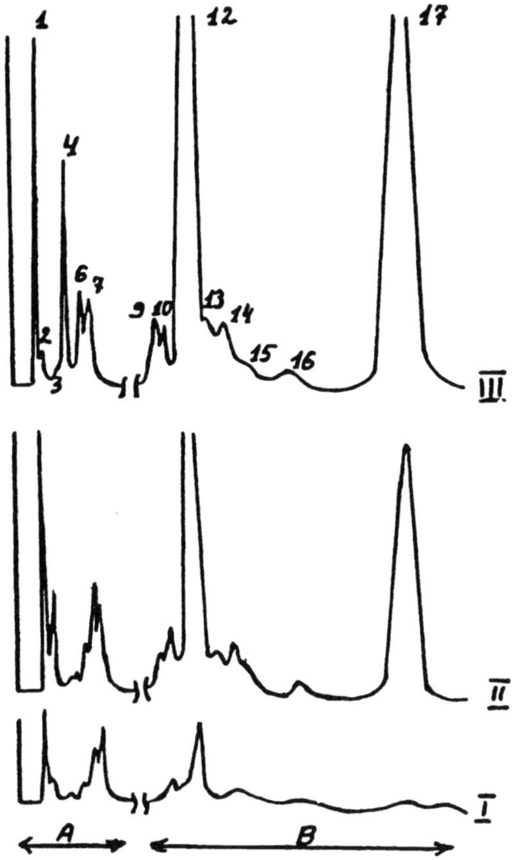

Fig.5. Chromatograms of volatile amines of lactobacilli Str. lactis670_6 at intervals 6(I), 24(II) and 30(III) hours in logarithmic phase of growth [10]. A - nitrogen flow rate - 26ml/min, B - 55ml/min.

1.Trimethylamine
2.Dimethylamine, methylamine
3.Ethylamine
Diethylamine, isopropylamine
5. Not identified
6.Diisopropylamine
7.Triethylamine
8.Not identified
9.Isobutylamine
10.Dipropylamine
11.n-Butylamine
12.Isopropyl-n-butylamine, pyrrolidine
13.Isopentylamine
14.Diisobutylamine, piperidine
15.n-Pentylamine
16.Tripropylamine
17.Not identified

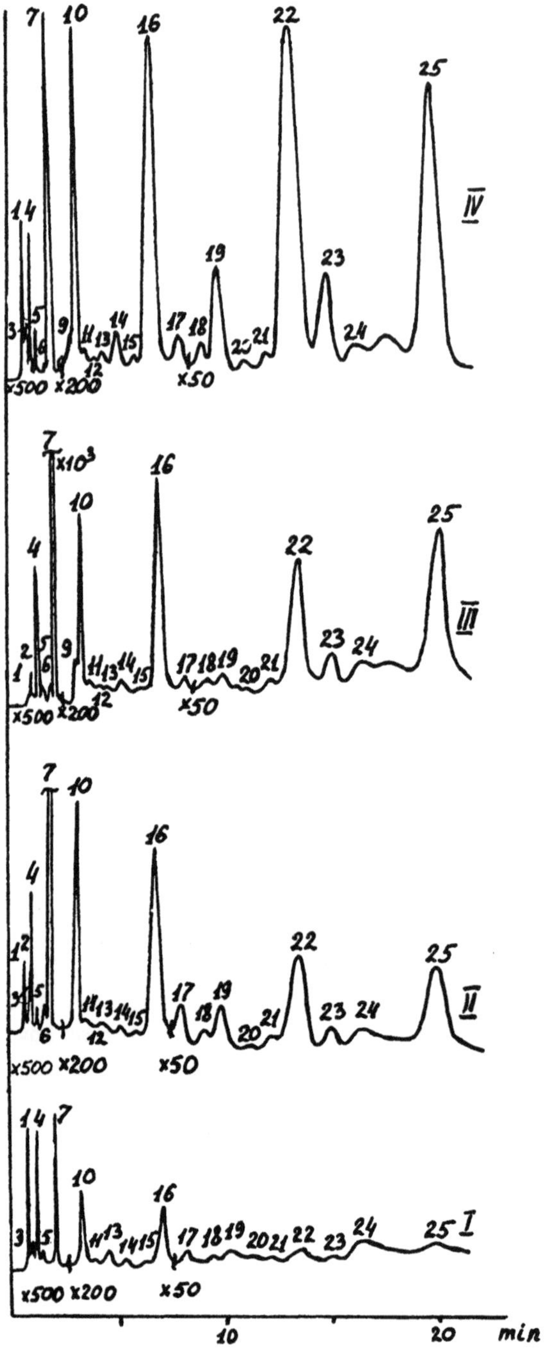

Fig.6. Chromatograms of volatile amines of strain Str. lactis-MGU at intervals 6(I), 9(II), 13(III) and 24(IY) hours [18]. Column with PEG-1000/Na_3PO_4, temperature - 100°C, gelium flow rate - 40ml/min.

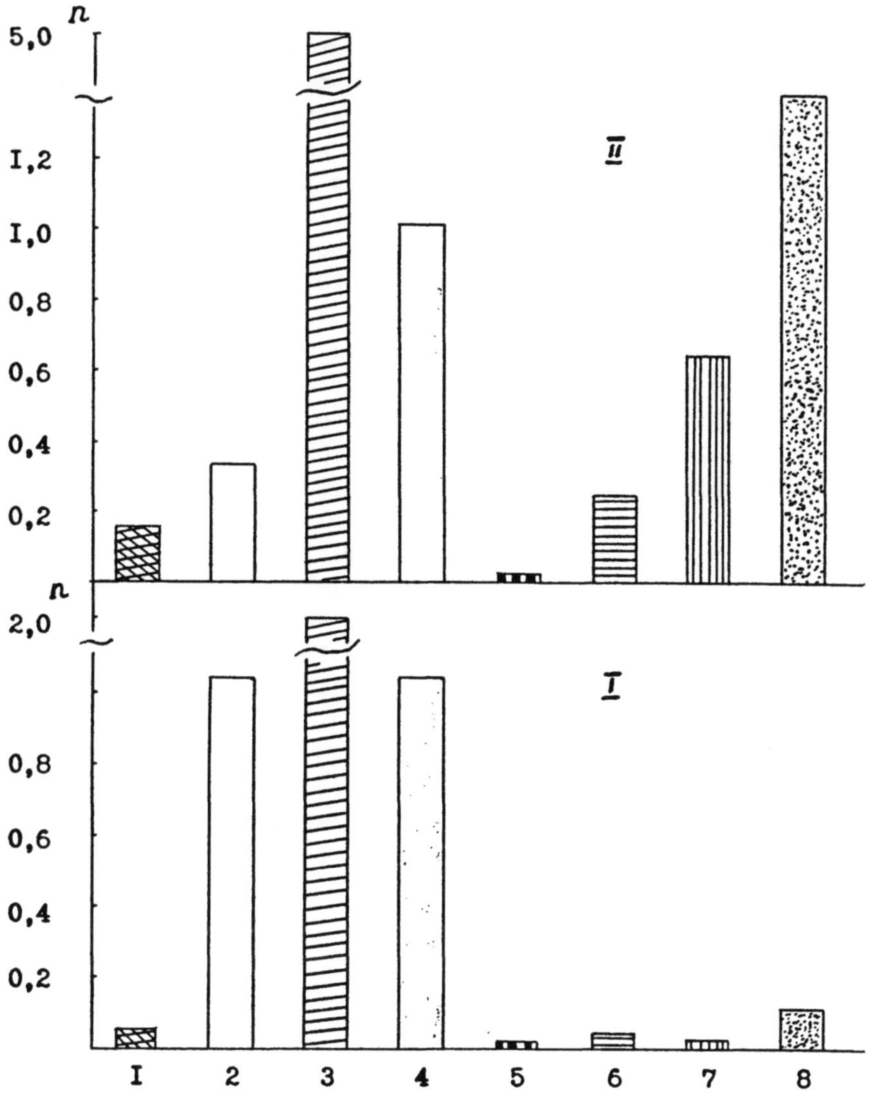

Fig.7. Quantitative correlation of selected volatile amines of strain lactobacilli Str. lactis-MGU in stationary phase (24h., I) and in decay stage (10 days, II) [18]. n- ratio of square of amine peak to square of peak of isobutyl-n-hexylamine (the square of isobutyl-n-hexylamine was accepted as 1).

1. Triethylamine, diethylamine
2. N-methylpyrrolidine
3. Diethylpropargylamine
4. Isobutyl-n-hexylamine
5. Isobutyl-cyclo-hexylamine
6. 2,5-and 2,6-dimethylpyrazine
7. 2,3-dimethylpyrazine
8. 2-methoxy-3-ethylpyrazine

The changes of the qualitative and quantitative composition of nitrogen-containing bases in the course of growth were typical for each of the five bacteria strains. As an illustration, the chromatographic profiles of volatile amines for three strains of lactobacilli, compared in the stationary stage of growth, were presented in a diagram in Fig. 8 [10,19]. The differences between the strains are clearly seen. It should be noted, that the cultures of bacteria used in production of "Biolact", valuable baby food, produce a minimal amount of volatile amines in growth period (Fig. 9) [20].

The chromatographic amine profiles can be used as a chemotaxonomic marker of lactobacilli and characterize culture purity.

They can also be used to determine the stages of the culture maximum productivity or its biomass accumulation. Thus, for example, the strain Str. lactis-MGU, nisin producer, releases the highest amount of N-methylpyrrolidine in the stage of maximum productivity of nisin, that can be seen from data given in Table 1. In this case the stage of maximum productivity of the strain Str. lactis-MGU coinciding with the maximum biomass accumulation (See the Table 1) can be determined from the amount of N-methylpyrrolidine. The dependence of amine composition on bacteria productivity during cultivation can be used for control of nisin production. N-methylpyrrolidine peak square can serve as an indicator of nisin accumulation [18].

Recently G.V.Morina and M.S.Umansky [22] investigated volatile amines in metabolites of 8 strains of microorganisms of series Str. lactis, Str. diacetilactis and Str. cremoris, cultivated on skimmed milk. It was found that all strains have a distinct difference in the chromatographic profile of organic bases (Fig. 10). The quantitative changes were significant in some of them. For example, dipropylamine ranged from 1,1 to 11%, isopentyl - from 1,9 to 18%, diethylpropyl- and isobutylamine - from trace amounts to 31,1%, tretbutylamine and diethylamine - from traces to 17,8%, and trimethylamine - from 24,7 to 84,8%, rel. The findings also show that the chromatographic profile of amines may serve as a chemotaxonomic marker of neighbouring microorganism cultures. The authors of the study [22] recommended to use the chromatographic profiles of amines in selecting strains as ferment components for cheese production.

Thus, total findings allow to recommend the chromatographic profiles of volatile amines as characteristics of microorganism cultures, since amine composition changes according to the proteolytic activity of microorganisms and the rate of protein containing source transformation during microbiological fermentation.

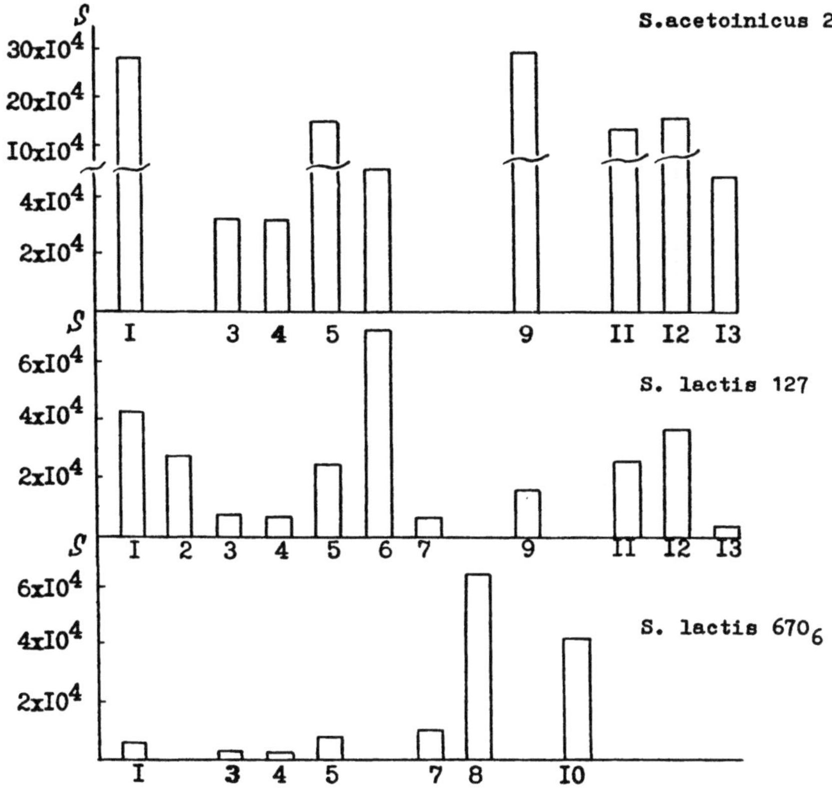

Fig. 8. Chromatographic profiles of amines in volatile metabolites of lactobacilli (the stationary phase) [10,19] . S - the value proportional to the square of peak on the chromatogram.

1.Trimethylamine	6. 6.n-Propylamine
2.Dimethylethylamine	7.Dipropylamine
3.Methylamine, dimethylamine	8.Isopropyl-n-butylamine
4.Ethylamine, isopropylamine, tret-butylamine	9.N-methylpyrrolidine
	10.Not identified
5.Diethylamine,triethylamine, diisopropylamine	11.Piperidine
	12.Pyridine

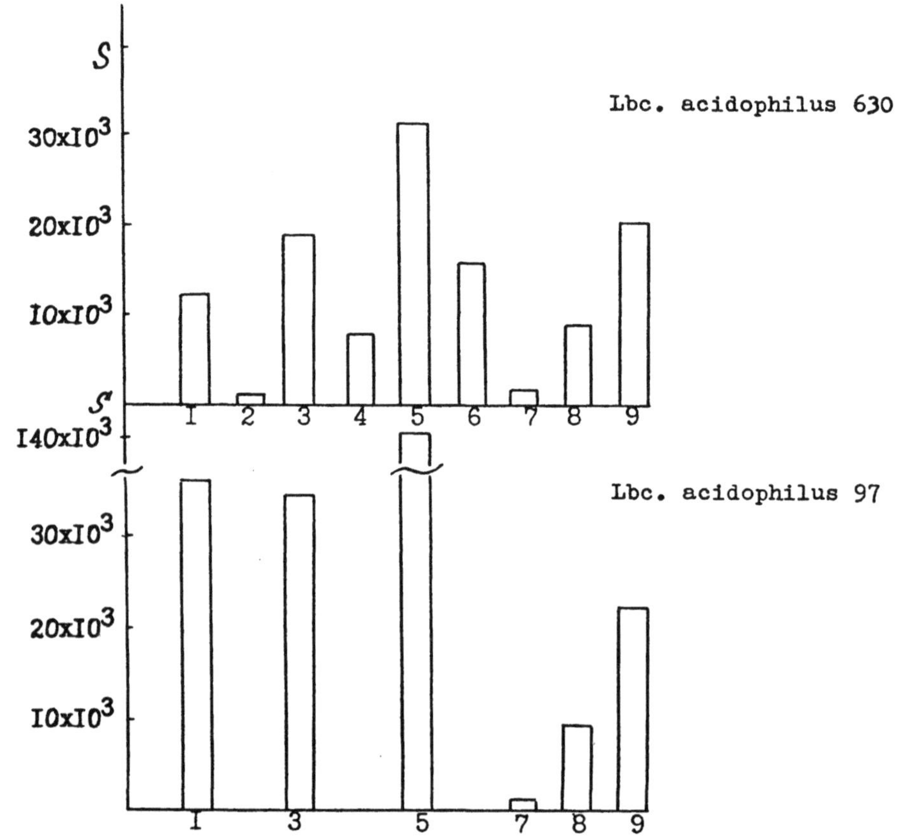

Fig. 9. Chromatographic profiles of amines in volatile metabolites of two strains of Lbc. acidophilus (the stationary phase) used in baby food "Biolact" [20]. S - the value proportional to the square of peak on the chromatogram.

1. Trimethylamine
2. Methylamine, dimethylamine
3. Not identified with I^{PEG}=849
4. Dimethyl-n-pentylamine
5. N-methylpyrrolidine
6. Isopropylisopentylamine, dimethyl-n-hexylamine
7. Piperidine
8. n-Hexylamine
9. Pyridine

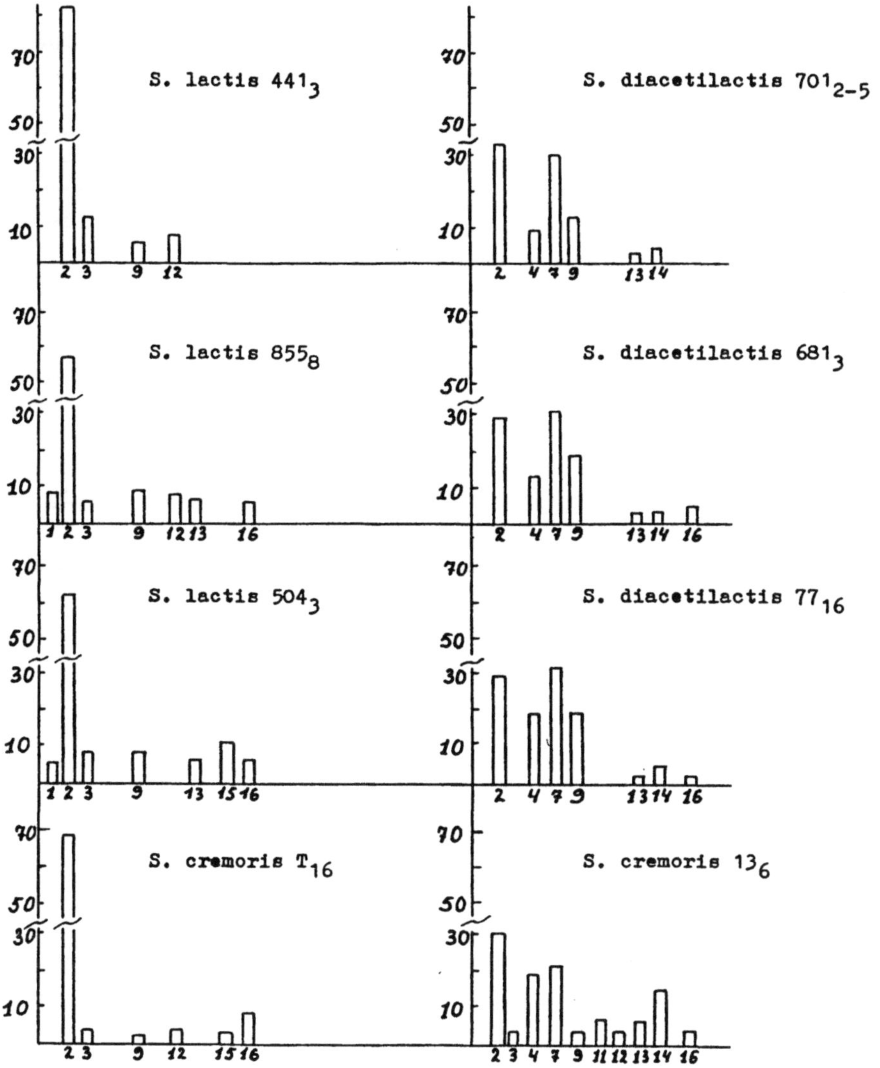

Fig.10. Chromatographic profiles of amines in volatile metabolites of 8 strains of lactobacilli (the stationary phase) [22]. The height of column is proportional to the contents of amines (rel.%)

Table 1
Dependence of nisin productivity by Str. lactis - MGU on the stage of growth indicated by quantity of N-methylpyrrolidine

Parameters	Stage of growth					
	Logarithmic phase				Stationary phase	Death period
	0 hours	6 h.	9 h.	13 h.	24 h.	10 days
Biomass (mg/l)	60	160	720	1330	1250	1100
Nisin ME/ml	-	-	1850	3300	2600	2200
N-methylpyrrolidine (rel.%)	23.2±2	24.7±4	38.5±2	48.5±5	17.1±2	5.2±2

The chromatographic profiles of volatile amines proved useful also for control of separate stages of technological procedures in production of various foodstuffs, including fermentation.

Wine-making

In wine-making, the enzymatic processes of some grape wine industrial production were studied. Thus, for sherry wines, the treatment with yeast races is the most responsible stage. Fermentation with sherry races of two wine materials ("Fetyaska" grape and white European varieties), alcoholized to 15-16 rev% and treated with races of aerobic sherry yeasts B-41 and 96-K, were studied [23]. Fermentation with sherry races lasts 6 months with the qualitative and quantitative composition of the volatile organic bases being changed.

In the volatile components of sherry wine materials 85 amines were identified, including 46 compounds which were first found in grape wines. A comparison of amines on chromatograms of the initial wine material and that one after 6 month fermentation with B-41 and 96-K sherry races allow to establish an increase in the number of highly volatile bases boiling at 110°, accomponied by a decrease in high-boiling amines (Fig. 11). The races differ in activity that was revealed by comparison of quantitative contents of individual components in chromatographic amine profiles. Irrespective of the initial wine material, the yeast race B-41 was the most active, that is illustrated by the diagrams of changes of the absolute content of nitrogen-containing bases scaled per 1 l of the sample (Fig.12). The sample, treated with the yeast race B-41, were also superrior to the samples with the race 96-K according to organoleptic estimation. With respect to the conclusion of the study [23] , the composition of the volatile bases can serve as a characteristic of yeast race and may be specific test of fermentation intensity with sherry races and formation of organoleptic wine characteristics.

Unlike fermentation with sherry races, champagne fermentation of wine with the culture of Saccharomyces vini yeast in industry carris out in anaerobic conditions. A comparison of the findings of the qualitative and quantitative composition of the volatile nitrogen bases at continuous champagne fermentation of wine materials allowed for the first time to identify over 60 amines in champagne. This results broadened our concepts of wine flavour and aroma formation [24]. Alongside with the primary, secondary and tertiary aliphatic amines, the pyridine and piperidine bases, pyrazines, diamines, aminoalcohols, aromatic amines, monoalkylacetamides were identified. The effect of gra-

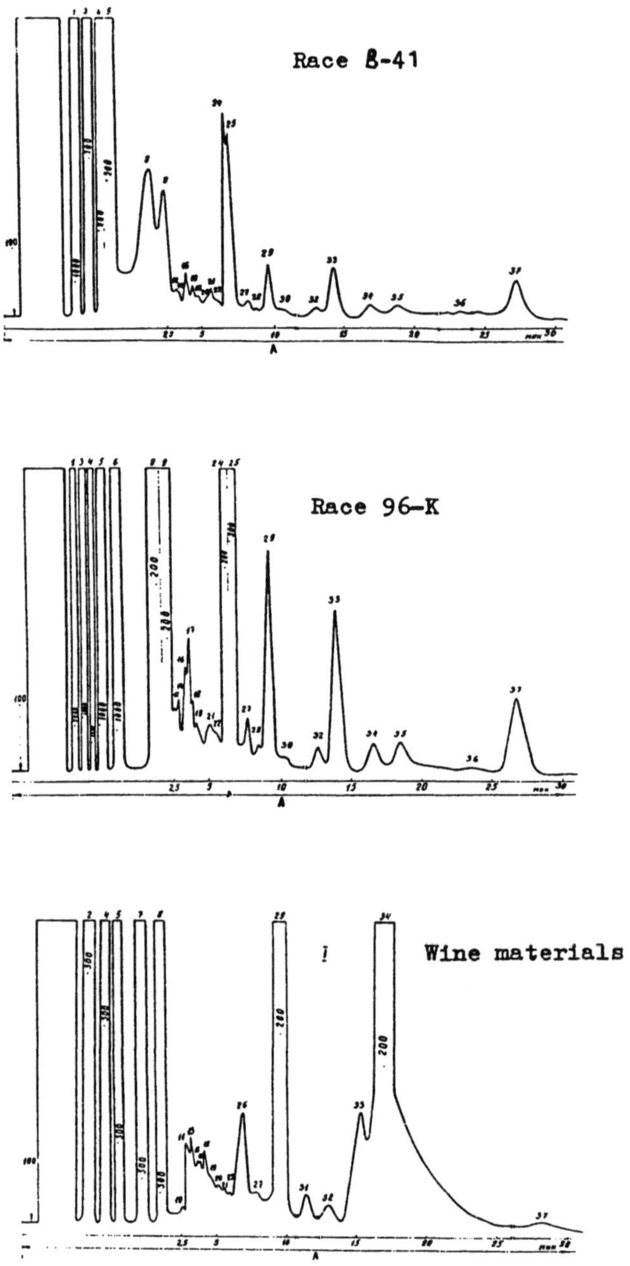

Fig.11. Chromatograms of volatile amines from wine materials after 6-month fermentation with sherry yeast races 96-K and β-41 [23]. Column with PEG-1000/Na_3PO_4, temperature - 150°C.

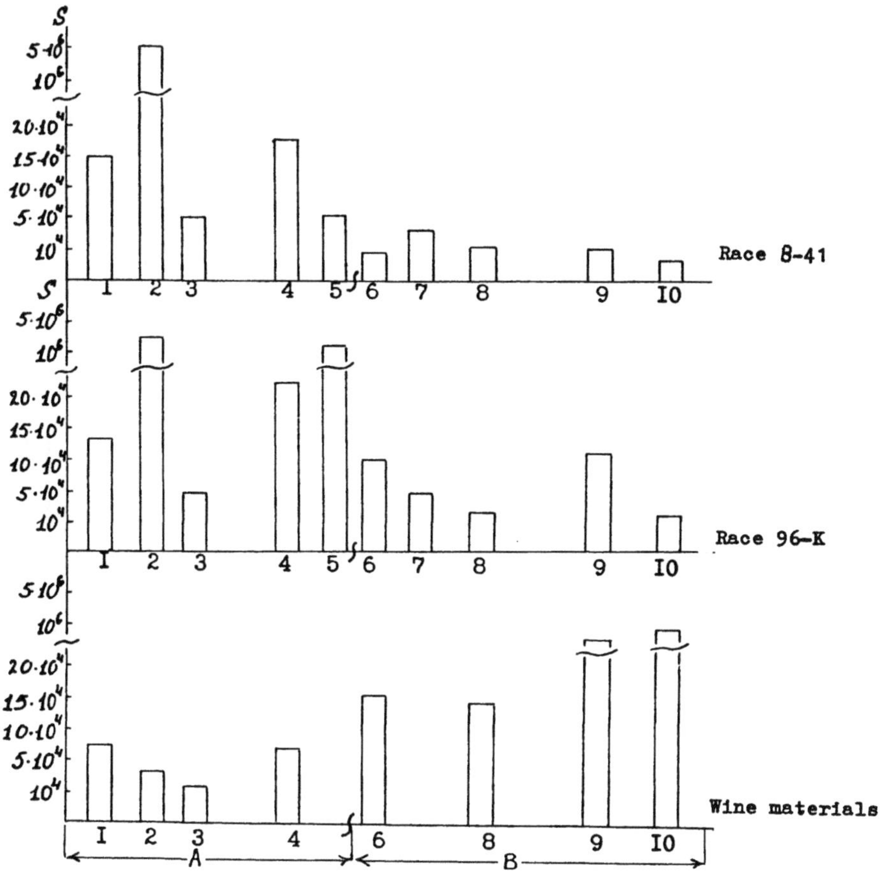

Fig. 12. Chromatographic profiles of volatile amines from wine materials after 6-month fermentation with sherry yeast races 96-K and β-41 [23]. S - the value proportional to the square of peak on the chromatogram.

1. Dimethylamine
2. Dipropylmethylamine, ethylpropylamine
3. Diethyl-sec-butylamine
4. N-ethylpyrrole, diisopentylamine
5. N-propylpyrrole, pyrazine
6. Not identified with $I^{PEG}=1276$
7. 2,6-Dimethylpyridine
8. 1,2-Diaminopropan
9. N-propylacetamide, isobutylacetamide
10. Not identified with $I^{PEG}=1959$

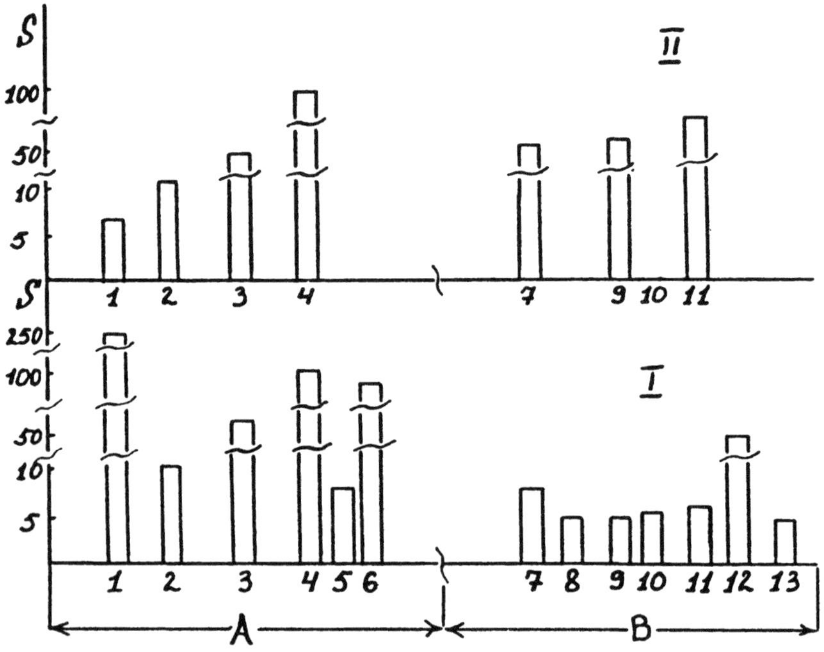

Fig.13. Chromatographic profiles of volatile amines of wine materials before and after champagne fermentation with yeast Saccharomyces vini [24]. A - temperature - 100°C, B - 150°C. S - the value proportional to the square of peak on the chromatogram.

1. Trimethylamine
2. Ethylamine
3. Isobutylamine
4. Methylbutylamine, isopentylamine
5. n-Pentylamine
6. Dibutylethylamine
7.
8. Ethanolamine
9. Putrescine
10. Isopropylacetamide
11. Cadaverine
12. Not identified with I^{PEG}=1582
13. Not identified with I^{PEG}=1623

pe variety on the composition of organic bases after champagne fermentation is insignificant indicating that the profile of amines is formed during fermentation only. The latter is illustrated with a diagram in Fig. 13 showing the changes in the content of some organic bases after 17 days of champagne fermentation at 10-15°. Unlike the aerobic fermentation with sherry races in the anaerobic champagne fermentation a tendency toward a decrease in highly volatile amines, such as methyl-, ethyl-, propyl-, n-pentyl-, dipropylamine and others is observed. On the other hand, such high-boiling compounds as ethanolamine, putrescine, N-isopropylacetamide are accumulated. The changes in volatile amine composition depends on complex biochemical transformations occurring in the course of growing of yeast cells responsible for the technological champagne fermentation.

The results of these studies allow to conclude that the qualitative and quantitative composition of the volatile nitrogen-containing bases could serve as a specific test of the intensity of champagne fermentation and fermentation with sherry races in industrial conditions. The findings allow to relate the presence and type of changes of nitrogen-containing bases to the trend of the technological process and enable its control for production of high quality wine-making products.

Bread baking

In bread baking, studies on the composition of volatile organic bases related to wheaten bread making were conducted.

It should be noted that formerly gas chromatography was used to study the volatile carbonyl compounds and changes of their composition in the course of wheaten bread baking, but no distinct relationship with technological stages was found [25, 26].

The study of volatile amines has shown that this group of compounds is essential for bread flavour formation [26-30]. Accumulation of aroma forming substances occurs basically in the stage of sponge and dough fermentations and develops finally during bread baking as a result of biochemical transformations of proteins, carbohydrates, amino acids under the effect of flour and yeast enzymes, and in the final stage during baking at high temperature. Proceeding from the concept that the organic bases can serve as specific indicators of intensity of biochemical and microbiological processes, we studied the composition of volatile organic bases and its changes in the course of wheaten bread baking using the traditional sponge dough method. The amines were determined in the following technological stages: after kneading and the end of fermentation sponge and dough, proofing of dough and bread baking [30]. The amine chromatograms illustrating only three points of the bread making process are given in Fig. 14, i.e. sponge and the dough after fermentation and the bread (separately for the crumb and crust). It can be seen that each stage is characterized by its own chromatographic profile. In the chromatogram of volatile dough amines the number of peaks is the highest at the end of fermentation. As compared to sponge, a number of new substances emerge: the peak 12 disappears, the peak 18 and a number of others emerge. The quantitative relationship of amines also undergoes substantial changes in all stages, especially in that of dough making (Table 2). It emphasizes the importance of the processes occurring in the course of dough fermentation and gives an opportunity to use the chromatographic amine profiles as indicators of the intensity of these processes. On the whole, about 100 organic bases were found by chromatography in all stages of bread baking, including 56 identified and 43 conventional ones. The duration of sponge and dough fermentation was found to affect the qualitative composition and quantitative relationship of amines in bread flavour condensates of crumb and crust.

Proceeding from the concept that the chromatographic amine profiles contain information about the processes occurring in bread baking, the chromatograms obtained in stages of traditional wheaten bread production were used as the standard and the bread sample as the control in development of a new compact technological process of wheaten bread production using the yeast pour half-finished product (HFP) [29, 30].

The chromatograms of volatile amines in control dough (III) and the dough on the pour half-finished product (I.) are presented in Fig. 15. It can be seen that the number of volatile amines on the chromatograms of the dough with HFP is much higher than on the chromatogram for control dough (compare No 18, 19, 25). Baked bread was of inferior quality to the control. The addition of 0,05% acetic acid to the pour half-finished product enabled normalization of the dough making stage. At the end of fermentation the amine chromatogram of the dough with HFP (II) was closer to the control (III). The organoleptic properties of bread made with HFP supplemented with acetic acid also improved and neared the control bread as it seen from Table 3.

The organic bases, product of protein substance decomposition, can serve as indicators of the processes occurring at dough fermentation It is recommended to use the results of amine analysis in the stage of dough fermentation to determine the optimal parameters of wheat

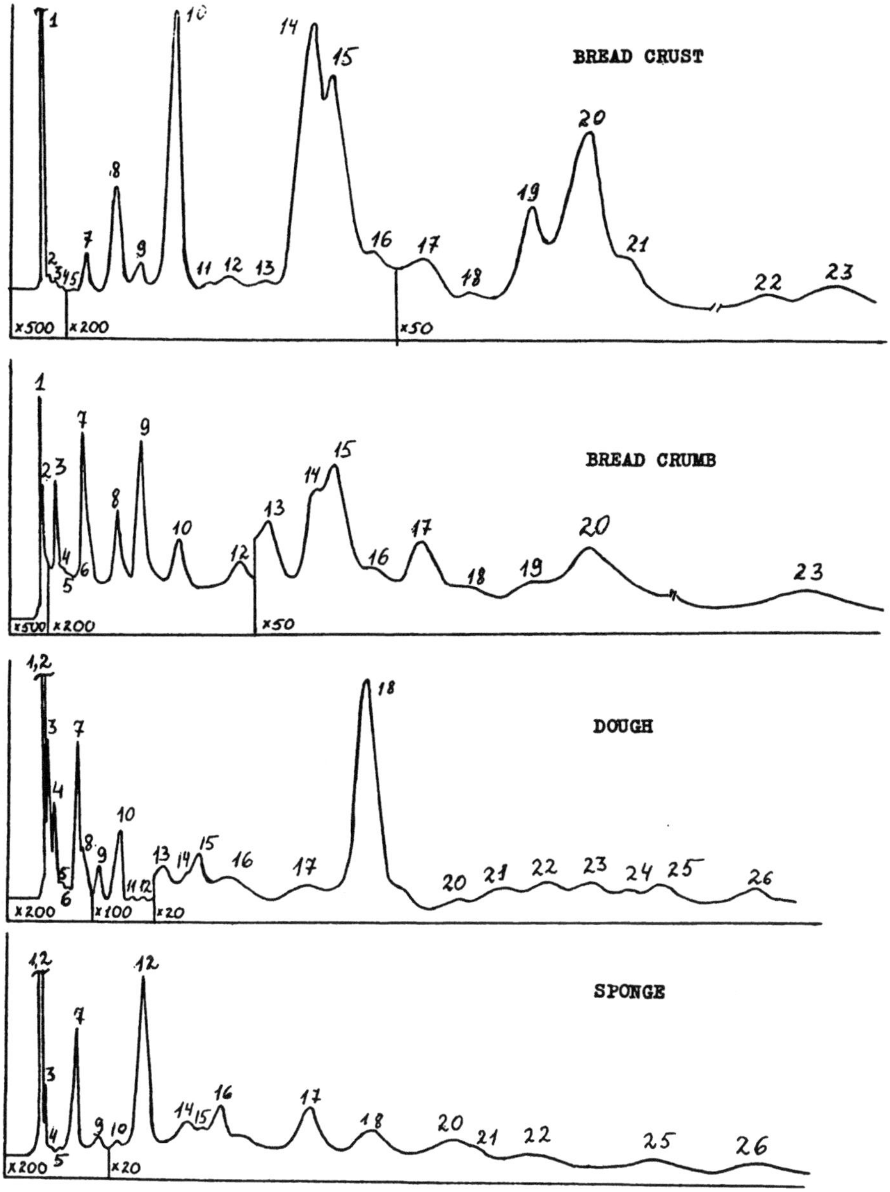

Fig. 14. Change of GC-profiles of volatile amines in the process of dough making and baking of white wheaten bread [30].

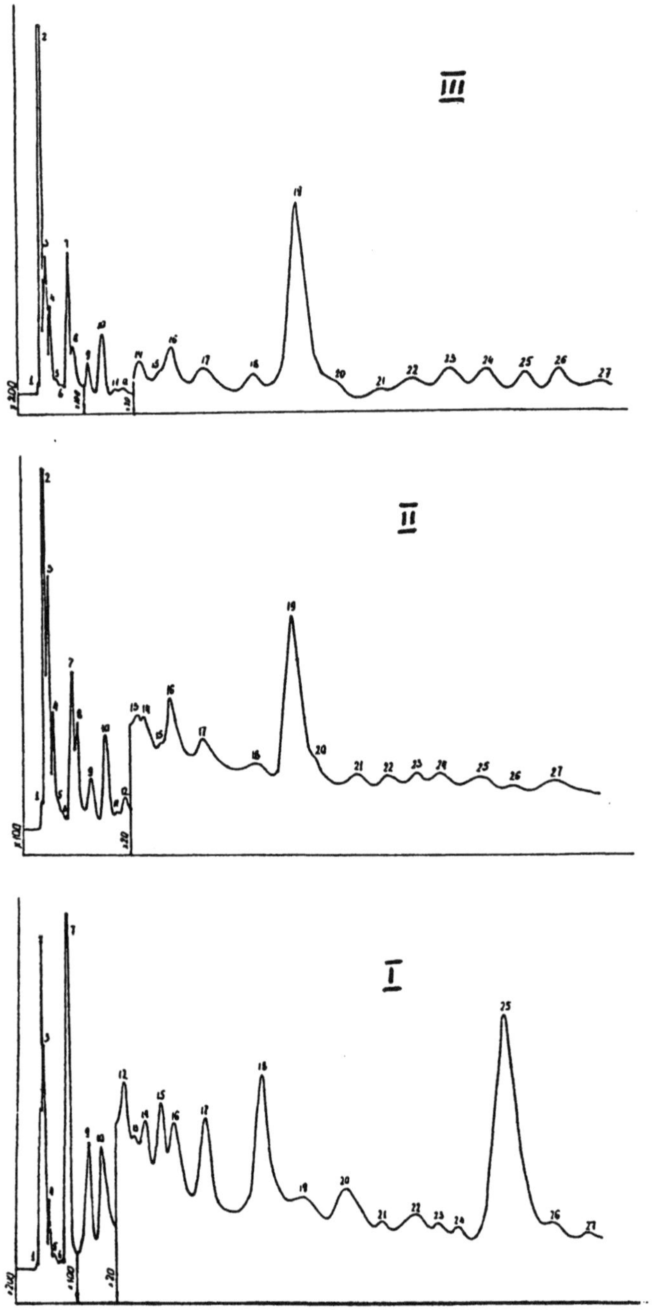

Fig.15. Volatile amines of wheaten dough prepered on pour half-finished product (H.F.P.) [30]. I - Dough on H.F.P., II - Dough on H.F.P. + 0,05% CH_3COOH, III - Dough prepered with traditionaly method (control).

Table 2
Quantitative relationships (% rel.) of some volatile organic bases of sponge, dough and wheaten bread (column with ApL at 100°)

AMINE	Sponge at the end of fermentation	Dough at the end of fermentation	Bread crumb	Bread crust
Methyl-, ethyl- dimethyl-, trimethyl- methylethyl-	48,5	30,1	24,7	21,5
Propyl-cyclo-pentyl- Trimethylpyrazine	1,5	1,0	0,9	5,1
2,3-dimethylpyrazine	1,4	0,4	4,3	9,6
Dimethylisopentyl- ethyl-n-butyl-	6,2	2,4	-	-
Methyl-n-pentyl- N-ethylpyrrole propyl-sec-butyl-	6,0	6,0	-	-
Diethyl-sec-butyl- di-sec-butyl- methyl-sec-heptyl-	2,6	0,6	-	-
Methyl-sec-butyl- dimethyl-cyclo-hexyl- dimethyl-n-heptyl-	1,4	11,6	-	-
Ethylisopentyl- α-picoline	-	-	5,0	14,9
Pyrrolidine isopentyl- diisobutyl-	-	-	22,6	1,8

Table 3
Organoleptic evaluation of white wheaten bread prepared from pour half-finished product (HFP)

PARAMETER	Dough from HFP	Dough from HFP + 0,05% CH_3COOH	Control dough (ordinary method)
Odour (w.f. = 0,8)	1,6	2,8	3,2
Taste (w.f. = 0,8)	1,6	2,8	3,2
Total organoleptic evaluation*	11,2	15,9	16,1

*evaluation was carried out on 8 parameters 20 point system taking into account weight factors(w.f.)

half-finished product ripening in developing new progressive techniques of bread, roll and bun production.
In conclusion it should be noted that, according to the findings, the research into the composition of volatile amines in foodstuff prodution, including fermentation, is promising. The chromatographic amine profiles used for control of traditional technological processes and purposeful search of new ones allow to ensure high organoleptic properties of the final foodstuffs. The thermodynamyc approach to flavour research of textured foodstuffs as a research of thermodynamically nonequilibrium systems of self-coordinated reactions, producing volatile substances is promising fruitful.

Literature

[1] Kolb, B.: Angewandte Gas-Chromatographie, 15, Perkin-Elmer, Überlingen, 1972.

[2] Vitenberg, A.G. and B.V. Ioffe: Gasovaya Ekstraktsiya in Khromatograficheskom Analize, Khimiya,Leningrad, 1982.

[3] Golovnya, R.V.: in Prikl. Kromatografii, K.I. Sakodynskii (ed.), Nauka, Moscow, 1984.

[4] Golovnya, R.V.: Talanta, 34 (1987) 1, 51-60.

[5] Talyzin, V.V., V.Ya. Anisimov, O.I. Yakovleva, T.A. Mischarina and R.V. Golovnya: J. Analit. Chimii, 42, (1987), 2076-2079.

[6] Mischarina, T.A., R.V. Golovnya and S.V. Vitt: Prikl. Biokhim. i Mikrobiol., 21 (1985), 107-110.

[7] Svetlova, N.I., Golovnya R.V,, I.L. Zhuravleva, D.N. Grigoryeva and A.L. Samusenko: Nahrung, 29 (1985), 143-151.

[8] Mischarina, T.A., S.V. Vitt and R.V. Golovnya: Biotechnologiya, 2 (1987), 210-216.

[9] Golovnya, R.V., I.L. Zhuravleva and S.G. Kharatyan: in Letuchie Biologicheski Aktivnya Soedineniya Biogennogo Proishozhdeniya, M. Telitchenko and A. Tambiev (eds.), MGU, Moscow, 1971.

[10] Golovnya, R.V., I.L. Zhuravleva and S.G. Kharatyan: J. Chromatog., 44 (1969), 262-268.

[11] Brooks, J.B. and W.E. Moore: Can. J. Microbiol., 15 (1969),433-437.

[12] Thacker, L. and J.B. Brooks: Infect. Immun., 9 (1974), 648-653.

[13] Dunn, S.R., N.L. Simenhoff and L.G. Wesson: Anal. Chem., 48 (1976), 41-46.

[14] Larsson, L., P.A. Mardh and G. Odham: Acta Pathol. Microbiol. Scand., 86B (1978), 207-213.

[15] Brooks, J.B., D.S. Kellogg, Jr. M. E. Shepherd and C.C. Alley: J. Clin. Microbiol., 11 (1980), 52-58.

[16] Davis, T.Y. and N.J. Hayward: J. Chromatog., 307 (1984), 11-15.

[17] Brondz, I. and I. Olsen: J. Chromatog., 379 (1986), 367-411.

[18] Golovnya, R.V., I.L. Zhuravleva, M.B. Terenina, A.N. Polin and V.A. Grushina: Biotechnologiya, 5 (1985), 51-58.

[19] Golovnya, R.V., M.B. Terenina and I.L. Zhuravleva: Prikl. Biokhim. i Mikrobiol., 22 (1986) 2, 217-225.

[20] Terenina, M.B., R.V. Golovnya and I.L. Zhuravleva: Biotchnologiya, 1 (1986), 35-40.

[21] Golovnya, R.V., M.B. Terenina and I.L. Zhuravleva: Abstr. II Wartburg Aroma Symosium, 1987.

[22] Morina, G.V. and M.S. Umansky: Prikl. Biokhim. i Mikrobiol., 27 (1987) 2, 275-280.

[23] Kishkovskii, Z.N., L.P. Palamarchuk, R.V. Golovnya and I.L. Zhuravleva: Prikl. Biokhim i Mikrobiol., 16 (1980) 3, 446-459.

[24] Volodarskaya, N.S., L.V. Palamarchuk, Z.N. Kishkovskii, I.L. Zhuravleva and M.B. Terenina: Tezisy v Vsesoyuznoi Konferentsii po Analiticheskoi Khimii Organicheskikh Soedinenii, Nauka, Moscow, 1984.

[25] Rothe, M.: Handbuch der Aromaforschung. Aroma von Brot. Akademie-Verlag, Berlin, 1974.

[26] Golovnya, R.V., N.G. Enikeeva, I.L. Zhuravleva and A.S. Zjuzko: Nahrung, 18 (1974) 2, 143-156.

[27] Zjuzko, A.S., N.G. Enikeeva, I.L. Zhuravleva and R.V. Golovnya: Izvestiya VUSOV, Pitshevaya Technologiya, 3 (1973), 161-164.

[28] Zjuzko, A.S., N.G. Enikeeva, R.V. Golovnya and I.L. Zhuravleva: Prikl. Biokhim. i Mikrobiol., 10 (1974) 3, 443-449.

[29] Kichaeva, T.T., N.G. Enikeeva, L.I. Pushkova, R.V. Golovnya, I.L. Zhuravleva and M.B. Terenina: Izvestiya VUSOV, Pitshevaya Technologiya, 4, (1984), 101-103.

[30] Kichaeva, T.T., N.G. Enikeeva, I.L. Zhuravleva and R.V. Golovnya: Prikl. Biokhim. i Mikrobiol., 23 (1987) 3, 418-425.

POTENTIAL MEANS FOR IMPROVING THE FLAVOUR OF ALCOHOLIC BEVERAGES

Lalli Nykänen[1)]

In the manufacturing of alcoholic beverages, the fermentable carbohydrates in the mash are converted by yeast to alcohol and carbon dioxide. Although the processes are well understood in the alcohol industry, some problems occasionally appear which demand special investigations. One of the most common problems is the occurrence of an off-flavour in highly rectified alcohol. The compounds responsible for the off-flavour generally are present in very low concentrations and, therefore, may be difficult to remove by means of distillation. An important question is, how effective must the spirit distillation be for the rectified spirit to be used in the production of the beverages of vodka types.. One must keep in mind that many vodkas contain an easily recognized grain flavour. Other vodkas, such as the Finlandia Vodka produced by the Finnish State Alcohol Company, however, are unflavoured and found to be free from all the flavour compounds typical of grain. In this case even an extremely small quantity of an off-flavour is easily detected. Therefore, in the alcohol beverage industry, the producers of grain spirits for top quality vodka have paid special attention to the rectifying process.

There is no doubt that the flavour is the most important feature of an alcoholic beverage and that many beverages are produced mainly for their palatable properties. Consequently, it is not surprising that many researchers have been interested in the composition of the chemicals responsible for flavour and tried to find relationships between the composition and gustatory properties. The formation of flavour compounds during fermentation, the occurrence of the most prominent compounds in distilled beverages, and the composition of the flavour have been elucidated in our older review (41). Recently, more than 1300 flavour compounds occurring in beer, wine and distilled alcoholic beverages have been reviewed (23).

Estimation of flavour by sensory analysis

Sensory analysis has been used for estimation of the flavour characteristics for a long time, and it seems likely to remain the most important tool for improving beverages in the future despite the development of many good instrumental methods. Usually the terms used in commercial

---

[1)]Research Laboratories of the Finnish State Alcohol Company, Alko Ltd. POB 350, SF-00101 Helsinki, Finland

degustation are somewhat different from those involved in determining the flavour profile in sensory studies. Although there are a great number of publications in the literature dealing with sensory analyses of alcoholic beverages and special terminology, I have taken only one example by Ann Noble (24). She made a wide inquiry about the terms used in American wineries and on the basis of the responses, she proposed a list of analytical terms which are shown in the wine aroma wheel in Fig. 1. Similar flavour terminology is used to describe the flavour characteristics in the beer (14) and whisky industries (37).

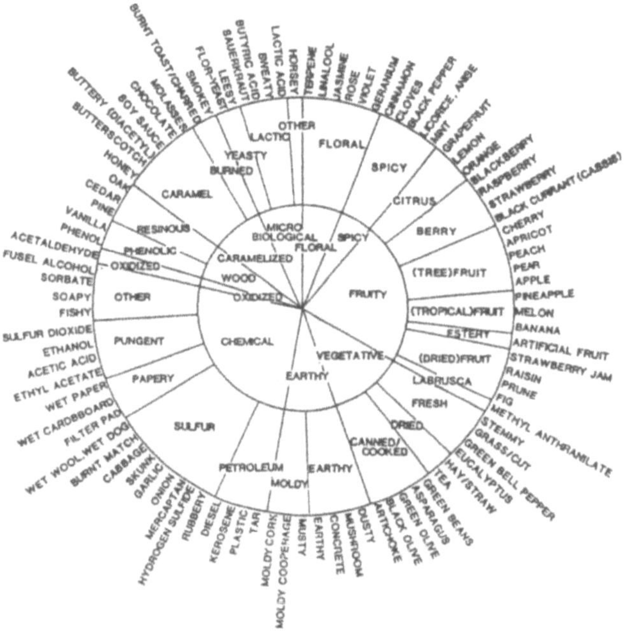

Fig. 1. Wine flavour wheel compiled by A. Noble (24).

Origin of flavour compounds

Since the 60's different instrumental methods, particularly, gas chromatography, infra red spectrometry and mass spectrometry, have been widely used to identify flavour compounds in alcoholic beverages. The investigations indicate that all the flavour compounds of the beverages are not formed at the same phase of the manufacturing process. Principally, the flavour compounds of the beverage originate from three sources. Some flavour compounds have been shown to occur in free or bounded form in the raw materials from which they are liberated during mashing and fermentation (39,53). A great number of the flavour compounds are formed during fermentation by yeast from sugars and other nutrients which are present in the medium. Particularly the yeast and the fermentation conditions may have a decisive effect on the composition of the flavour (18,19,40,43,44). With regard to distilled beverages, the distillation procedure naturally influences the composition of the flavour by stripping both low boiling and high boiling compounds from the distillate. In many beverages the final flavour composition is achieved only after maturation. During the maturation process, some new compounds are formed in slow chemical reactions, while some other compounds disappear (6,32). Furthermore, a number of the compounds are liberated by alcohol from the wooden barrels used for the maturation (18,20,28,32,33). In fact, there are many possibilities to improve the flavour by changing the manufacturing conditions. By tasting and smelling different beers, wines, whiskies, cognacs and rums, one can easily find out that there are differences in flavours between the different brands.

Grapes as a source of terpenic compounds

The flavour of wine is known to involve a vast number of terpenic compounds. It has been shown that many volatile compounds originally present in grapes contribute varietal characteristics of the grapes to wines, thus enabling wines made from individual cultivars to be differentiated by analytical methods (31,34-36,52-54). Many of these grape compounds are monoterpene alcohols and oxides which seem to be transferred more or less unchanged into the wine. An interesting finding is that only a portion of terpenes occur in a free form while others are bound in the grapes as glycosides. Williams (39,53,54) was able to isolate glycosidic derivatives from Muscat grapes and confirm their structure by analytical methods. They found that the precursors in both Muscat of Alexandria and Riesling grapes were not simply β-D-glucosides, but a complex mixture of β-rutinosides and β-O-α-L-arabinofuranosyl-β-D-glucopyranosides of several monoterpene alcohols. The pattern and concentration of free terpene compounds in grapes and wines are dependent on the mechanism by which the compounds are formed. In an acidic medium, glycosides are hydrolyzed by the acids to sugars and the terpene alcohols, but enzymes that can accomplish the hydrolysis in grape juice also occur in grapes. Because substantial amounts of grape monoterpenes seem to be tied up in the fruit as glycosides or as flavourless polyols, they form a potential reserve of flavour in wine making.

Flavour contribution of fermentation

Most of the compounds formed during fermentation are volatile. Although the term "volatile compound" is a rather diffuse one, generally speaking all the compounds occurring in alcoholic beverages can be grouped according to whether or not they are distilled along with the alcohol and water steam. Among the volatile compounds there are aliphatic carbonyl compounds, alcohols, monocarboxylic acids and their esters, nitrogen and sulphur containing compounds, hydrocarbons, terpenic compounds, and heterocyclic and aromatic compounds. The major part of the non-volatile extract of the alcoholic beverages consists of unfermented sugars, di- and tribasic carboxylic acids, coloured substances, and tannic and polyphenolic substances (23).

Because esters are numerically the largest group of flavour compounds, their formation has been thoroughly investigated (17-19,40,41,43,44). A possible mechanism used for illustrating the formation of esters in fermentation is based on investigations performed by Nordström (15).

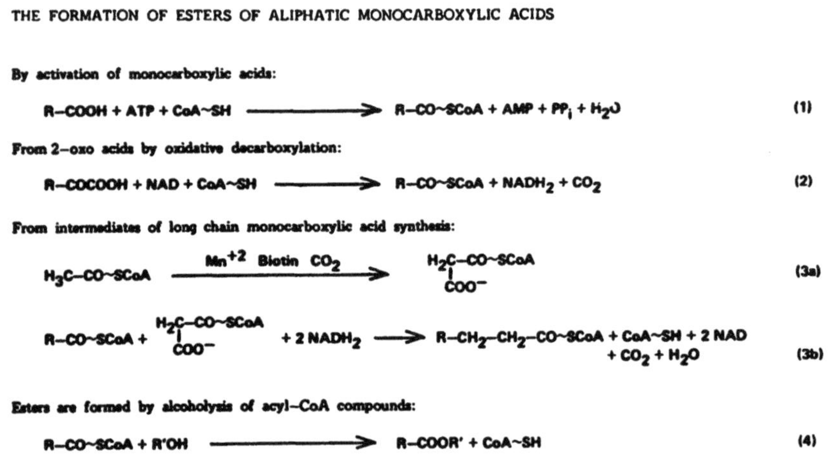

Fig. 2. Scheme for the fatty acid ester synthesis (15).

According to the mechanism described in Fig. 2, acyl-coenzyme-A functions as the key compound in the biosynthesis of esters. During fermentation yeast produces acyl-CoA in cells either through the activation of fatty acid or through the oxidative decarboxylation of keto acid. These two mechanisms mainly differ in their capacity to utilize ATP, the former requiring ATP and the latter not. Furthermore, in the ester synthesis as well as in the fatty acid synthesis, the lengthening of the carbon chain of the acid moiety occurs so that malonyl-CoA bonds with acyl-CoA in the enzyme complex, bringing two more carbon atoms into the chain of the acid. The final step is the cleavage of the enzyme from the complex. In the presence of alcohol, the reaction produces an ester, whereas when water is present, the result is a free fatty acid. The metabolic pathway in yeast cells beginning from acetyl-CoA, therefore, leads to esters whose acid moiety has an even number of carbon atoms. In fact, esters having an even carbon number regularly appear as primary components in the ester fractions of wines and distilled alcoholic beverages.

Particular attention has also been given to the ester formation by different yeasts. *Hansenula anomala* and *Candida krusei* yeasts, for example, have been found to produce more ethyl acetate than do *Saccharomyces cerevisiae*, *Schizosaccharomyces pombe* and *Pichia membranaefaciens* (38). On the other hand, of these five yeasts, *Hansenula anomala* and *Candida krusei* have been found to form the lowest amounts of the ethyl esters of octanoic, decanoic and lauric acids (25). We investigated the ability of strains of *Saccharomyces cerevisiae* and *S. uvarum* yeasts to produce acetates of isopentyl and phenethyl alcohol and ethyl esters of the C_6 - C_{12} fatty acids in semiaerobic sugar fermentations, performing the fermentations with 57 strains of *Saccharomyces cerevisiae* and three strains of *S. uvarum* yeasts (18,21). The yeast strains of *S. cerevisiae* were found to produce more isopentyl acetate, ethyl hexanoate, ethyl octanoate, ethyl decanoate and phenethyl acetate than do *S. uvarum* yeasts. Furthermore, we were able to show that the distribution of the fatty acid esters between the yeast cells and the fermentation medium is dependent on the chain length of the acid part of the esters. Both of the acetates, isoamyl acetate and phenethyl acetate, and ethyl hexanoate appear solely in the medium, whereas ethyl octanoate and decanoate are present in both the medium and the cells. Ethyl laurate was found only in the cells. The distribution of esters between the medium and the yeast cells seem to be dependent upon the yeast species. The *S. uvarum* strains appear to retain relatively more of the total amount of ethyl octanoate and decanoate in the cells than do the *S. cerevisiae* strains. The esters are liberated from the yeast cells by distillation and, hence, the quantity of the esters amounts in distilled beverages can be regulated by performing the distillation in the presence or absence of the yeast.

The effect of oxygen on the formation of esters

Aerobic conditions are known to have a significant effect on the formation of esters. Nordström (15) concluded that aerobic conditions for a fermentation solution restrict the formation of isopentyl acetate and ethyl hexanoate in fermentations by *S. cerevisiae*. According to Bertrand (7) aerobic conditions cause the formation of differing amounts of esters in fermentations performed with different yeasts. Bertrand reported that *S. ellipsoideus* and *S. oviformis* in a strictly anaerobic fermentation produce eight times more isopentyl acetate and five times more ethyl hexanoate than in a semi-aerobic fermentation, whereas *Schizosaccharomyces liquefaciens* produces isopentyl acetate and ethyl hexanoate approximately in the same amounts under anaerobic and semi-aerobic conditions. The formation of isopentyl acetate by *Saccha romycodes ludwigii* declines to a third when shifting from semi-aerobic to anaerobic fermentation. Of the yeasts used by Bertrand (7), *Kloeckera apiculata* was found to produce the highest levels of

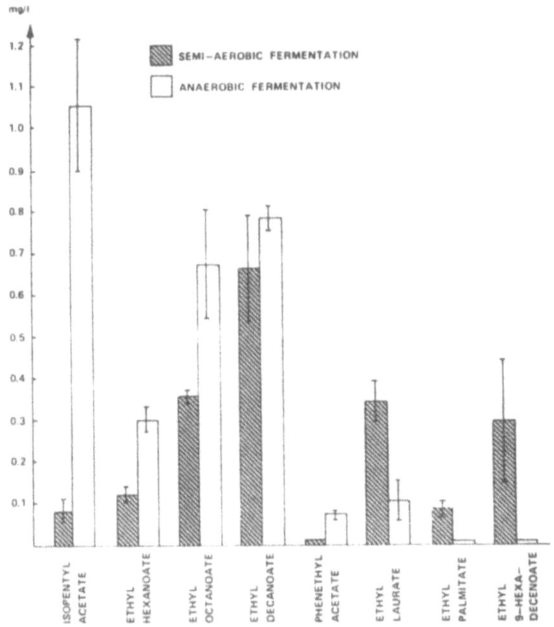

Fig. 3. The formation of esters in semi-aerobic and anaerobic sugar fermentation by Saccharomyces cerevisiae yeast.

isopentyl acetate but only in semi-aerobic fermentation; the yeast does not produce it at all in anaerobic fermentation.

We found that the fermentation by a S. cerevisiae wine yeast, obtained from Geisenheim, produced under strictly anaerobic conditions considerably more of isopentyl acetate than in semi-aerobic conditions (18,21). Furthermore, we showed that twice as much ethyl hexanoate and octanoate as well as phenethyl acetate was formed in anaerobic fermentation than in semi-aerobic fermentation. It is also worth mentioning that clearly less ethyl laurate is produced in a fermentation performed under anaerobic conditions. Considerably less ethyl palmitate and an unsaturated ester, ethyl 9-hexadecenoate, also are produced by this wine yeast under anaerobic conditions than under semi-aerobic conditions.

The effect of alcohols on the formation of esters

While investigating the effect of added alcohols on the formation of esters, we found that the addition of 2-methylpropanol and 3-methylbutanol to the fermentation mixtures interrupted the synthesis of fatty

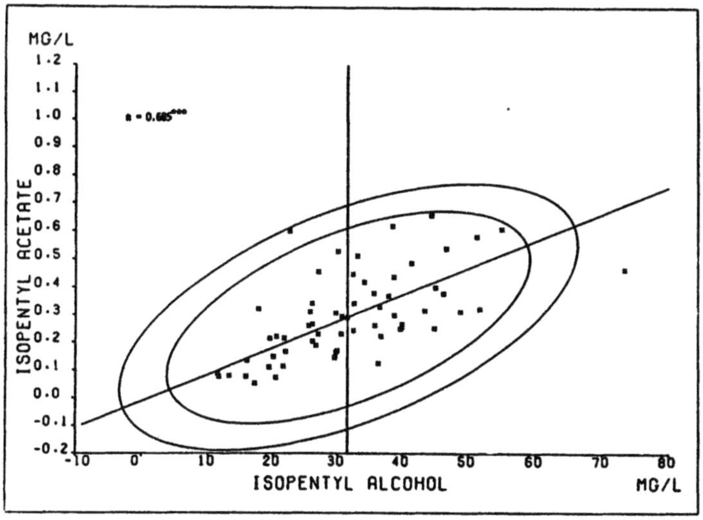

Fig. 4. Relationship between the amounts of iso-pentyl alcohol and acetate formed by S.cerevisiae yeast in sugar fermentation.

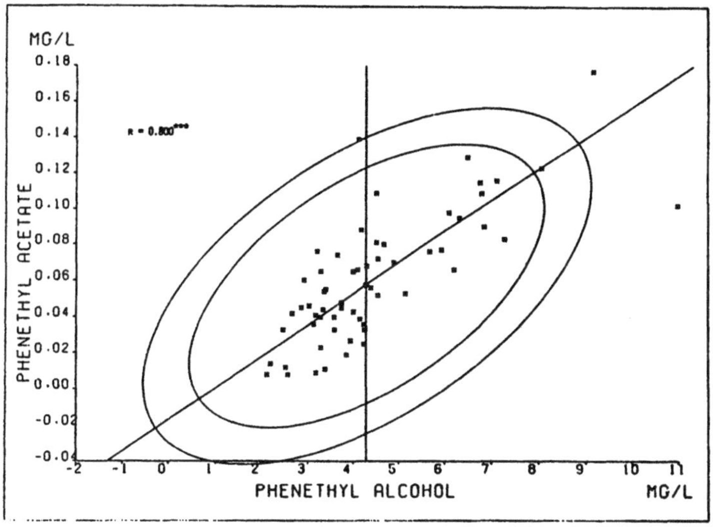

Fig. 5. Relationship between the amounts of phenethyl alcohol and acetate formed by S. cerevisiae yeast in sugar fermentation.

acids so that no acids higher than acetic acid were formed. Thus, only 2-methylpropyl acetate and 3-methylbutyl acetate were formed from the added alcohols. This raised the possibility that the formation of isopentyl acetate depends upon the formation of isopentyl alcohol (Figs. 4 and 5) and, correspondingly, phenethyl acetate upon the formation of phenethyl alcohol in sugar fermentation (17). The gas chromatogram in

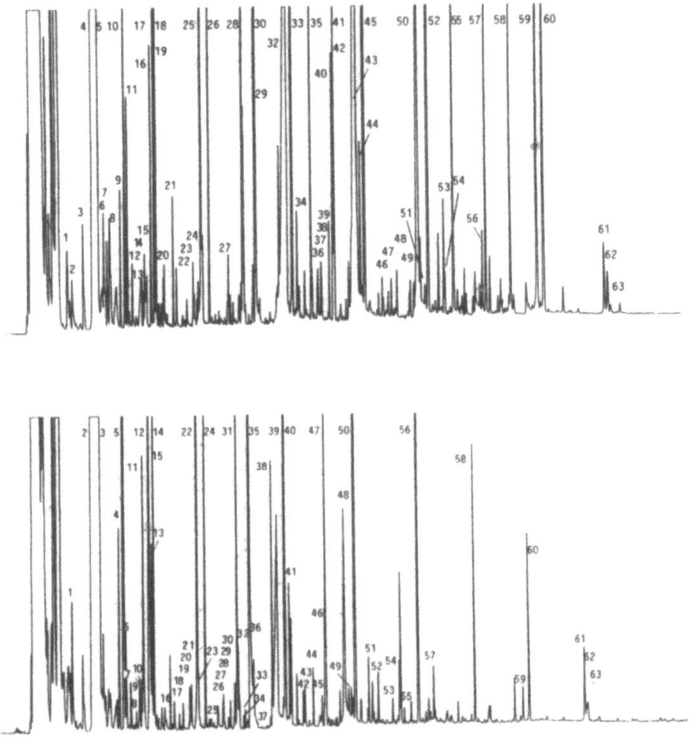

Fig.6. The flavour compounds (see text) of Scotch whisky (upper) and French cognac (lower) gas chromatographed on a 50 m OV-101 fused silica capillary column (17).

Fig.6 illustrates the composition of an extract from whisky. Ethyl esters having even carbon atoms in the acid moiety are prominent. The occurrence of palmitoleic acid ethyl ester (peak 59) and palmitic acid ethyl ester (peak 60) in about equal concentrations is typical of Scotch whisky. The large amounts of the unsaturated ethyl palmitoleate in Scotch whisky is probably the result of fermentation of the wort under semiaerobic conditions. The occurrence of the long chain carboxylic acid esters, such as ethyl laurate (peak 52) and ethyl myristate (peak 57), in addition to the long chain alcohols, tetradecanol (peak 55) and hexadecanol (peak 58), is conspicuous. The fatty flavour of these compounds may be responsible for the stearinelike smell that is characteristic of Scotch malt whisky. The the flavour composition determined in French cognac (22) can be found to be qualitatively similar to that for Scotch whisky.

The role that yeast plays in establishing the basic flavour constituents of alcoholic beverages is notable; flavour compositions of beverages produced from different raw materials are similar if the same

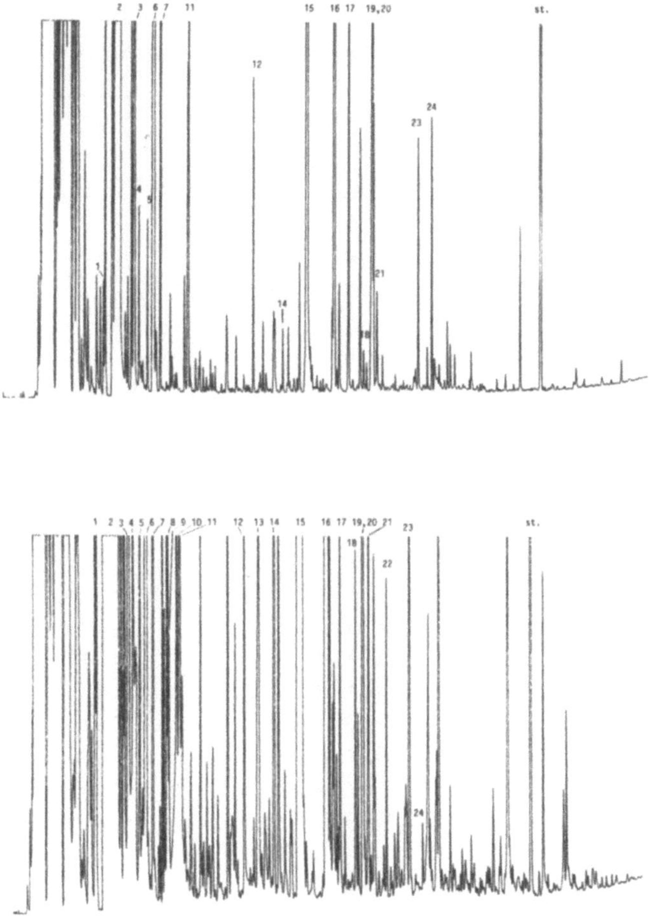

Fig. 7. Gas chromatograms of the aroma extract of Finnish berry wine of a sherry type (upper) and that of a Spanish sherry (lower). The compounds are listed in Table 1. Column: A 50 m OV-101 fused silica capillary column (17).

yeast is used. Some twenty years ago we compared the flavour compositions of Spanish sherry to that of Finnish berry wine of a sherry type. When both beverages are produced by a solera method, using a Jerez yeast, the compositions are remarkably similar even though the sherry was made from grapes and the Finnish berry wine was made from currants. Recently, high resolution gas chromatography was used for a similar study (17). The chromatograms (Fig 7) indicate that although some clear quantitative differences can be found, the qualitative compositions are similar. It is interesting to note that among the compounds, excluding aliphatic fusel alcohols, phenethyl alcohol is the largest component in both the Spanish sherry and Finnish berry wine of sherry type. Diethyl succinate, ethyl lactate and 2,3-butanediol are

Table 1. Contents (mg/L) of flavour compounds in a genuine Spanish sherry and a Finnish berry wine of sherry type (17).

Peak no	Compound	Spanish sherry	Finnish berry wine
1	Ethyl propionate	0.268	0.040
2	Isopentanol	not measured	
3	2,3-Butanediol	1.306	0.602
4	2,3-Butanediol	0.563	0.068
5	Ethyl butyrate	0.382	0.066
6	Ethyl lactate	3.804	2.862
7	2,3-Pentanediol (tent.)	1.324	0.556
8	Ethyl isovalerate	0.294	traces
9	3-Hexen-1-ol	0.238	n.d.
10	Hexanol	1.076	n.d.
11	Isopentyl acetate + gamma-butyrolactone	0.795	0.277
12	Ethyl hexanoate	0.558	0.108
13	Benzyl alcohol	0.717	n.d.
14	Ethyl 2-hydroxyisohexanoate	0.830	0.024
15	Phenethyl alcohol	26.640	2.636
16	Diethyl succinate	11.488	1.628
17	Ethyl octanoate	0.259	0.240
18	Lactone (MW 146)	0.147	0.016
19	Lactone (MW 146)	1.031	traces
20	Hydroxy acid ethyl ester	-	0.637
21	Lactone	0.269	0.047
22	5-Butyl-4-methyl-dihydro-(3-H)-2-furanone	0.166	traces
23	Lactone	0.615	0.084
24	Ethyl decanoate	0.034	0.096

n.d. = none detected

also prominent components in both beverages, even though the amounts in Spanish sherry are higher than in the Finnish berry wine. Consequently, sherry yeast produces the same aroma components irrespective of whether grapes or berries are used as the raw material.

The role of carbonyl compounds

The most volatile aroma fraction of alcoholic beverages has been found to be composed of carbonyl compounds. While investigating the occurrence of carbonyl compounds in completely fermented medium, Suomalainen

(40) showed that a certain relationship can be shown to exist between keto acids and the structurally corresponding amino acids, aldehydes and alcohols which are listed in Table 2.

Table 2. Key compounds in the formation of alcohols and amino acids by yeast (40).

Alcohols	Aldehydes	Keto acids	Amino acids
Ethanol	Acetaldehyde	2-Ketopropionic acid	Alanine
Glycol	Glyoxal	3-Hydroxy-2-keto-propionic acid	Serine
1-Propanol	Propionaldehyde	2-Ketobutyric acid	2-Aminobutyric acid
1-Butanol	Butyraldehyde		
2-Methyl-1-propanol	Isobutyraldehyde	2-Ketoisovaleric acid	Valine
2-Methyl-1-butanol	2-Methylbutyr-aldehyde	2-Keto-3-methyl-valeric acid	Isoleucine
3-Methyl-1-butanol	Isovaleraldehyde	2-Ketoisocaproic acid	Leucine
1-Hexanol	Hexanal		
Phenethyl alcohol		3-Phenyl-2-keto-propionic acid	Phenylalanine
Tyrosol		3-(4-hydroxyphenyl)-2-ketopropionic acid	Tyrosine
Tryptophol			Tryptophan

Suomalainen with his co-workers (40,42-45) was able to show that aldehydes and keto acids are essential for the amino acid synthesis as well as for the formation of fusel alcohols. After the compounds have been formed in the yeast cells, they transfer to the medium. Consequently, the keto acids which were found to be present in the yeast cells could be found in the fermentation solutions, although the yeast is able to metabolize keto acids during fermentation.

Aldehyde content

There are several reasons for the variation of aldehyde levels in alcoholic beverages. While investigating the formation of carbonyl compounds, Radler (29,30) found that the highest aldehyde level is reached during fermentation when the action of yeast is in the most vigorous phase. The total aldehyde content seems to depend upon the yeast used. In assays performed with 300 yeasts (49,50), the aldehyde content was found to vary from 6 to 190 mg per litre. It is interesting to note that baker's yeast, wild-type yeast and brewer's yeast differ distinctly in their aldehyde production due to their respiratory capabilities. For a given yeast strain, the amount of acetaldehyde it produces is related to the activity of its pyruvate decarboxylase. The nutrient composition may also have an effect on the carbonyl formation. A deficiency of nutrient substances in the fermentation of grapes gives rise to increased levels of aldehydes in the wine; this is because the formation of ethyl alcohol is delayed (55,56). With a deficiency of amino acids, the path from the carbon source to the fusel alcohols is clearly diverted to keto acids. With fermentation conditions in which amino acids are present in abundance, amino acids are converted instead to fusel alcohols by deamination and decarboxylation. The precursors of aldehydes, the 2-keto acids, are formed as intermediates in both processes (1).

In addition to the biochemical reactions described above, oxidation of alcohols, oxidative Strecker degradation of amino acids, and autoxidation of fatty acids may also produce aldehydes in alcoholic beverages. In sherries, for example, ethyl alcohol is slowly oxidized by air into acetaldehyde; this reaction takes place during the solera process, although the oxidation of ethyl alcohol generally takes place slowly. Of course, the aldehyde level in wines increases during aging to some extent. The presence of vicinal di- and tri-hydroxy phenols in model wine solutions has been shown to accelerate the oxidation of ethyl alcohol to acetaldehyde (51). It is supposed that the autoxidation of phenols produces a strong oxidant, probably hydrogen peroxide, which then initiates a number of oxidation reactions involving the oxidation of ethyl alcohol to acetaldehyde. These findings show that the formation of aldehydes in wines and distilled beverages during aging is a much more complex phenomenon than merely the oxidation of alcohols by air.

Acetaldehyde is frequently the major component among the carbonyls and generally constitutes more than 90 % of the total aldehyde content. Because of the low boiling point (21 ^{o}C) acetaldehyde is easily distilled together with water and alcohol into distillates and, therefore, it is ubiquitous. The total aldehyde content has been reported to vary from 0.1 to 15 mg/litre in most beers, and up to 300 mg/litre in distilled beverages. The unpleasant smell of large quantities of acetaldehyde in beverages can be recognized without difficulty. The sensory threshold for acetaldehyde is 25 ppm (13). Rectification decreases the aldehyde level in distilled beverages to some extent. Because of the low aldehyde content of the rectified alcohol used for the production of vodka, the aldehyde content of different vodkas generally is some 10 mg per litre or less. Relatively large aldehyde amounts can be found in whisky, cognac, brandy and rum. According to Guymon (2,3) the aldehyde content of commercial brandy distillates are generally low, the mean being some 11 mg/l at 50 % alcohol, but acetaldehyde concentrations as high as 264 mg/l have been found in a low quality brandy (5). For comparison, some aldehyde contents in distilled beverages are listed in Table 3.

Table 3. Aldehyde content in distilled alcoholic beverages (23).

Beverage	mg/l (50 % alcohol)
American whisky	av. 43
Bourbon whisky	20 - 60
Canadian whisky	10 - 36
Irish whisky	20 - 70
Scotch blended	20 - 110
Wine distillate	19 - 55
Brandy	63 - 308
Cognac	av. 105

In addition to saturated aldehydes, a potential effect of some unsaturated aldehydes on the flavour of alcoholic beverages should be taken into consideration. When Kahn et al. (10) determined by gas chromatography the low boiling compounds in a head fraction distilled from a Canadian beer still product, they found acrolein. They suggested that acrolein, which has a pungent odour and a lachrymatory property, is responsible for the "peppery" smell in some whiskies. Probably acrolein is produced either by bacteria from glycerol, or glycerol is dehydrated on hot surfaces of the distillation column and converted into acrolein.

Table 4. Unsaturated aldehydes identified in cognac (8).

Acrolein	2,4-Heptadienal, trans, trans-,
2-Methyl-2-propenal	2-Hexenal, trans-,
2-Methyl-2-butenal, cis-,	2-Heptenal, trans-,
3-Methyl-2-butenal	2-Octenal, trans-,
2-Pentenal, trans-,	2,4-Nonadienal, trans, trans-,
2-Methyl-2-pentenal	2-Nonenal, trans-,
1-Cyclohexene-1-carbaldehyde	2,4-Decadienal, trans, trans-,
1,2,3-Trimethyl-3-cyclo-pentenyl-1-acetaldehyde	

Acrolein reacts easily with three alcohol molecules to form 1,1,3-triethoxypropane. Recently, we found 1,1,3-triethoxypropane in a molasses spirit distilled from mash containing glycerol. Our experiments showed that the acrolein content quickly decreases in the spirit, whereas 1,1,3-triethoxypropane seems to be fairly stable. Acrolein has been found to be present also in cognac and rum. A number of unsaturated aldehydes identified in cognac (8) are listed in Table 4.

Vicinal diketones

The vicinal diketones, 2,3-butanedione (diacetyl) and 2,3-pentanedione, are ubiquitous flavour components of wine and distilled alcoholic beverages. Both diketones are prominent flavour compounds; diacetyl has a particularly special position in beverages because of its low sensory threshold, and therefore its formation and occurrence should be discussed briefly.

The mechanism by which 2,3-butanedione is formed in a fermentation solution was settled only recently. The metabolic pathway originally suggested for 2,3-butanedione involves the formation of acetolactate and its degradation to 2-hydroxy-3-butanone, but this mechanism has not been confirmed. On the contrary, it has been shown (26) that 2,3-butanedione is not formed by oxidation of 2-hydroxy-3-butanone in yeast fermentation. According to Suomalainen and Ronkainen (46), 2,3-butanedione is formed outside the yeast cell by spontaneous decomposition of 2-acetolactate which is formed in the yeast cells. The 2-acetolactate is an intermediate in the biosynthesis of 2-ketoisovalerate, which is a precursor of valine, leucine and 2-methylpropanol formation in the yeast cell. A plausible alternative mechanism for the formation of

Table 5. Content of 2,3-butanedione in wines and distilled beverages (23).

Beverage	2,3-Butanedione mg/l
White wine	0.05 - 3.40
Rose wine	0.11 - 1.24
Red wine	0.02 - 4.06
Brandy	0.07 - 3.0
Cognac	0.10 - 0.31
Scotch whisky	0.09 - 0.32
Rum	0.03 - 4.4
Vodka	0.01 - 0.10

2,3-butanedione has been also presented (9). Its key step involves an oxidative decarboxylation of 2-acetolactate to 2,3-butanedione. The concentrations of vicinal diketones, 2,3-butanedione and 2,3-pentanedione, in wines and distilled beverages vary widely (Table 5). Postel and Güvenc (27) reported that the average concentration of 2,3-butanedione was 0.42 mg/l in wines, with no differences between the quality classes. The average content of 2,3-butanedione in red wines of various European origins is 1.46 mg/l. Leppänen et al. (12) investigated the occurrence of vicinal diketones in wines. They found that some 50 % of the total 2,3-butanedione in white wines and 25 % of its total amount in red wines is probably bound with sulphur dioxide. Scotch whisky and cognac contain an average of 0.16 mg of 2,3-butanedione per litre. A Martinique rum was found to contain as much as 4.4 mg/l.

The sensory threshold for 2,3-butanedione has been reported to be 0.15 ppm in beer (13). The threshold for it in wine and distilled beverages such as whisky, cognac and rum may be somewhat higher than that in beer. The observed concentrations seem to exceed this limit in some cases to such an extent that the butterscotch flavour of 2,3-butanedione should be recognisable in the samples. Also 2,3-pentanedione is frequently found in wine and distilled beverages. However, it is of minor importance to the flavour because of its fairly low concentrations and high sensory threshold.

The effect of maturation on the flavour of beverages

Beverages contain also a number of minor compounds which appear only after fermentation. Maturation is usually a necessary stage for improving the flavour of alcoholic beverages. It has been shown that the amounts of aldehydes, esters, acids, furfural as well as non-volatile tannins, colouring matter and solids increase (6) in the whisky distillate during ageing. In addition to the formation of these compounds which are progressively produced in slow chemical reactions during ageing, a number of degradation reactions take place in the wall of the cask. Approximately 2 % of the water and alcohol pass yearly through the inner wall of the cask; while doing so they initiate the hydrolysis of hemicelluloses and large lignin compounds. The constant flow of water and alcohol through the interstices and cells in the hardwood of oak effectively extract these hydrolysis products into the distillate. The picture (Fig. 8) made by means of scanning electron microscopy shows that the oak has been changed to a completely dead hard wood (20). Therefore, it cannot transport water in the stem, and is an excellent material for casks. We attempted to elucidate this slow process by investigating the compounds liberated from oak chips by alcohol (20).

It is interesting to see that the content of alcohol has a marked effect on the amount of total extract. The maximum efficiency was reached with a solvent which contained 60 % alcohol, which is a

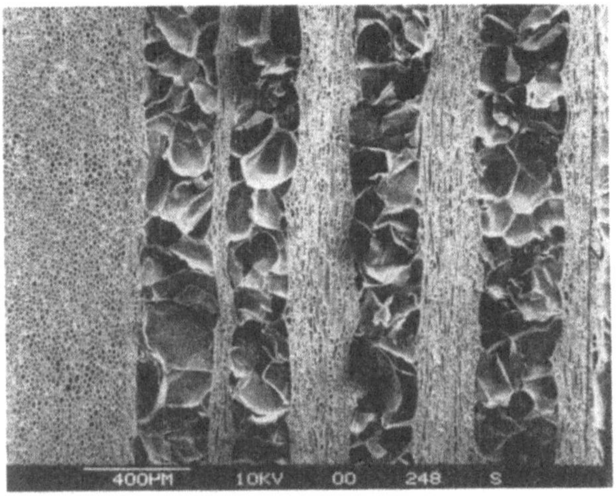

Fig 8. Scanning electron microscopy view of oak chip (20).

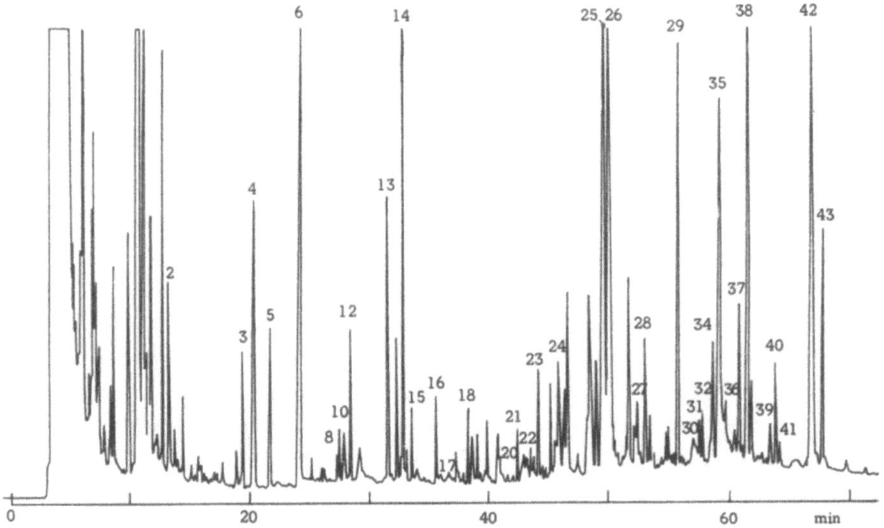

Fig. 9. Gas chromatogram of volatile compounds liberated by alcohol from oak chips at 60 % alcohol. A fused silica capillary column of 50 m were used; see the compounds in Table 6.

usually alcohol strength during maturation. When the composition of the extract was analysed in detail by means of high resolution gas chromatography and gas chromatography combined with mass spectrometry (Table 6), some acidic compounds were found, which probably catalyze the formation of acetals, esters and lactones. Ethyl esters of the dicar-

Table 6. Carboxylic acids and some non-acidic components liberated from oak chips by alcohol (20).

Peak	Compound	Peak	Compound
1	3-Hexanol	23	Cinnamic acid
2	Valeric acid	24	Undecanoic acid
4	Hexanoic acid	25	Lauric acid
5	1,1-Dimethoxyethane	26	Azelaic acid
6	2-Furancarboxylic acid	27	Tridecanoic acid
7	Unknown	28	Sebacic acid
8	Fumaric acid	29	Tetradecanoic acid
9	Heptanoic acid	30	Undecanoic acid
10	Succinic acid	31	Branched Pentadecanoic acid
11	Ethyl hexanoic acid	32	Branched hexadecanoic acid
12	Limonene	33	Pentadecanoic acid
13	Benzoic acid	34	Benzenetricarboxylic acid
14	Mesaconic acid	35	Ferulic acid
15	Octanoic acid	36	Branched hexadecanoic acid
16	Phenylacetic acid	37	Palmitoleic acid
17	2-Hydroxybenzoic acid	38	Palmitic acid
18	Nonanoic acid	39	Branched heptadecanoic acid
19	5-Butyl-4-methyl-2(3_H_)-furanone _cis_	40	Phthalic acid
		41	Heptadecanoic acid
20	5-Butyl-4-methyl-2(3_H_)-furanone _trans_	42	Oleic acid
		43	Stearic acid
21	Decanoic acid	44	Hexadecanedioic acid
22	3,4-Dimethoxyphenol		

boxylic acids, fumaric, succinic and azelaic acid, were also identified. In addition to acids, _cis_- and _trans_- isomers of 3-methyl-4-hydroxyoctanoic acid gamma-lactone (whisky lactone) were detected in the alcoholic extract of oak chips. Webb (11) has reported the occurrence of the whisky lactones in Cabernet Sauvignon wines, and Guymon and Crowell (4) detected them in brandies.

Flavoured alcoholic beverages

Essential oils or alcoholic extracts of spices and herbs are widely used for flavouring alcoholic beverages. On a manufacturing scale, the isolation of flavour compounds is carried out by percolating the spices with an alcohol-water mixture followed by distillation of the solution, usually under reduced pressure. The composition of the flavour extract is dependent upon the alcohol concentration and the equipment. We have

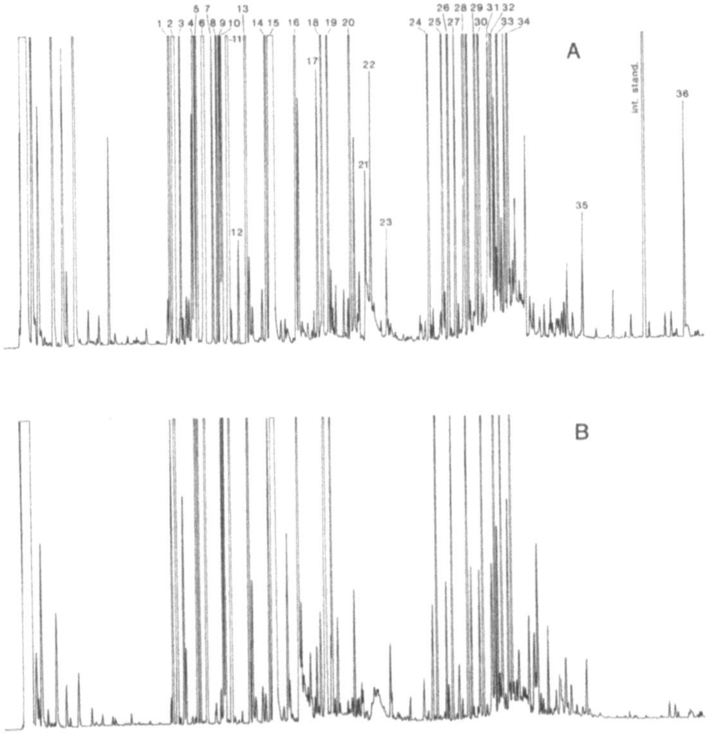

Fig. 10. High resolution gas chromatograms of the flavour compounds determined in a Finnish (upper) and an English (lower) gin: 1 α-thujone, 2 α-pinene, 3 camphene, 4 sabinene, 5 β-pinene, 6 β-myrcene, 7 α-phellandrene, 8 3-carene, 9 α-terpinene, 10 *p*-cymene, 11 limonene, 12 *trans*-ocimene, 13 γ-terpinene, 14 terpinolene, 15 linalool, 16 camphor, 17 4-terpinonyl acetate, 18 4-terpineol, 19 *trans*-piperitol ethyl ether, 20 4-ethoxy-1-*p*-menthene, 21 nerol, 22 *cis*-piperitol ethyl ether, 23 bornyl acetate, 24 α-cubebene, 25 α-copaene, 26 β-elemene, 27 α-gurjunene, 28 β-caryophyllene, 29 β-sequiphellandrene, 30 α-humulene, 31 γ-muurolene, 32 δ-cadinene 1, 33 α-guaiene, 34 δ-cadinene 2, 35 γ-elemene / germacrene B, 36 ethyl palmitate.

been especially interested in the flavour compounds occurring in juniper berries and angelica root (47,48) because both herbs are used for flavouring gin. Among the compounds of a juniper berry distillate analysed by a gas chromatographic method, α-pinene, sabinene, myrcene and α-terpinene were found in fairly large quantities. When the same method was used to determine the flavour compounds in an ether-pentane extract of angelica root, a total of 64 compounds were found, with the major ones being α-pinene, 3-carene, α-phellandrene, limonene, β-phellandrene and p-cymene.

The gas chromatograms in Fig. 10 demonstrate the flavour composition of a sample of Silver Gin produced in Finland and that of an English gin. In this case the flavour was separated and concentrated using a liquid-liquid extraction technique and gas chromatographed on a fused silica capillary column. A Total of 35 compounds were identified and determined quantitatively. The same compounds which were found in the juniper berry and angelica root extracts were confirmed to be present in the Finnish product. When the chromatogram is compared to the next one which presents the flavour composition of an English Gordon Gin sample, the similarity is remarkable, although some differences in the concentrations of the compounds may appear. These gin analyses provide a good example how successful sensory analyses made by a well trained professional panel can be confirmed by means of instrumental methods.

LITERATURE

1. Äyräpää, T.: On the formation of higher alcohols by yeasts and its dependence on nitrogenous nutrients. Kem. Tidskr. 83 (1971) 79-90.

2. Guymon, J. F.: Composition of California commercial brandy distillates. Amer. J. Enol. Viticult. 21 (1970) 61-69.

3. Guymon, J. F.: Brandy distillates. Composition and quality. Wines and Vines 52 (1971) 27-29.

4. Guymon, J. F. and E. A. Crowell: GC-separated brandy components derived from French and American oaks. Amer. J. Enol. Viticult. 23 (1972) 114-120.

5. Guymon, J. F. and D. L. Wright: Collaborative study of the direct bisulfite method for the determination of aldehydes in brandy and wine spirits: American Society of Enologists - AOAC joint report. J. Assoc. off. analytic. Chemists 55 (1972) 566-569.

6. Baldwin, S. and A. A. Andreasen: Congener development in bourbon whisky matured at various proofs for twelve years. J. Assoc. off. analytic. Chemists. 57 (1974) 940-950.

7. Bertrand, A.: Utilisation de la chromatographie en phase gazeuse pour le dosage des constituants volatils du vin. Thèse, Faculté des Sciences de l'Université de Bordeaux 1968.

8. ter Heide, R., P. J. de Valois, J. Visser, P.P. Jaegers, and R. Timmer: Concentration and identification of trace constituents in alcoholic beverages, in "Analysis of Food and Beverages. Headspace Techniques". G. Charalambous, ed., Academic Press, New York, 1978 pp. 249-281.

9. Inoue, T: Influence of temperature on formation of acetohydroxy acids by brewer's yeast. Rep. Res. Lab. Kirin Brewery 17 (1974) 25-27.

10. Kahn, J. H., E. G. Laroe, and H. A. Conner: Whiskey composition: identification of components by single-pass gas chromatogra-

phy-mass spectrometry. J. Food Sci. 33 (1968) 395-400.

11. Kepner, R. E., A. D. Webb, and C. J. Muller: Identification of 4-hydroxy-3-methyloctanoic acid gamma-lactone (5-butyl-4-methyldihydro-2-(3H)-furanone as a volatile component of oak-wood-aged wines of Vitis vinifera var. "Cabernet Sauvignon". Amer. J. Enol. Viticult. 23 (1972) 103-105.

12. Leppänen, O., P. Ronkainen, T. Koivisto, and J. Denslow: A semi-automatic method for the gas chromatographic determination of vicinal diketones in alcoholic beverages. J. Inst. Brewing 85 (1979) 278-281.

13. Meilgaard, M.C.: Aroma volatiles in beer: Purification, flavour, threshold and interaction. In: Drawert, F. ed, Geruch- und Geschmackstoffe. Internationales Symposium aus Anlass des Einhundertjährigen Bestehens der Firma Haarmann & Reimer GmbH, Holzminden, Bad Pyrmont, Germany, 2-4 October, 1974, Verlag Hans Carl Nürnberg 1975 pp. 211-254.

14. Meilgaard, M.C., C.E. Dalgliesh, and J.F. Clapperton: Progress towards an international system of beer flavour Terminology. Am. Soc. Brew. Chem. 37 (1979) 42-52.

15. Nordström, K.: Studies on the formation of volatile esters in fermentation with brewer's yeast. Svensk kem. Tidskr. 76 (1964) 510-543.

16. Nykänen, L.: Aroma compounds liberated from oak chips and wooden casks by alcohol. In: Nykänen, L. and P. Lehtonen eds, Flavour Research of Alcoholic Beverages. Instrumental and Sensory Analysis. Proceedings of the Alko Symposium on Flavour Research of Alcoholic Beverages, 13-15 June, 1984, Helsinki, Finland, Foundation for Biotechnical and Industrial Fermentation Research, Vol 3, Helsinki, Finland 1984 pp. 141-148.

17. Nykänen, L.: Formation and occurrence of flavour compounds in wine and distilled alcoholic beverages. Amer. J. Enol. Viticult. 37 (1986) 84-96.

18. Nykänen, L. and I. Nykänen: Production of esters by different yeast strains in sugar fermentation. J. Inst. Brewing, 83, (1977) 30-31.

19. Nykänen, L. and I. Nykänen: Rum flavour. In: Piggott, J.R. ed, Flavour of Distilled Beverages: Origin and Development, Ellis Horwood Ltd., Chichester, 1983 pp. 49-63.

20. Nykänen, L., I. Nykänen, and M. Moring: Aroma compounds dissolved from oak chips by alcohol, in "Progress in Flavour Research 1984". Proceedings of the 4th Weurman Flavour Research Symposium, Dourdan, France. 9-11 May 1984, J. Adda ed., Elsevier Science Publishers B.V., Amsterdam 1985 pp. 399-346.

21. Nykänen, L., I. Nykänen and H. Suomalainen: Distribution of esters produced during sugar fermentation between the yeast cell and the medium. J. Inst. Brewing, 83, (1977) 32-34.

22. Nykänen, L., P. Savolahti and I. Nykänen: First experiences with HRGC-FT-IR analysis of flavour compounds in distilled alcoholic beverages. In: Berger, R.G., S. Nitz and P. Schreier, eds. Topics in Flavour Research. Proceedings of the International Conference, Freising-Weihenstephan April, 1-2 1985, H. Eichorn, Marzling-Hangenham, Germany 1985 pp. 109-123.

23. Nykänen, L. and H. Suomalainen: Aroma of Beer, Wine and Distilled Alcoholic Beverages. In: Rothe, M. ed, Handbuch der Aroma Forschung, Akademie-Verlag, Berlin and D. Reidel Publishing

Company, Dordrecht, Holland / Boston, U.S.A. / London, England 1983 pp. 1-413.

24. Noble, A., R.A. Arnold, B.M. Masuda, S.D. Pecore, J.O. Schmidt, and P.M. Stern: Research note. Progress towards a standardized system of wine aroma terminology. Am. J. Enol. Vitic. 35 (1984) 107-109.

25. Parfait, A., M. Namory and P. Dupois: Les esters éthyliques des acides gras supérieurs des rhums. Annales de Technologie Agricole 21 (1972) 199-210.

26. Portno, A. D: The influence of oxygen on the production of diacetyl during fermentation and conditioning. J. Inst. Brewing 72 (1966) 458-461.

27. Postel, W. and U. Güvenc: Gaschromatographische Bestimmung von Diacetyl, Acetoin und 2,3-Pentandion in Wein. Z. Lebensm. Unters. Forsch. 161 (1976) 35-44.

28. Puech, J.-L.: Characteristics of oak wood and biochemical aspects of armagnac aging. Am. J. Enol. Vitic. 35 (1984) 77-81.

29. Radler, F.: Über die Bildung von Carbonylbildungen durch Hefen während der Gärung. 3. Symp. Techn. Mikcobiol. Berlin (1973) 391-396.

30. Radler, F. : Bedeutung und Möglichkeiten der Verwendung von Reinkulturen von Hefen bei der Weinbereitung. Weinberg u. Keller 20 (1973) 330-350.

31. Rapp, A.: Terpene compounds in Wine. In: Nykänen, L. and P. Lehtonen. eds, Flavour Research of Alcoholic Beverages. Instrumental and Sensory Analysis. Proceedings of the Alko Symposium on Flavour Research of Alcoholic Beverages, 13-15 June, 1984, Helsinki, Finland, Foundation for Biotechnical and Industrial Fermentation Research, Vol 3, Helsinki, Finland 1984 pp. 255-274.

32. Reazin, G.H.: Chemical analysis of whisky maturation. In: Piggott, J.R. ed, Flavour of Distilled Beverages: Origin and Development, Ellis Horwood Ltd., Chichester 1983 pp. 225-240.

33. Reazin, G.H., S. Baldwin, S.H. Scales, H.W. Washington and A.A. Andreasen: Determination of congeners produced from ethanol during whisky maturation. J. Assoc. Off. Anal. Chem. 59 (1976) 770-776.

34. Schreier, P.: Formation of Wine aroma. In: Nykänen, L. and P. Lehtonen eds, Flavour Research of Alcoholic Beverages. Instrumental and Sensory Analysis. Proceedings of the Alko Symposium on Flavour Research of Alcoholic Beverages, 13-15 June, 1984, Helsinki, Finland, Foundation for Biotechnical and Industrial Fermentation Research, Vol 3, Helsinki, Finland 1984 pp. 9-37.

35. Schreier, P., F. Drawert, A. Junker, und L. Reiner: Anwendung der multiplen Diskriminalyze zur Differenzierung von Rebsorten anhand der quantitativen Verurteilung flüchtiger Weininhaltsstoffe. Mitt. Rebe, Wein, Obstbau und Früchteverwertung 25 (1976) 225-234.

36. Schreier, P., F. Drawert und A. Junker: Gaschromatographisch-massenspektrometrische Differenzierung der Traubenaromastoffe verschiedener Rebsorten von Vitis vinifera. Chem. Mikrobiol. Technol. Lebensm. 4 (1976) 154-157.

37. Shortread, G.W., P. Richards, J.S. Swan, and S. Burtles: The flavour terminology of Scotch Whisky. Brewers Guardian Nov. 2-6 (1979).

38. Sponholz, W.-R. and H.H. Dittrich: Die bildung von SO_2-bindenden Gärungs-Nebenprodukten, höheren Alkoholen und Estern bei einigen Reinzuchthefestämmen und bei einigen für die Weinbereitung wichtigen "wilden" Hefen. Wein-Wiss. 29 (1974) 301-314.

39. Strauss, C.R., P.J. Williams, B. Wilson and E. Dimitriadis: Formation and identification of aroma compounds from non-volatile precursor in gapes and wine. In: Nykänen, L. and P. Lehtonen eds, Flavour Research of Alcoholic Beverages. Instrumental and Sensory Analysis. Proceedings of the Alko Symposium on Flavour Research of Alcoholic Beverages, 13-15 June, 1984, Helsinki, Finland, Foundation for Biotechnical and Industrial Fermentation Research, Vol 3, Helsinki Finland 1984 pp. 51-60.

40. Suomalainen, H.: The Structure and function of the yeast cell. In: Aspects of Yeast Metabolism, eds. A.K. Mills and H. Krebs. Guinnes Symposium, 1967, Blackwell Scientific Publications, Oxford and Edinburg 1968, 1-30.

41. Suomalainen, H., O. Kauppila, L. Nykänen and R.J. Peltonen: Branntweine. In: Schormüller, J. ed., Handbuch der Lebensmittelchemie. Band VII, Alkoholische Genussmittel, Springer Verlag, Berlin-Heidelberg-New York 1968 pp. 496-653.

42. Suomalainen, H. and A.J.A. Keränen: Keto acids formed by baker's yeast. J. Inst. Brewing 73 (1967) 477-484.

43. Suomalainen, H. and M. Lehtonen: Ursprung des Aromas in alkoholischen Getränken. Die Nahrung,24 (1980), 49-61.

44. Suomalainen, H. and L. Nykänen: Investigations on the aroma of alcoholic beverages. Naeringsmiddelindustrien, 23 (1970) 15-30.

45. Suomalainen, H. and P. Ronkainen: Keto acids in baker's yeast and in fermentation solution. J. Inst. Brewing 69 (1963) 478-483.

46. Suomalainen, H., and P.Ronkainen. Mechanism of diacetyl formation in yeast fermentation. Nature (London) 220 (1968) 792-793.

47. Taskinen, J. and L. Nykänen: Chemical composition of Angelica root oil. Acta Chem. Scand. B 29 (1975) 757-764.

48. Taskinen, J. and L. Nykänen: Volatile constituents of an alcoholic extract of juniper berry. Int. Flavours Food Addit. 7 (1976) 228, 233.

49. Then, R. and F. Radler: Regulation der Acetladehydkonzentration im Medium während der alkoholichen Gärung durch Saccharomyces cerevisiae. Arch. Mikrobiol. 72 (1970) 60-67.

50. Then, R. and F. Radler: Vergleichende Untersuchung der Acetaldehydbildung bei der aeroben Vergärung von Glucose bei verschiedenen Stämmen von Saccharomyces cerevisiae und Saccharomyces carlsbergensis. Mschr. Brauerei 24 (1971) 127-130.

51. Wildenradt, H.L. and V.L. Singleton: The production of aldehydes as a result of oxidation of polyphenolic compounds and its relation to wine aging. Am. J. Enol. Viticult. 25 (1974) 119-126.

52. Williams, A.A: Recent developments in the field of wine flavour research. J. Inst. Brewing, 88 (1982) 43-53.

53. Williams, P.J., C.R. Strauss, B. Wilson and E. Dimitriadis: Recent studies into grape terpene glycosides. In: Adda, J. ed, Progress in Food Flavour Research 1984. Proceedings of the 4th Weurman Flavour Research Symposium, Dourdan, France, 9-11 May, 1985 Elsevier Science Publishers B.V. Amsterdam-Oxford-New York-Tokyo, Printed in The Netherlands 1985 pp. 349-357.

54. Williams, P.J., C.R Strauss, B. Wilson, R.A. Massy-Westropp: Novel monoterpene disaccharide glycosides of Vitis vinifera grapes and wines. Phytochemistry 21 (1982) 2013-2020.

55. Wucherpfennig, K. and G. Semmler: Acetaldehydbildung im Verlauf der Gärung in Abhängigkeit von Wuchsstoffgehalt des Gärsubstrates. I. Mitteilung. Z. Lebensmittel-Unters. u. Forsch. 148 (1972) 77-82.

56. Wucherpfennig, K. and G. Semmler: Über den SO_2-Bedarf der Weine aus den verschiedenen Weinbaugebieten der Welt und dessen Abhängigkeit von der Bildung von Acetaldehyd in Verlauf der Gärung. Dtch. Weinbau 28 (1973) 846, 851-855.

CHANGES IN COFFEE AROMA DURING STORAGE

Ferenc Örsi[1)]

In recent years coffee consumption in Hungary evenly increased, and Hungary entered the rank of nations consuming considerable quantities of coffee. Coffee is sold in the roasted packed form, mostly in 250 g packages, and is ground by the consumer. In our study on the time of quality preservation our aim was to investigate changes in the organoleptic properties of coffee, and in conjunction with it chemical changes occurring in coffee, particularly the rancidification of fats present in substantial quantities in coffee and containing also in considerable quantities unsaturated fatty acids.

Experimental storing conditions provided for rapid changes were chosen, while at the same time, by determining the parameters of changes and following process kinetics, the formulation of a mathematical model of the processes was attempted which made also possible to conclude from the results obtained on normal storing conditions.

Materials and methods

The roasted coffee blend used for the investigations, consisting of 30 % Minas, 40 % Santos and 30 % Columbia varieties, was made available by the factory ZAMAT for Biscuit and Wafer.

Using 5 kinds of wrapping materials, packages of 100 g were made from the coffee. Description of the wrapping materials and their permeability properties are given in Table 1.

Moisture permeability was measured with gravimetric method between species of 100 and 0 % relative humidity, respectively. Oxygen permeability was measured by manometric method using moist oxygen.

The packages had a volume of 230 cm^3 and a surface of 240 cm^2. A void fraction of 124 cm^3 was measured between the coffee seeds by aqueous filling.

An air-conditioning equipment operating at 40 ^oC was used for the storage experiments. In this 0 % relative humidity and 100 % oxygen concentration were provided by gaseous oxygen passed through a sulfuric acid

[1)]Institute of Biochemistry and Food Technology, Technical University of Budapest

gas scrubber, while 100 % moisture content and 100 % oxygen by free water surface on large trays and by oxygen bubbled through an aqueous scrubber.

Table 1. Construction and characteristics of wrapping materials

Series number	Denomination, construction	Moisture permeability g/m^2 day	Oxygen permeability ml O_2/m^2 day
1.	Parchment	2300 ± 76	94 ± 5.6
2.	Cellophane	1050 ± 28	23 ± 0.4
3.	9 μm alufoil 70 g/m^2 paper	5.5 ± 1.4	1.4 ± 0.05
4.	14 μm alufoil 40 g/m^2 paper polyethylene 15 g/m^2	1.7 ± 0.4	0.11 ± 0.05
5.	30 μm alufoil 35 μm polyethylene	1.0 ± 0.1	0.0082 ± 0.0009

Each week the moisture content of the samples was determined, organoleptic tests were made, fat was recovered and peroxide number and carbonyl number of the fatty substances were determined.

Organoleptic tests were performed by 10 trained experts on coffee (10 cm^3/g) made on an espresso-type machine, and evaluated according to the system shown in Table 2. In each case coffee made of a sample stored in a refrigerator was also served. This represented with respect to every property a sample of point number 5.

Moisture content was caluclated from the loss in mass of a sample of 4 - 6 g, comminuted on an electric coffee grinder after drying at 105 °C for 4 hours.

The oil content of coffee was extracted for 4 hours with a 2 : 1 chloroform : methanol mixture in a Besson's flask, and evaporating to dryness one part of the oil, the quantity of oil was determined by weighing again. The peroxide number was determined from the residual solution by the spectrophotometric method of Stine and Hatland, based on the oxidation of Fe^{2+} ions into Fe^{3+} ions, the quantity of which was determined on the basis of its colour reaction with ammonium rhodanide. The quantity of carbonyl compounds was determined according to the prescription of Heinrich with 2,4-dinitrophenyl reagent, similarly by photometry.

The fatty acid composition of coffee oil was determined by the gas chromatographic analysis of methyl esters.

Table 2. Coffee qualifying table

Property	Grade	Point number
Odour	excellent	5
	good	4
	not unpleasant	3
	mild foreign odour	2
	unpleasant odour	1
Sour taste	unperceptible	5
	perceptible	4
	mildely sour	3
	sour	2
	strongly sour	1
Bitter taste	pleasant	5
	mildely bitter	4
	bitter	3
	strongly bitter	2
	unpleasantly bitter	1
General impression	excellent	5
	good	4
	satisfactory	3
	acceptable	2
	bad	1

Results

The oil content of the coffee tested ranged from 17 to 19 % decreasing somewhat during storage. The fatty acid composition is shown in Table 3.

It can be seen that the ratio of unsaturated fatty acids is more than 50 %.

Table 3. Fatty acid composition of raw and roasted coffee

Fatty acid	Raw coffee	Freshly roasted coffee
Myristic acid	4.42	2.51
Palmitic acid	35.76	31.75
Stearic acid	8.92	8.95
Oleic acid	9.92	10.22
Linoleic acid	40.98	46.87

The sorption isotherm in coffee, shown in Fig. 1, was taken at 40 °C in an atmosphere of 20, 40, 50, 70, 90 and 100 % relative humidity, following the establishment of equilibrium by mass measurement.

The change in moisture content of the samples is shown in Figs. 2 and 3. It may be noted that in a space of 0 % relative humidity packages suffered a loss in weight while in a space of 100 % relative humidity they revealed an increase in weight as a function of permeability.

Change in peroxide number during storage is shown in Figs. 4 and 5. It can be seen that peroxide number increases in both cases during storage.

The quantity of oxo compounds is shown in Figs. 6 and 7.

The changes in sensory properties, moisture content, peroxide number and aldehyde content, further the change in total oxygen quantity bound were investigated together by principal component analysis.

Data of changes observed at 0 and 100 % relative humidity, respectively, are summarized in Tables 4 and 5, while in Table 6 the significant correlation coefficients of the characteristics are summarized.

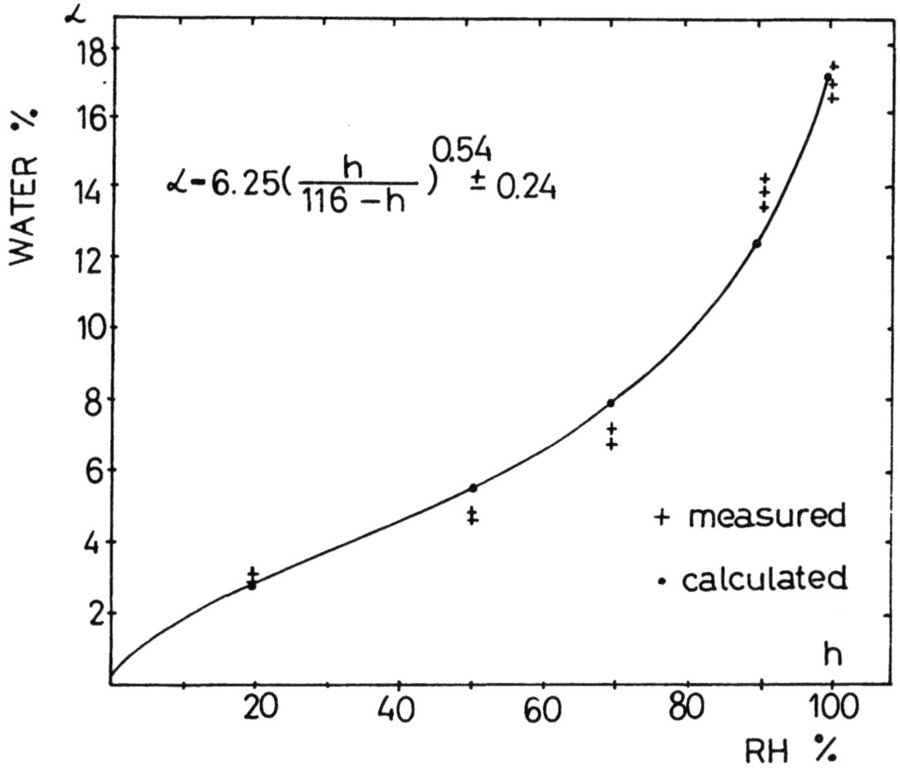

Fig. 1. Sorption isotherm of roasted coffee

Since there are significant close correlations between the characteristics it was to be expected that the system can be described by principal component analysis with relatively few principal component variables. It was actually found that the variance of 8 variables can be described to 92 % by two principal components. Table 7 shows the variable components of the two principal components.

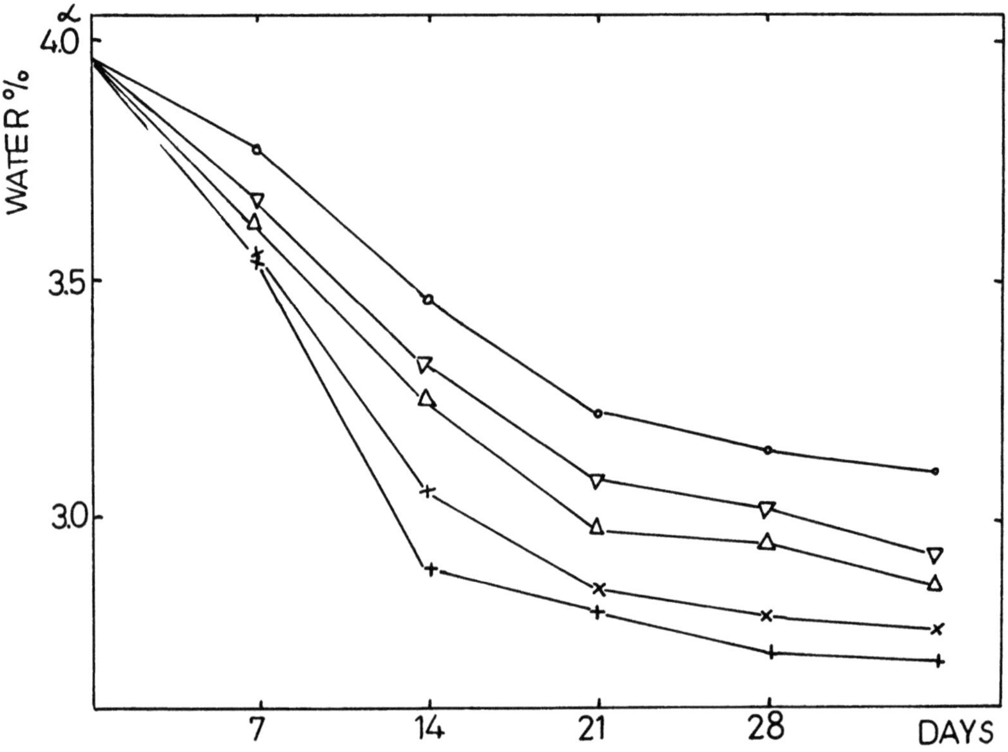

Fig. 2. Change of water at 0 % relative humidity (RH)

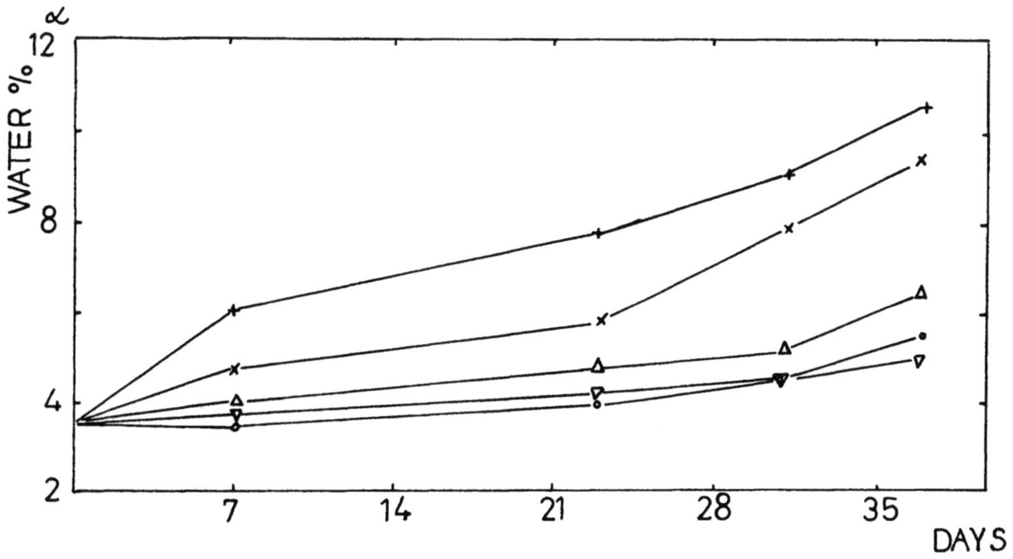

Fig. 3. Change of water at 100 % relative humidity (RH)

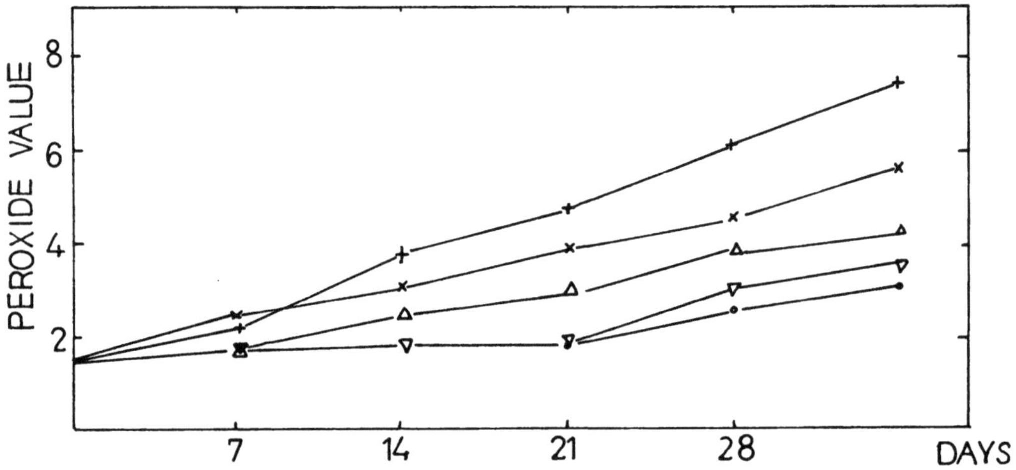

Fig. 4. Change of peroxide value at 0 % relative humidity (RH)

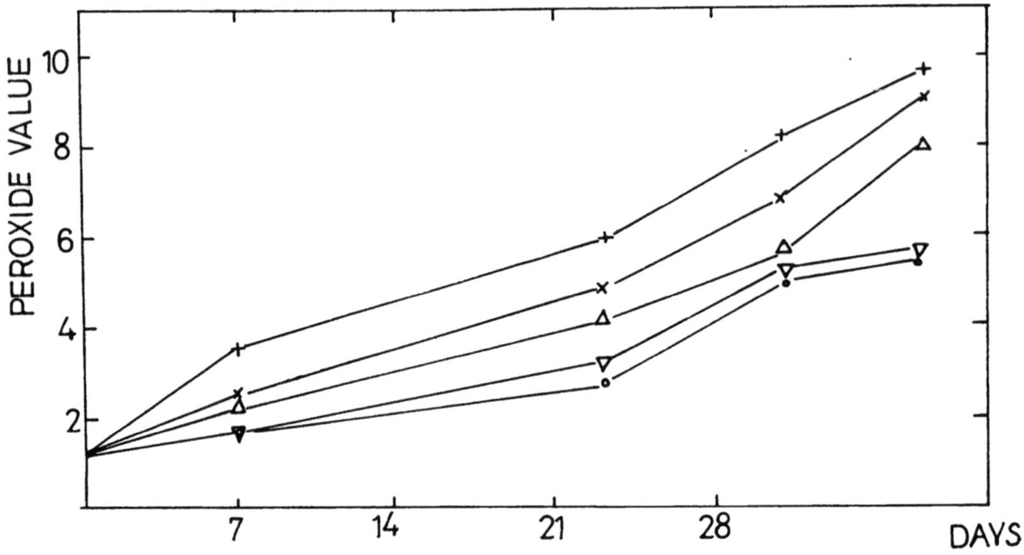

Fig. 5. Change of peroxide value at 100 % relative humidity (RH)

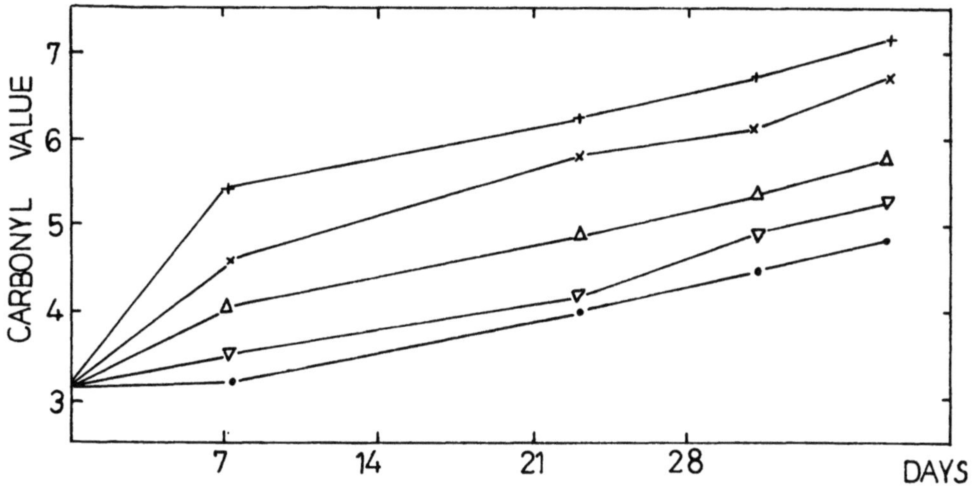

Fig. 6. Change of carbonyl value at 0 % relative humidity (RH)

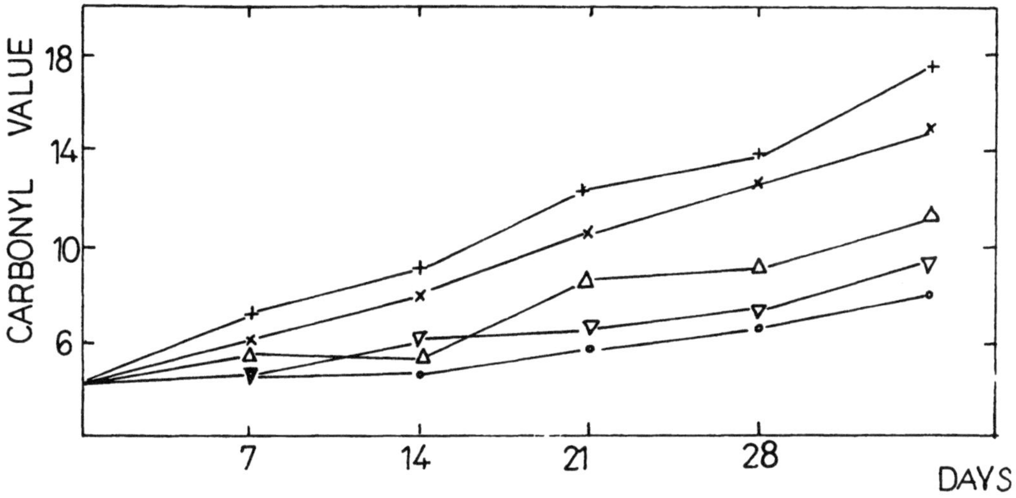

Fig. 7. Change of carbonyl value at 100 % relative humidity (RH)

Table 4. Coffee characteristics at 0 % relative humidity (RH)

Storage time	Wrapping material	Water content (%)	Odour	Bitter taste	Sour taste	Total impression	Peroxides (μmol O_2/g lipids)	Oxo compounds (μmol O_2/g lipids)	Total oxygen bound
0	1.	4.0	4.3	4.6	4.5	4.5	1.48	4.28	5.76
7		3.6	3.7	3.5	3.6	3.5	2.19	7.19	9.83
14		2.9	3.3	3.1	3.0	3.1	3.92	9.10	13.02
21		2.8	2.8	3.3	3.0	2.4	4.77	12.50	17.33
28		2.7	2.4	2.9	2.0	1.9	6.15	13.96	20.11
35		2.7	2.6	2.6	2.7	2.3	7.41	17.57	24.98
7	2.	3.6	3.5	3.6	3.7	3.7	2.44	6.21	8.65
14		3.0	3.4	3.4	3.5	3.6	3.05	8.00	11.05
21		2.9	3.4	3.3	2.8	3.1	3.88	10.76	14.64
28		2.8	3.0	3.1	2.8	2.3	4.63	12.48	17.11
35		2.8	3.0	3.1	3.3	2.8	5.59	14.96	20.55
7	3.	3.6	3.9	3.8	3.9	3.7	1.80	5.37	7.17
14		3.2	3.8	3.5	3.8	3.3	2.47	6.23	8.70
21		3.0	3.5	3.6	3.6	3.1	2.98	8.59	11.55
28		3.0	3.6	3.2	3.1	2.9	3.90	9.08	12.98
35		2.9	3.5	3.3	3.6	3.5	4.23	11.31	15.54
7	4.	3.7	3.5	3.7	4.0	3.2	1.72	4.75	6.47
14		3.3	4.2	4.0	3.8	3.7	1.80	5.31	7.11
21		3.1	4.1	3.8	3.8	3.6	1.86	6.62	8.48
28		3.0	3.4	3.6	3.6	3.1	3.09	7.60	10.69
35		2.9	3.6	3.1	3.5	3.4	3.62	9.48	12.10
7	5.	3.8	3.8	3.6	3.7	3.5	1.74	4.85	6.59
14		3.5	4.2	4.2	3.9	4.2	1.86	5.09	6.95
21		3.3	4.2	3.9	3.8	3.8	1.83	5.98	7.81
28		3.2	3.3	3.6	3.6	3.4	2.51	6.59	9.10
35		3.1	3.8	3.6	3.6	3.6	3.04	8.10	11.14

Table 5. Coffee characteristics at 100 % relative humidity (RH)

Storage time	Wrapping material	Water content (%)	Odour	Bitter taste	Sour taste	Total impression	Peroxides (μmol O_2/g lipids)	Oxo compounds (μmol O_2/g lipids)	Total oxygen bound
0	1.	3.7	4.1	4.6	4.5	4.5	1.20	3.15	4,35
7		6.1	4.0	3.7	4.2	3.7	3.55	5.43	8.98
21		7.8	2.6	2.5	2.4	2.6	5.90	6.21	12.11
31		9.1	2.2	2.5	2.3	2.0	8.20	6.69	14.89
37		10.5	2.0	1.9	2.0	1.7	9.60	7.12	16.72
7	2.	4.9	3.8	3.5	3.8	3.4	2.50	4.48	6.98
21		5.8	3.2	2.9	2.6	2.8	4.85	5.81	10.66
31		8.0	2.8	2.9	2.8	2.6	6.85	6.13	12.98
37		9.4	2.4	2.5	2.5	2.3	9.05	6.70	15.75
7	3.	4.0	4.3	4.1	4.3	4.4	2.25	4.01	6.26
21		4.8	3.5	3.4	3.2	3.3	4.10	4.85	8.95
31		5.2	3.2	3.4	3.0	3.3	5.60	5.30	10.90
37		6.5	3.1	3.2	3.1	3.3	6.95	5.75	12.60
7	4.	3.8	4.3	4.5	4.6	4.5	1.70	3.50	5.20
21		4.1	3.9	4.1	3.7	4.2	3.20	4.20	7.40
31		4.6	3.7	4.0	3.7	3.7	5.25	4.90	10.15
37		5.2	3.7	4.1	4.1	3.9	5.70	5.25	10.95
7	5.	3.5	4.3	4.5	4.5	4.4	1.75	3.45	5.20
21		4.0	4.1	4.2	4.0	4.5	2.75	4.00	6.75
31		4.5	4.1	4.2	4.1	4.2	5.00	4.50	9.50
37		5.5	4.0	4.3	3.8	4.3	5.55	4.83	10.38

Table 6. Correlation coefficients ($100 \cdot r_{xy}$)

Number of variables	1 5	2 6	3 7	4 8
1	1 - 35*	- 49* 68*	- 46* - 30*	- 43* 9
2	 93*	1 - 83*	91* - 57*	92* - 80*
3	 92*	 - 74*	1 - 56*	91* - 75*
4	 91*	 - 77*	 - 55*	1 - 77*
5	1	- 70*	- 64*	- 79*
6		1	38*	74*
7			1	89*

* significantly differs from 0

Principal component I, describing 75 % of the variance, contains with near identical loading the sensory characteristics, the peroxide number and total oxygen bound. In principal component II moisture and aldehydes have the highest loading factor.

Conditions are clearly shown in Fig. 8 where the projections of the original variables are plotted in the plane of the first two principal components.

The nearly opposite course of peroxide number and oxygen bound and of the sensory properties can be readily seen as well as the strict correlation of sensory characteristics. Change in moisture content and quantity of aldehydes proceed in opposite direction and have little connection with the other characteristics.

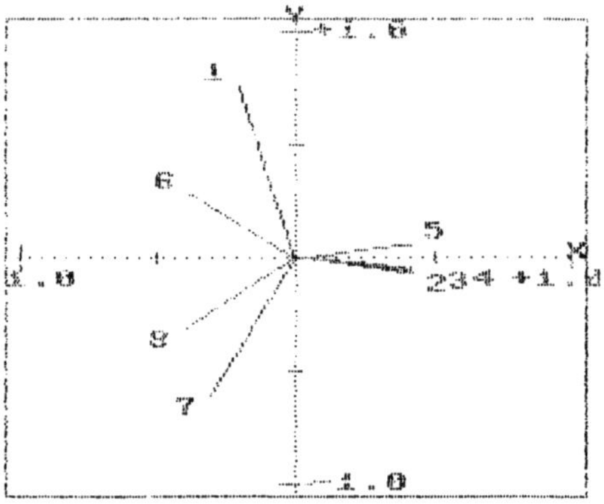

Fig. 8. Projection of characteristics in the plane of principal components

x = 1. principal component y = 2. principal component

1 = water content

2 = odour

3 = bitter taste

4 = sour taste

5 = total impression

6 = peroxides

7 = oxo compounds

8 = total oxygen bound

In Figs. 9 - 10 samples are plotted in the plane of the first two principal components. Fig. 10 shows a magnified part of the right half-field.

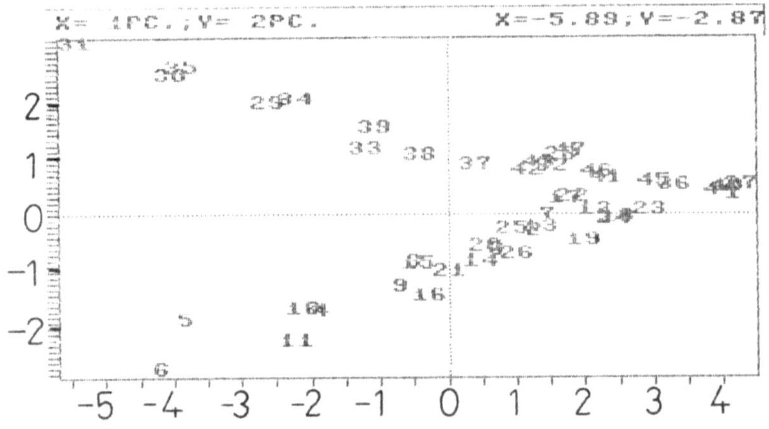

Fig. 9. Coffee samples presented in the plane of the first and second principal component

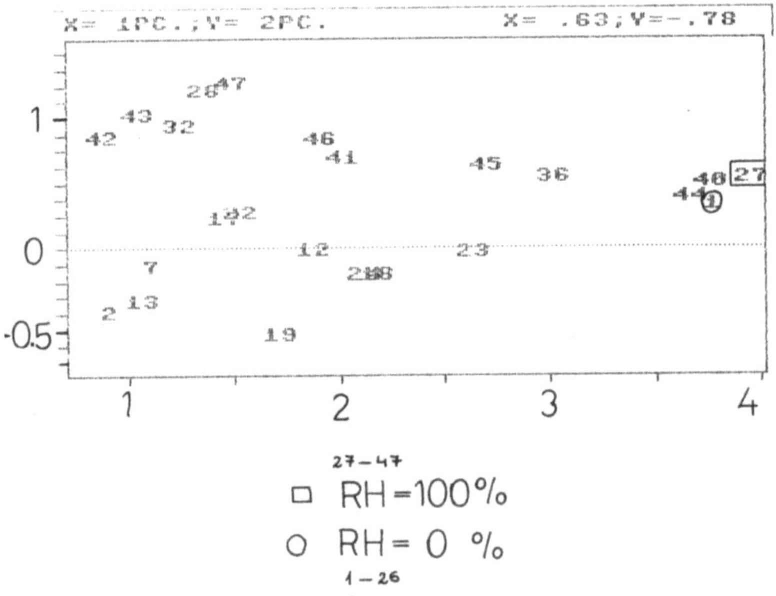

Fig. 10. Magnified part of the right half-field

Table 7. Composition of the two principal components (factor loading)

Variable	Principal component I.	II.
1. Moisture	3.4	50
2. Odour	16.0	0.5
3. Bitter taste	15.0	0.3
4. Sour taste	15.0	0.2
5. Total impression	15.0	0.2
6. Peroxides	13.0	6.6
7. Aldehydes	7.8	34
8. Oxygen bound	13.6	8.8
Self-value %	73	19

It can be seen that the change from starting values is nearly linear, but it determines different straight lines according to the relative atmospheric humidity. The distance measured from the starting point is an unequivocal function of deterioration, the farther we proceed the larger are peroxide number and the quantity of bound oxygen, and sensory characteristics worsen, too. The figures show the relationship between peroxide number and bound oxygen and sensory characteristics as a function of peroxide number and bound oxygen, respectively, The fact that the quantity of aldehydes formed during rancification shows no direct relationship with sensory characteristics, though its correlation coefficient is significant, indicates that the deterioration process is not to be attributed to changes in aroma or taste produced by them (Fig. 11 - 14).

The effect of storage time involved in the correlation coefficients was investigated by the calculation of the partial correlation coefficients (assuming that t = const.). Partial correlation coefficients of the characteristics are summarized in Table 8.

It can be seen that significant correlation coefficients are maintained, indicating that the case is not a simple co-variation in time but a true relationship.

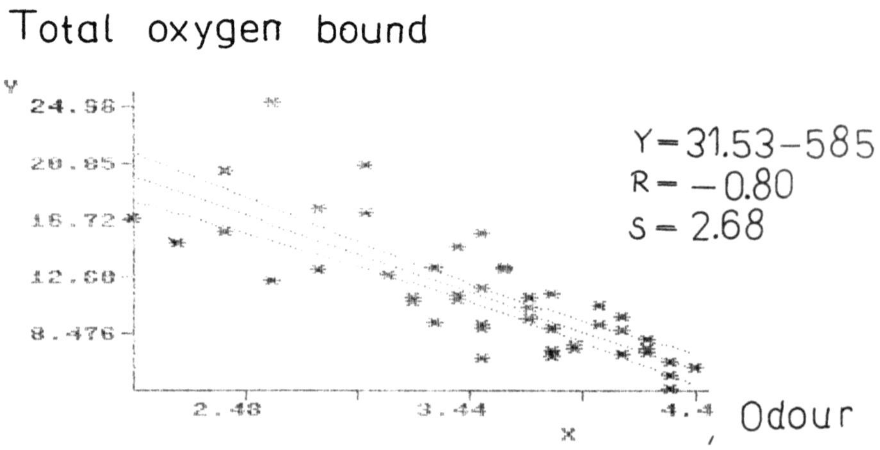

Fig. 11. Relationship between the total oxygen bound and odour

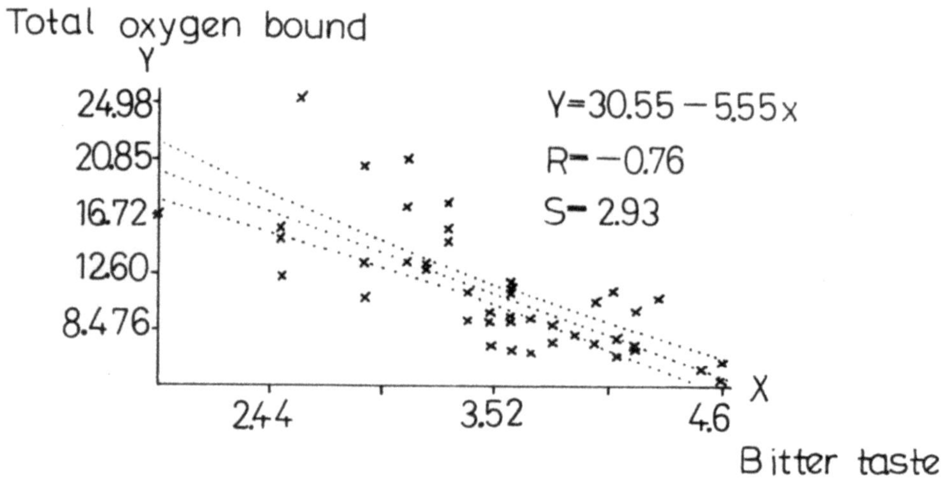

Fig. 12. Relationship between the total oxygen bound and bitter taste

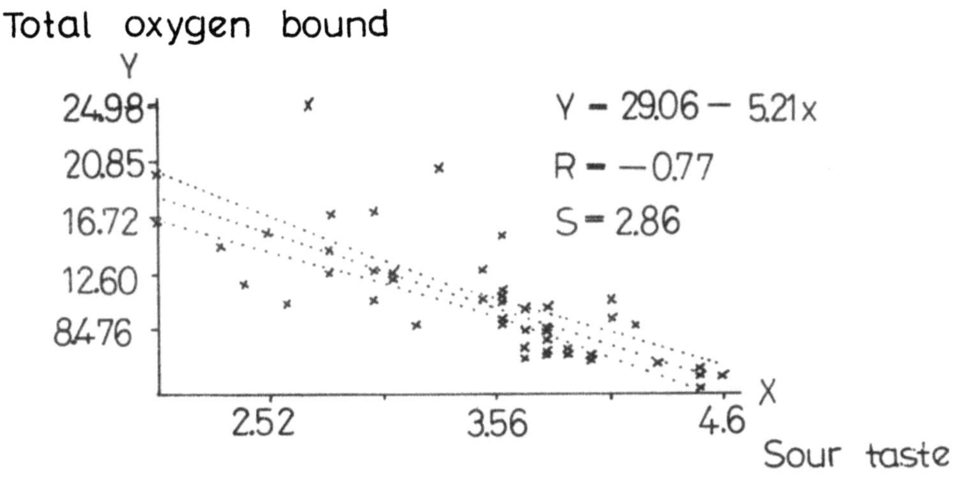

Fig. 13. Relationship between the total oxygen bound and sour taste

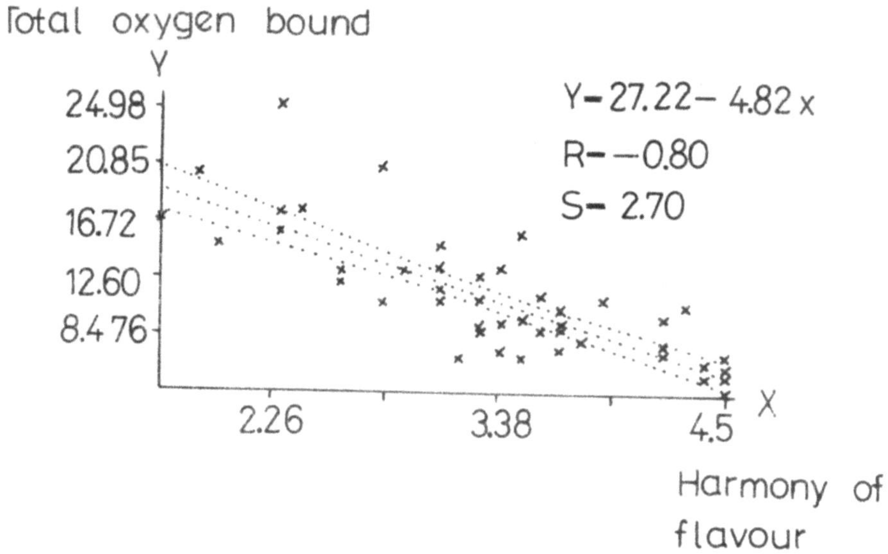

Fig. 14. Relationship between the total oxygen bound and total impression

Table 8. Partial correlation coefficients between the pairs of variables investigated excluding the effect of time (100 rxy • t)

Variable	2	3	4	5	6	7	8
1	-42^{*}	-38^{*}	-34^{*}	-25^{*}	74^{*}	-51^{*}	-14
2		89^{*}	89^{*}	91^{*}	-76^{*}	-76^{*}	-69^{*}
3			89^{*}	91^{*}	-64^{*}	-44^{*}	-65^{*}
4				89^{*}	-65^{*}	-42^{*}	-64^{*}
5					-59^{*}	-55^{*}	-73^{*}
6						8	48^{*}
7							91^{*}
8							

From the relationship between total oxygen bound and number of sensory point we attempted the determination of limit values causing minimal sensory changes. A change of a magnitude corresponding to standard error along the curve means already a significantly perceivable deviation from the initial value (Y axis section), and values belonging to it give those limit values, and from their relationship with time those storage times during which these changes take place.

Table 9. Wrapping materials and preservation time

Sensory characteristic	Standard error along the straight line	Maximum oxygen bound 1)	Minimal preservation time 2) (days) 0 % RH		100 % RH	
Odour	2.68	0.34	10	37	17	32
Sour taste	2.68	0.38	11	40	18	34
Bitter taste	2.93	0.36	11	41	19	35
Total impression	2.70	0.41	10	37	17	33
		Foil	1.	5.	1.	5.

1) decreasing sensory point number from 5 to 3

2) the time until the bound oxygen exceeds 2times the standard error

Thus results show that changes in taste and aroma are directly connected, even if it cannot be proved that this is to be attributed to the presence of aldehydes and ketones formed.

Results make also possible the determination of the kinetic parameters of the changes permitting to draw conclusions on storability under normal storage conditions.

Considering the size of the packages and the quantity of oxygen enclosed in the package (about 60 μmol O_2/g fat) if the coffee is not packed under vacuum, the permeability of packing material is of no importance from the aspect of storability (foil 1.: 52 μmol/g in 10 days, foil 5.: 0.016 μmol/g in 40 days).

Comparing storability times, this is nearly equal for all the characteristics; at low moisture content the rate of deterioration is higher in films of high permeability while high humidity seems to hinder deterioration in such packing material. However, in films of low permeability storage is limited by oxygen passing through it.

Summary

Change in aroma of roasted packaged coffee was followed under intensive storage conditions (40 ^{o}C, 0 to 100 % RH and 100 % oxygen) by sensory method.

Sensory changes can be directly connected with lipid changes during storage though the exclusive role of rancidification products cannot be proved.

A practical consequence of the results is that oxygen enclosed in the package is already sufficient to induce the deterioration process.

Literature

1. Heinick, A. S.: Handbuch der Lebensmittelchemie. IV. 880, 1950
2. Stine, M. C. and H. A, Harland: J. Dairy Sci. 37 (1954) 202
3. Wada, K. et al.: Agric. Biol. Chem. 51 (1987) 1745
4. Wada, K. et al.: Agric. Biol. Chem. 51 (1987) 1753

PRODUCTION AND USE OF HERB AROMAS MADE WITH ESSENTIAL OILS

R. J. Thomann, U. Bräuer and P. Kretschmer[1)]

The dot on the i in preparing quite a lot of dishes is to add a small amount of freshly cutted herbs.
These herbs not only contribute to an attractive appearance of the food, but also complete the taste in a special way. Several dishes have a long tradition of being used together with some herbs. I want to remind you of cooked fish and dill cream.

The utilization of herb aromas for the industrial food producers was the aim we formulated some years ago.
Special interest was laid on dill, parsley, lovage and celery. The flavour carrier in these herbs are the essential oils. At the usual stage of harvest of these herbs, the contents of essential oils are as follows:

Table 1: Parameters of herbs at the point of harvest in conventional processing

	essential oil (1) (ml/100 g)	(l/t)	green weight (t/ha)	essential oil (l/ha)
dill	0,06	0,6	20	12,0
parsley	0,05	0,5	21	10,5
celery	0,05	0,5	17	8,5
lovage	0,2	2,0	4	8,0

Table 1 gives the theoretically possible values (100 %). Under practical conditions we got 30 - 80 % of these values.
Regarding the yields involved, we found that, except lovage oil, which is very expensive, the distillation is not profitable.

1) Institute of Cereal Processing, A.-Scheunert-Allee 40/41, Bergholz-Rehbrücke, GDR-1505

With the progressive growing of the dill plant the oil content increases continuously without reducing the typical herb flavour.

Figure 1 Content of essential oil of dill HERKULES during growing 1987

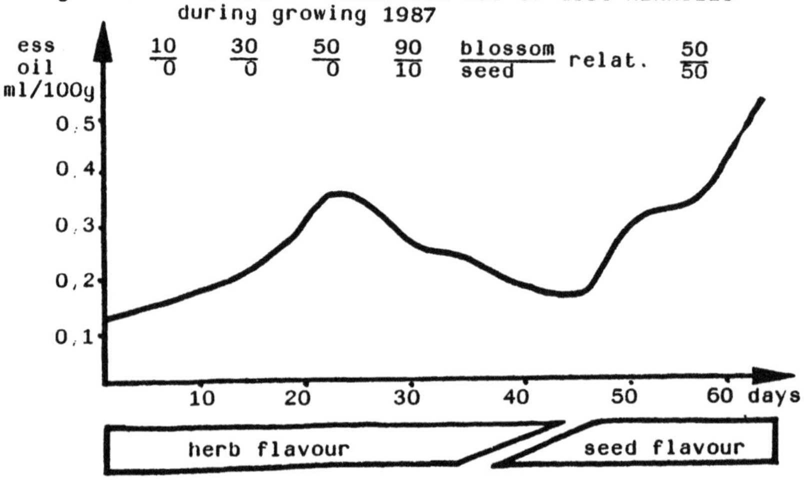

The distillation of dill at the optimum point yields under practical conditions about 45 kg of essential oil per ha with the following composition:

Table 2: Composition of essential dill oils

	before blossom (3)	before blossom (4)	during blossom	after the optimum point of harvest
carvon	2,4	0,5	11	17,5
α-phellandrene	25	44,2	30	30,8
limonene	10	7,5	35	31,2
dill-ether*)	31	21,9	14	6,7

*)dill-ether = 3,6-dimethyl-2,3,3a,4,5,7a-hexahydro-benzofurane (5)

Investigations on the composition of the essential oil in the stage of bud forming by HUOPALAHTI (3) are in conformance with our results (4).

In accordance with GÜNTHER (2) the content of α-phellandrene is regarded as the main component.
Distillation after the optimal point of harvest showed, however, that even a content till 45 % of α-phellandrene is no guarantee for giving the dill oil the typical herb flavour.
By GÜNTHER a maximum content of 35 % of carvone does not yet impair the herb flavour. We found 20 % as an acceptable upper limit.

Parsley and celery belong to those herbs, which form their seeds only in the second year of growing. In the stage of seed formation the essential oils have a strong, herb-like flavour and may be produced with good economic results.

Table 3: Yields of essential oil of herbs during seed formation

	essential oil (ml/100 g)	(l/ha) (theoretically)
dill	0,92	138
parsley	0,54	135
celery	0,22	110

The real yield of essential oil to be obtained depends on two facts: how you manage harvesting and transporting the herb to the distillation unit in a short time and whether you are able to distill with high efficiency.

Some parameters to characterize the oils.

Table 4: Chemical, physical and sensoric parameters of essential oils

	density (g/cm^3)	refractive index	richness (ml)
dill oil	0,866	1,476	$2,5 \cdot 10^{-4}$
parsley oil	0,945	1,519	$7 \cdot 10^{-5}$
celery oil	0,871	1,486	$5 \cdot 10^{-5}$
lovage oil	0,939	1,490	$1 \cdot 10^{-5}$

Richness means that volume of essential oil in 100 ml of a 0,5 % NaCl-solution which allows the consumer to determine the kind of herb.

Of course, the variety of the used herb influences the taste of oil. E. g. the dill oil of 'HERKULES' is weaker, but also fresh and mint-like, on the other hand the oil of 'GEWÖHNLICHER DILL' is stronger and more spicy.

The distillation is done in the following steps:

- full mechanized harvest
- cutting the herbs at the place of the distillation unit,
- distillation in a stainlesss steel container,
- heating by direct steam injection.

In distilling lovage we add chopped straw during the filling process in order to guarantee an equal steam penetration.

charge:	about 1 t
steam temperature:	120 °C
distillation time:	2 h
condensate temperature:	30 °C
	(at parsley 50 °C to prevent stiffening)

The essential oils are collected in Florentine receivers, dried over sodium sulphate and are ready for use after analytical evaluation and corrections.

Application

In applicating the herb aromas we have to differ into three main directions:

1. Adding the essential oil to a fat or oil containing component of the final product, e. g.
   - fish in flavoured vegetable oil
   - fish in flavoured cream or mayonnaise
   - fatty curd preparations.

   Due to the good miscibility with oil technological problems do not arise.

2. The use of liquid aromas. They contain about 2 % of essential oil and ethanol and acetic acid as solvents.

The enterprise 'Aromatic Leipzig' was gained as the producer of these products. Because of the very good miscibility with water and dosage quality the producers of

- pickled cucumbers,
- canned fish in sour brine or aspic jelly,
- poor fat curds,

are the favourite users of these herb aromas.

3. The third and most complicated kind of application are dry micro-encapsulated herb aroma granulates.
   Our investigations to fix the essential oils on sodium chloride by the aid of additives didn't result in the necessary storage stability.
   In using fluid bed granulation it was possible to form granulates by spraying an emulsion of
   - essential oil,
   - gum arabicum and
   - dextrin

   on dextrin, so that we obtained granulates with up to 10 % of essential oil.

These granulates are fully free running and soluble in warm water. The storage stability is very good. The losses of essential oil during the granulation process are smaller than usual storage losses.

Figure 2 Loss of essential oil during fixing process and storage

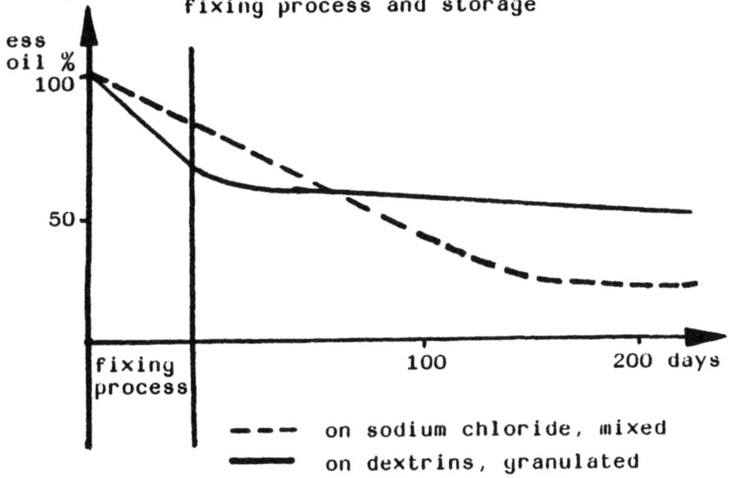

The reason is the internal structure. Micrographes of dill aroma granulates show

1. enlargement 290 x the agglomerate
2. enlargement 1350 x the closed surface of the granulates
3. enlargement 1350 x at the spot of rupture the foam-like structure under the closed surface.

The dill-type products were immediately accepted by the processing industry. For parsley, lovage and celery products recipe developments for soups and broth are finished.

Summary

Following the wishes of consumers and food processing industry the process of distillation of essential oils from dill, parsley, lovage and celery was introduced at the agricultural cooperative in BRÜNDEL (GDR). Liquid and granulated forms for application were developed so that since 1987 a suitable herb aroma for nearly every kind of application may be available.

We express our thanks to the enterprise Kosmetik Kombinat Berlin and to the cooperative Bründel for years of good collaboration and their assistence in the distillation of essential oils and in finding out the measurement data.

References

1. GDR-Standard, TGL 37785/02
'Bestimmung des Gehaltes an ätherischem Öl mit der Apparatur nach Brückner'.

2. Günther, E.: 'The Essential Oils'. Van Nestrand, New York, 1960, vol. IV, 619.

3. Huopalahti, R. and R. R. Linko: 'Composition and content of aroma compounds in dill ...'. J. Agric. Food Chem. 31 (1983) 2, 331 - 333.

4. Thomann, R. J. and P. Kretschmer. Unpublished results.

5. Göckeritz, D. Pharmazie 34 (1979) 12, 846.

FIRST RESULTS OF EXTRACTION OF MARJORAM WITH SUPERCRITICAL GASES

U. Tietz[1)] and K.-P. Roethe[2)]

What is the reason for selecting extraction with supercritical gases for the production of marjoram extracts?

The essential oils - aroma of spices and herbs - are obtained by water steam distillation. The distillation of the volatile flavour compounds is relatively simple to realize, cheap and gives products of an excellent quality.

The essential oil of marjoram, obtained by water steam distillation does not, however, represent the typical herb of marjoram. Often it is not suitable for application in food industry. Therefore marjoram extracts are used for products of meat processing and soup industry.

At present spice extracts are produced by using low boiling organic solvents. Disadvantages of this procedure are the removal of solvent residues from the extract as well as the environmental pollution by the organic solvents. Off-flavours may also arise.

In the sixties first research works began to use the special solving properties of fluid gases. The high pressure extraction developed to a method for isolation of valuable and sensible natural substances, more and more interesting also for spice industry.

Compared with water steam distillation and normal solvent extraction high pressure extraction needs a relatively complicated and expensive technical equipment.

A modern spice processing industry cannot, however, ignore the advantages of high pressure extraction:

- extraction with non-toxic gases like carbon dioxide,
- therefore the method does not involve environmental pollution,

[1)] Institute for Cereal Processing, Arthur-Scheunert-Allee 40/41, Bergholz-Rehbrücke, GDR-1505

[2)] Central Institute of Physical Chemistry, Academy of Sciences of GDR, Berlin, GDR

- many variations by the use of gas mixtures, that means purposeful extraction,
- depending on the task to be fulfilled the production of the full extract or the selective isolation of single components are possible (for instance: production of hop extracts - isolation of coffeine from coffee),
- gentle treatment of sensible substances at low temperatures is possible,
- there are no residues of organic solvents.

Experiments were carried out to obtain optimum conditions for the extraction of marjoram. Ground Thuringian marjoram (Majorana hortensis Moench) was used for extraction. The marjoram had a content of essential oil of 0.7 %.

The experiments were carried out in a laboratory at the Academy of Sciences. This plant was constructed for extraction with supercritical carbon dioxide.

As known from literature[1] and according to own results, the following conditions were used for finding the specific solvent amount:

pressure of extraction	15 MPa and
temperature of extraction	between 45 and 58 °C.

For evaluation of the extractions the nonextracted amount of flavour of the extraction residues was determined. To that end the marjoram residues were extracted a second time with dichlormethane. Quality especially was estimated by organoleptic analyses in determining the yield of extracts and the aromatic flavour characteristics.

Only 50 % of the aromatic substances extractable with dichlormethane were obtained by high pressure extraction with carbon dioxide at the mentioned conditions. The amount of essential oils obtained, however, was 96 %.

Variations of the specific solvent amount for 3 times do not considerably decrease the residue, extractable with dichlormethane. At the separation valve the separation of solid white waxes from the extract was observed.

The CO_2-extract has marjoram-like sensoric properties. The influence of oxygen, however, immediately led to an off-flavour.

From the literature (1) the influence of added substances on separation factors in supercritical systems is known. This change of selectivity was used for the extraction of marjoram. Pressured ethanol at an amount of 6 - 10 % by weight was added to the supercritical carbon dioxide. The extraction now was subdivided into two steps: first the extraction with CO_2/ethanol mixture, followed by drying of the extracted marjoram. The following table shows the conditions of extraction:

Table 1: Extraction of marjoram with CO_2/ethanol

extraction	p	15 MPa
	T	50 °C
separation	p	5 - 6 MPa
	T	40 °C
ethanol in gas mixture		6 - 10 Ma %
specific solvent amount		2 - 35 g CO_2/g
time of extraction		0.5 - 5 h
time of drying		1 h
initial matter		50 g

It was very important to improve the separation of the extract from the CO_2/ethanol mixture. For this purpose the construction of the separation unit and the temperature of the cooling receiver were varied.

In this way a highly concentrated ethanolic marjoram extract was collected at a temperature of -10 °C. In a second collector ethanol was separated at a temperature of -40 °C. This ethanol had only traces of flavour components.
The obtained extract is easy to disperse in water and is suitable for sensoric estimation.

Compared with pure carbon dioxide, the decrease by weight of the initial matter is higher but clearly influenced by the specific solvent amount. It was found that an optimum sensoric quality of the extracts is only achieved by a limited specific solvent amount of 6 g CO_2/g and 0.6 g ethanol/g for extraction and of 6 g CO_2/g for drying.

The extraction time is of special importance to the sensoric properties of marjoram extract.

It was found that an extraction of 1 h followed by drying of 1 h is optimum. It is supposed that a longer time of extraction as well as higher specific solvent amounts lead to an increase of undesired flavour components in marjoram extracts.

A correlation of analytical and sensoric results was not markably detected. Furthermore it was striking that also 1.5 to 2.0 % of lipids were extracted. The natural marjoram contains 2.7 to 3.5 % of lipids (determined by extraction with petroleum ether).

Dried ground marjoram has a water content of 8.2 %. Only traces of 0.6 % of this water are extracted. The marjoram extract has an essential oil content of 8 - 13 %.

The components of marjoram flavour thoroughly were investigated by Brosche (2), Oberdieck (3) and Fischer (4). cis-Sabinenhydrate, terpinen-4-ol, α-terpineol, linalool, α - and γ -terpinene and sabinene mainly contribute to the marjoram flavour. But none of these compounds may be considered as a key component of the marjoram flavour.

The following table shows the composition of marjoram flavour according to Brosche, Oberdieck, Tietz and Schmidt. Only main components are mentioned.

Table 2: Composition of essential oil of marjoram - main components

Compound	Brosche (%)	Oberdieck (%)	Tietz/Schmidt (%)
sabinene	5.6	3.8	
α -terpinene	7.3	3.4	
terpinolene	2.3	1.9	
γ -terpinene	11.4	7.1	11.4
trans-sabinenhydrate	4.3	4.7	4.6
cis-sabinenhydrate	21.1	13.9	11.6
linalool	2.5	4.4	3.0
terpinen-4-ol	23.5	22.1	25.0
α -terpineol	3.7	4.3	6.8
linalyl acetate	3.3	8.1	

The influence of various extraction methods on the composition of marjoram flavour was investigated by Fischer.

Table 3 shows the composition of marjoram flavour obtained by different distillation pressures.

Table 3: Composition of marjoram flavour, obtained by water steam distillation (Fischer)

Compound	reduced pressure (%)	normal pressure (%)
sabinene	0.9	6.5
α-terpinene	0.8	7.5
γ-terpinene	1.8	12.4
trans-sabinenhydrate	13.9	7.9
terpinolene	0.3	2.6
linalool/ cis-sabinenhydrate	64.2	34.9
terpinen-4-ol	14.8	24.7
α-terpineol	3.3	3.5

Comparing the composition, an increase of terpenes for 10 times is visible, obtained by heat treatment during water steam distillation at normal pressure. It is assumed that the change of flavour composition, especially the increase of terpinen-4-ol and of the terpinenes, is responsible for a considerable alteration of the sensory effect of cooked marjoram compared with the fresh herb.

The CO_2-extracts contain 8 - 13 % of essential oil as well as 90 % of non-volatile compounds. The evaluation of this non-volatile part is a task of the next time. Therefore the marjoram extracts and the essential oil of marjoram were evaluated not only by gas chromatography but also by thin layer chromatography.

The eluant was a mixture of benzene - chloroforme - acetic acid in a proportion of 50 : 50 : 4. A vanillin-sulphuric acid solution was used for detection.

The comparison of the thin layer chromatograms of CO_2-extract, dichlormethane extract and essential oil reveals significant differences in the composition.

Identification of the components (except α-terpineol and terpinen-4-ol) will follow in the next time, this research just being in the beginning.

Sensory analysis is of great importance for the quality evaluation of the extracts. The lowest concentration of marjoram extracts in a salt solution (0.2 % of NaCl in water) was determined which is required to perceive a taste of marjoram.

At a concentration of $5 \cdot 10^{-2}$ mg extract/100 ml the typical flavour of marjoram clearly is tasted. The taste value for essential oil of marjoram was found to be $2 \cdot 10^{-4}$ ml oil/100 ml.

Seasoning with CO_2-extracts and natural marjoram was compared to determine the effect of the extraction by carrying out a duo-trio-test in a neutral soup. The aroma of extracts obtained after 1 h achieve an intensity of 80 - 85 % as compared with natural marjoram. Extracts obtained after 3 h only show an intensity of 60 - 65 % for seasoning.

For the description of the flavour of marjoram aromatic flavour characteristics were established. The flavour can be described by the following taste components:

Table 4: Aromatic flavour characteristics

Taste components	Intensity
mint-fresh	medium-strong
marjoram-like	strong
herb-like green	strong
hay-like	no aroma
bitter	weak

A good aroma of marjoram combines the single taste components in the mentioned intensity. The herb-like green taste of the extracts is not so clearly percepted as in the fresh herb. Sometimes also a perfumic flavour was described in the extracts.

As compared with the essential oil, the extracts of marjoram represent the herb in a good quality.

The suitability of the new aromas as well as their good flavouring quality were proved in selected products in the laboratory.

Marjoram extracts are a favourite ingredient for seasoning soups, meat, sausages and legume dishes.

References

1. Brunner, G. and S. Peter: Chem.-Ing.-Techn. 53 (1981) 7, 529 - 542

2. Brosche, Th. and O. Vostrowsky: Z. Naturforsch. 36c (1981) 23 - 29

3. Oberdieck, R.: Deutsche Lebensmittel-Rundschau 77 (1981) 2, 63 - 74

4. Fischer, N., S. Nitz and F. Drawert: Chem. Mikrobiol. Technol. Lebensm. 9 (1985) 87 - 94

BIOTECHNOLOGICAL PRODUCTION OF AROMAS AND FLAVOURINGS

H. Ruttloff and M. Rothe[1)]

Introduction

Already since the turn of the century aroma compositions and flavourings have been produced for commercial purposes. Their industrial production has gained particular importance during the last 20 years (1). The reasons for this development are:

- increasing interest of consumers in aroma-intensive products,
- substitution of special raw materials or food ingredients owing to enhancement of prices or shortage of traditional resources,
- mechanization of harvesting processes in the agrarian primary production,
- rationalization of food technologies,
- arising of new consumer desire and expectation,
- low aroma quality of healthy, dietetically significant or energy-reduced foods (e. g. reduction of fat, sugar, sodium chloride).

Flavours consist of a great number of single aroma components which are present within the special food in different proportions. The food-specific perception of sense is realized only when the qualitative and quantitative composition corresponds with this pattern (2). The variety of flavour notes demonstrates the problems modern flavour research and production are confronted with, because as a first step these very complex systems have to be enlightened by sensory and instrumental methods.

Flavours are synthesized by biogeneration either during natural growing and ripening (e. g. in fruits and vegetables, nuts, herbs and spices) or as a result of fermentation processes (e. g. alcoholic beverages, pickled vegetables, fermented foods, special milk and dairy products). They can be developed also by physical influences (above all heat) or special enzymatically and/or microbially produced aroma precursors (e. g. baking products, roasted meat, cooked mushrooms, spirits, coffee, cocoa) or by other factors (e. g. smoking) (Tab. 1).

[1)] Zentralinstitut für Ernährung Potsdam-Rehbrücke, Akademie der Wissenschaften der DDR, Arthur-Scheunert-Allee 114/116, Bergholz-Rehbrücke, DDR-1505

Table 1: Aroma formation during ripening, processing and refining at the production of food (3)

Type of reaction	Food
Biochemical (animal or plant enzymes)	Fruit, vegetable, spices, ripened fish or ripened meat
Biochemical (microbial enzymes, partially with enzymes inherent in the substrates)	Butter, cheese, curdled milk products, beer, wine, fermented foods
Biochemical (plant or animal enzymes + microbial enzymes)	Tea, tobacco, ripened dry sausage
Biochemical and chemical (enzymes and physical influences, [e. g. heat, distillation processes])	Bread, cocoa, coffee, malt, alcoholic beverages
Chemical (partially + enzymes)	Cooked meat, cooked fish

Application of biological systems in flavour production

The production of aroma concentrates and flavourings for commercial purposes traditionally uses aromatic raw materials of natural origin as well as the composition of chemically synthesized single compounds (if key substances of natural aroma are known) (Tab. 2). Commercial aroma concentrates often contain 80 and more chemically synthesized or isolated single components. However, frequently the quality of the genuine natural product is not reached completely. Furthermore the international increasing environmental and nutritional awareness of people has effected worldwide an increasing turn to a 'biogeneration' of commercial aroma concentrates and flavourings or of their production by application of biological systems, respectively (4, 5). In consequence of this development US-American food legislation makes a difference only between 'natural' and 'artificial' aromas and flavourings while the term 'nature identical' is not used. This instruction is based on the fact that natural products possibly contain suspect substances only in traces or very low quantities, which can be toxic only in higher concentration. This danger is supposed to be higher if such substances are produced industrially. In Europe indeed this restriction is not usual.

Table 2: Traditional aroma production

- Production out of natural substances (e. g. fruits, vegetables, herbs, spices, sea animals)
  . extracts
  . juices
  . macerates, pulps
  . concentrates
  . dry products (pastes, powders)
  . distillates
  . fractions

- Composing of aroma concentrates on the basis of isolated or chemically synthesized compounds
  . without additional admixtures
  . admixture of products out of natural substances

In the application of commercial aroma concentrates and flavourings of course economical aspects are of particular importance. Nevertheless, the consumer is quite willing to accept a higher price if he can receive 'natural' aroma admixtures in foods. This means, that economical restrictions are not always standing in the fore when 'natural' flavour compositions are planned to be introduced into food production to a higher extent. These considerations have caused a more intensive employment of biotechnological processes in the production of commercial aroma concentrates and flavourings. However, up to now only the first steps have been gone.

The strategy of this intention requires the following premises:

- the biotechnological production of flavour compositions for commercial purposes is based on the course of the natural metabolism,

- consequently all those intermediate and end products of biogeneration are synthesized which are produced in vivo, too,

- the complexity of diverse substances for flavour impression and the proportion between the single components of them must be guaranteed and no deviations in genetic material or in regulation of the system may appear.

Then from biogeneration the following advantages should be expected:

- the quality is identical with that of natural flavour,
- the chemical synthesis of a great number of single compounds is not necessary or can be reduced,
- the economy in special (not in all) cases may be more favourable.

With regard to this conception the biological systems and biochemical reactions must be transferred into a technically practicable form, i. e. the biochemical pathways of aroma production have to be transferred from an in-vivo into an in-vitro-system (6). Under economic aspects a considerable acceleration of the biosynthesis rate should be realized. The following 'biological systems' are found to be suitable for this purpose:

- enzymes,
- microorganisms,
- plant material.

Enzymes

World literature indicates that up to now the application field of enzymes used for aroma production has not yet been very broad (Tab. 3). In our research group investigations on the production of a cocoa powder substitute and a butter fat lipolysate are to be mentioned.

Cocoa powder substitutes

In natural fermentation of cocoa beans after harvesting proteins and starch are splitted to low-molecular compounds (peptides, amino acids, reducing sugars or gluco-oligosaccharides, respectively) owing to the action of plant and microbial enzymes. These degradation products are precursors for the nonenzymatic browning reaction during the roasting step of cocoa beans releasing flavour substances (Fig. 1). Considering these facts a laboratory procedure aimed at a directed browning reaction between protein partial hydrolysates and reducing sugars has been developed. Suitable proteins are partially hydrolyzed by a protease and formed degradation products react with glucose under a defined temperature-time-regime. A product with characteristics similar to cocoa powder is obtained (7).

Other protein-containing raw materials can also be used. For instance wheat germs in mixture with reducing sugare are suitable for this purpose (8).

Table 3: Flavour generation by enzymes

Enzymes	Effect
Lipases	Production of butterfat lipolysates Shortening of the ripening time) special cheese types Increasing of aroma intensity)
Enzyme extracts from ripening flora of . cheese . dry sausage	Shortening of the ripening time Increasing of aroma intensity
Proteases	Increasing of flavour of fermented foods Shortening of the ripening time (special cheese types) Production of protein hydrolysates followed by heat treatment with reducing sugars . meat aroma concentrates . cocoa powder substitutes Elimination of off-flavour by plastein reaction
Naringinase	Debittering of citrus juices
'Aroma enzymes'	Reconstitution of aroma after food processing

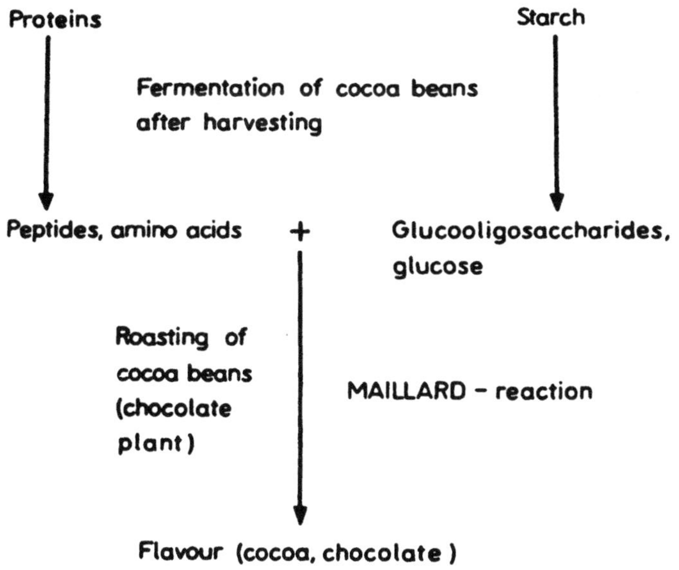

Fig. 1: Generation of flavour substances at fermentation and roasting of cocoa beans

Butter fat lipolysate

Commercial butter aromas contain a multitude of different chemically synthesized single components (e. g. diacetyl, acetoin, lactones, carbonyl compounds, S-substances, fatty acids and others). In general, enzymatically treated butter fat can be used as a basic material for such compositions.

The main problem of the production is to find such experimental parameters which guarantee a highly intensive hydrolysis of the substrate. The enzymatic release of the fatty acids has to be realized in such a way that a suitable quality of the final product results. The hydrolysis value of the enzymatic lipolysis should be high enough to limit the dosage to 0.1 - 0.3 % of the food to be flavourized at considering the fact that further aroma additives have to be included in this 'basic aroma'. A high quality of the lipolysate depends not only on the grade of hydrolysis but also on the composition of the free fatty acids released. In cooperation with the 'Kombinat Öl- und Margarineindustrie/ Magdeburg' such a product was developed (9), based on the lipolysis in a 2-phase-system (Fig. 2). The released fatty acids remain within the fatty phase. Their composition is shifted in the direction of middle and short chain fatty acids.

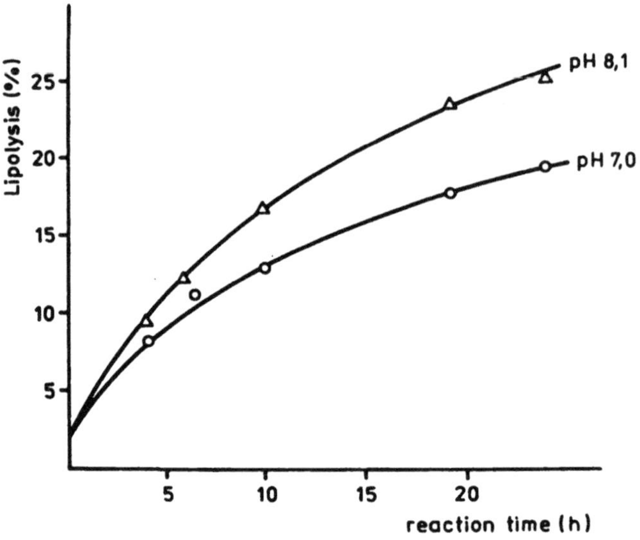

Fig. 2: Lipolysis of butter fat by pancreatic lipase (34 °C; ratio lipid: water = 1 : 2; 10 lipase units/ml)

Butter fat lipolysates can be used as basic flavourings for diverse food products (Tab. 4) or as partial components of commercial aroma compositions.

Table 4: Foods conveniently flavoured by addition of lipolyzed products (10)

Bakery/cereal products
Cake and cookie mixes,
chemically leavened bakery formulations
sweet doughs
cheese cake mixes
pancake mixes
cereals
Candy/confectionary products
Milk chocolate
creams and cream centers
toffee and caramel fudges
Dairy products
Cheese dips
coffee whiteners
Miscellaneous products
Margarines, imitation dairy products
popcorn oils
salad dressings
sauces, soups, snack foods

Microorganisms

Microorganisms are capable of producing flavour substances in various ways. A lot of foods are produced traditionally by the help of microorganisms. In the course of such fermentation processes ripening, refinement, chemical conversion and other changes as well as an intensive generation of flavour substances occur leading to an improvement in quality (3).

Tab. 5 gives a general view about the principal possibilities of microbial ability to synthesize or transform food raw materials which can be used in flavour production (11 - 24). Microorganisms are able to synthesize a lot of single components as well as mixtures of volatiles which in total may be responsible for the sensory impression of a special flavour note. Biotransformation has also to be mentioned here which may be a useful tool in the field of aroma production (12, 24).

Table 5: Flavour generation by microorganisms

	De novo synthesis	Biotransformation
Single compounds	Mixtures of several or many components	
Acetic acid	Mushroom flavours	Oxydation
Lactic acid	Fruity notes	Reduction
Citric acid	Citrus notes	Isomerization
Gluconic acid	Coconut notes	Splitting
Acetoin + diacetyl	Strawberry notes	Carboxylation
Ribonucleotides	Banana notes	Derivatisation (e. g. of terpenes)
Phenylalanine		
Aspartic acid	Roquefort note	

Blue cheese aroma concentrate

Aroma concentrates obtained from natural cheese are low in intensity. For flavour purposes, however, commercial cheese aroma concentrates and flavourings must have high intensity. To get such concentrates a biotechnological process for the production of commercial Roquefort aroma was developed in cooperation with the 'Kombinat Milchwirtschaft/Dresden'. Butter fat - as a fundamental ingredient of a suitable nutrient medium - is metabolized in an aerated stirred reactor by a combined action of pancreatic lipase preparation and a culture of Penicillium roqueforti. As a first step butter fat is split by enzymatic hydrolysis. The released fatty acids - especially the middle-chain ones - are transferred into 2-alkanones (Fig. 3). 2-Heptanone and 2-nonanone are discussed as key substances for the formation of the Roquefort note.

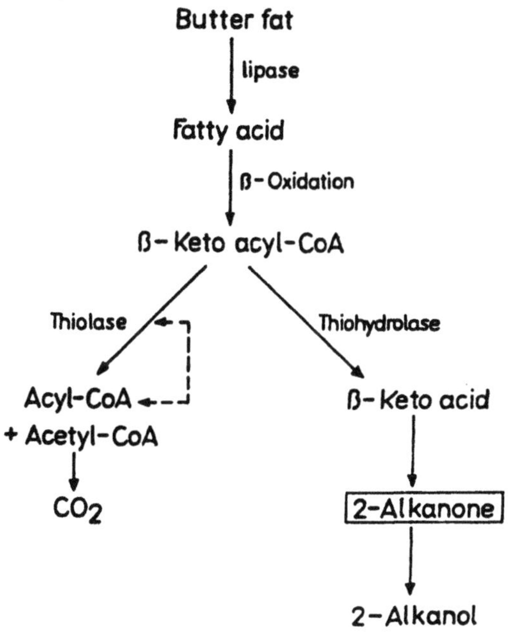

Fig. 3: Formation of methyl ketones by Penicillium roqueforti (25)

In the first steps of fermentation the content of middle-chain fatty acids increases rapidly followed by degradation to methyl ketones. Under technical conditions the production of methyl ketones approximates to a maximum value after a fermentation time of 40 - 60 h (Fig. 4). The

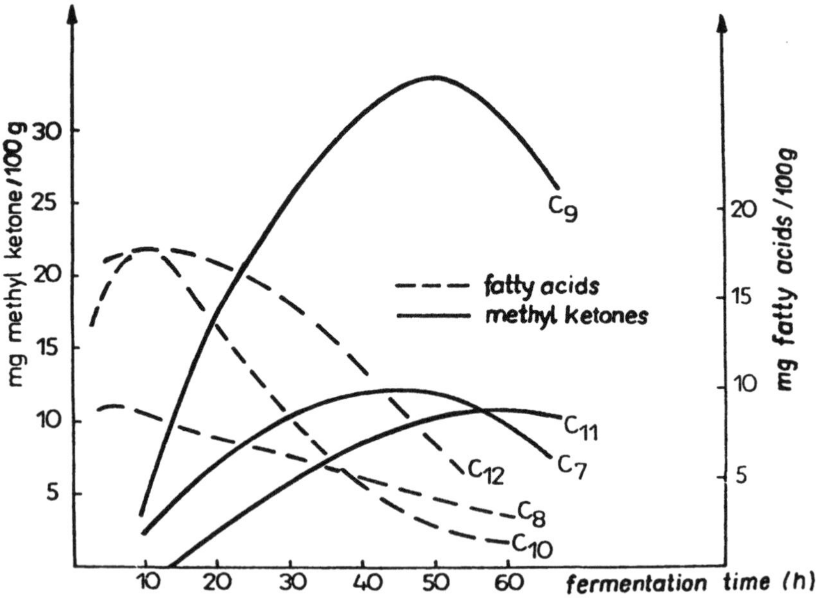

Fig. 4: Free fatty acids and methyl ketones during biotechnological production of Blue cheese aroma concentrates (26, 27)

content of methyl ketones and fatty acids can be used as an index for controlling the fermentation process. According to the distribution coefficient of the formed methyl ketones and further products of fat metabolisms the main part of aromatic substances is accumulated in the fatty phase.

The main problem of technical fermentations is the supposed fact that in such complicated systems many influence and goal parameters have to be considered. For optimizing such process a combination of sensory and instrumental methods is necessary. Statistical plans as well as the evaluation of data by means of multivariate analyses have to be applied. At the end of the cultivation the fermentor content is stabilized by heat and separated.

The separated aroma concentrate can be used in milk products in a dosage of 0.75 - 1.5 %. The application field includes the aromatization of low-flavoured processed cheeses as well as of fresh cheese sorts and of cheese cremes. As to their sensory quality the products with aroma additives are similar to natural Roquefort cheese. Intensity and quality of application products are controlled by sensory profile diagrams (Fig. 5).

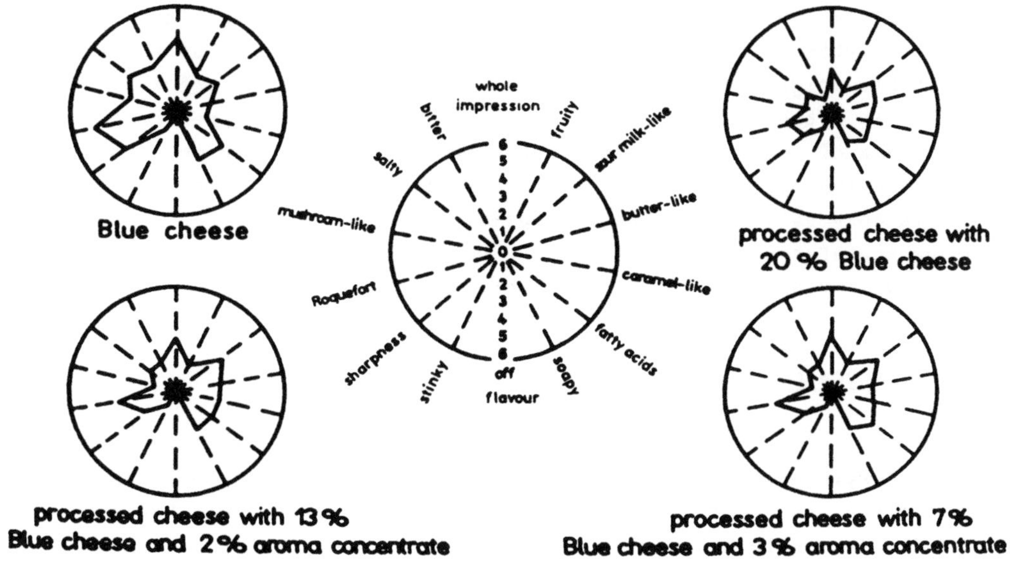

Fig. 5: Effect of the exchange of Blue cheese against aroma concentrate in processed cheese shown by sensory profiles

Plant material

In principle there are several possibilities of aroma production by application of plant material:

- extracts (so-called 'aroma enzymes'),
- macerates (enzymatically produced),
- cell and tissue cultures
  . biotransformation,
  . de novo-synthesis,
    .. simple systems (e. g. flavourings),
    .. complex systems (e. g. fruity notes).

Plant extracts

Naturally the enzyme system of the plant cell which is generally highly differentiated is able to synthesize flavour volatiles of high quality most effectively. Therefore numerous investigations were made to use plant systems for an aimed aroma production.

The application of plants for this purpose has been investigated already in the sixties. According to HEWITT (28) numerous aromas of fruit and vegetable origin are volatilized and lost during technological processing or in kitchen preparation. On the other hand aroma precursors which often are of nonvolatile character overcome such thermic processes without changing. By addition of a fresh fruit or plant extract (so-called 'aroma enzymes') out of the same plant the precursors are discussed as being transferred to regenerate the aroma. In practice, however, the expectations seem not to be realized.

Plant macerates

Mechanically or enzymatically produced macerates on principle can catalyze biosynthetic processes and are also discussed in connection with flavour production. The aroma intensity reached, however, is low. Thus a commercial utilization of these results seems not to be profitable.

In the seventies GUADAGNI et al. (29, 30) as well as VAMOS-VIGYAZO et al. (31) have used crushed fruit tissues or waste products of fruit processing (e. g. apple peels, apple tresters) for aroma production in such a way. The Hungarian authors have developed a laboratory process in which waste products of industrial apple manufacturing are incubated in a closed vessel. The aroma originated by influence of the fruit-own enzymes on the precursors is removed by a stream of inert gas and condensed in a cooling trap. But also this interesting principle solution up to now has not been realized in industrial scale.

Plant cell and tissue cultures

In future also plant cell and tissue cultures may be applied for stuff production ('phytoproduction') in increasing amount (32 - 35). The advantages of such possibilities are:

- plant flavours are secondary plant metabolites,
- each plant cell possesses the genetic material of the species ('totipotency'),
- the synthesis of numerous secondary plant metabolites is not possible by microorganisms,

- independence of imports,

- independence of climatic conditions and world-wide economic problems,

- absence of pesticides,

- cultivation of tissues of rare plants.

Therefore the production of secondary plant metabolites must be possible on principle. In general, the attention is concentrated on substances which are relatively expensive due to their production by traditional chemical technologies or by extraction from plants. In this case a technical production via fermentation seems to be justified from the economic point of view (e. g. shikonin). This method is also particularly suitable for biotransformation purposes. Up to now, however, only single compounds are produced in this way and no multi-component mixtures (i. e. natural aromas).

There are several disadvantages with regard to in-vitro-cultivation of plant cells and tissues:

- in spite of their existence no expression of numerous genes for secondary stuff production,

- morphological and functional despecialization (dedifferentiation),

- the great number of aroma components within most food aromas makes the problem more difficult,

- high requirements for development of suitable screening methods,

- extended cultivation time compared with the effect of microorganisms,

- increased shearing sensity in fermentation processes,

- disintegration of plant material for obtaining intracellular secondary stuffs is necessary,

- economy of aroma production via phytoproduction is questionable.

First of all the biochemical research is asked for solving the problem of despecialization and dedifferentiation, respectively (36).

However, during the last 10 years remarkable methodical advances have been obtained:

- improved knowledge of regulating metabolic processes,

- improved methodical efficiency (fermentor design, instrumentation),

- discovery and application of phytohormones for stimulation of secondary metabolism,

- increased application of biotransformation,

- application of 'elicitors',

- separation into the phases growth and secondary stuff production during fermentation,

- improvement of plant cells as to their ability for secondary stuff production,

- increasing of cell wall permeability,

- occuring of 'somaclonal variation',

- cultivation of 'hairy roots'.

From the point of view of health, nutrition physiology and food quality the use of plant cell and tissue cultures for production of aroma concentrates and flavourings seems to be justified. In the near future single components of high commercial value may actually come to the fore (e. g. flavourings, key substances of expensive aroma compositions, biotransformation). According to international emphasis of investigation as to biogeneration of flavour compounds via plant cell and tissue cultures activities can be expected in the following fields:

- sweetening substances,
- bitter substances,
- spice flavours,
- colouring substances,
- terpenes, essential oils (biotransformation),
- fruit flavours,) first cautious steps are gone in this direction.
- vegetable flavours)

From the examples shown it is evident that - considering the risk - some groups of substances have a chance to be produced via in-vitro-cultivation by plant cell and tissue cultures one day. On the other hand complex systems as fruit or vegetable aromas or even secondary flavour substances cannot be produced by this technology until a deepened insight and knowledge into the metabolic regulation is gained.

Literature

1. Ruttloff, H. und M. Rothe: Produktion von Aromen und Geschmacksstoffen unter besonderer Berücksichtigung biotechnologischer Verfahren. 1. Mitt. Ernährung 11 (1987) 395 - 399.

2. Rothe, M.: Introduction to Aroma Research. Akademie-Verlag, Berlin 1988.

3. Ruttloff, H.: Impact of biotechnology on food and nutrition. In: Knorr, D. (Ed.): Food Biotechnology. Marcel Dekker, Inc., New York and Basel 1987, pp. 37 - 94.

4. Ruttloff, H.: Biotechnology and aroma production. Nahrung 26 (1982) 575 - 589.

5. Parliment, T. H. and R. Croteau: Biogeneration of aromas. 180th meeting of the Amer. Chem. Soc., Chicago/Ill. (1985). American Chemical Society, Washington, DC 1986.

6. Ruttloff, H. und M. Rothe: Produktion von Aromen und Geschmacksstoffen unter besonderer Berücksichtigng biotechnologischer Verfahren. 2. Mitt. Ernährung 11 (1987) 466 - 473.

7. Rödel, W. und H. Ruttloff: Verfahren zur Herstellung eines Kakaoaromastoffes. DD-PS 239 942 (1986).

8. Walpuski, J. and H. Ruttloff: Principle possibilities for the production of cocoa powder substitutes, compare Poster Abstracts 2nd Wartburg Aroma Symposium, Eisenach, November 16 - 19, 1987.

9. Rothe, M., C. Kornelson, W. Rödel, H. Ruttloff und W. Paul: Verfahren zur Herstellung einer Butteraromakomposition auf der Basis der lipolytischen Spaltung von Butterfett. DD-PS, angemeldet am 14. 1. 1987.

10. Shahani, K. M., R. G. Arnold, A. Kitara and B. K. Dwivedi: Role of microbial enzymes in flavor development in foods. Biotechnol. Bioeng. 18 (1976) 891 - 907.

11. Kempler, G. M.: Production of flavor compounds by microorganisms. Adv. Appl. Microbiol. 29 (1983) 29 - 51.

12. Latrasse, A., J. R. Degorce-Dumas and J. Y. Leveau: Production d'arômes par les microorganismes. Sci. Aliment. 5 (1985) 1 - 26.

13. Martin, A. M.: A review of fundamental process aspects for the production of mushroom mycelium. J. Food Proc. Eng. 8 (1986) 81 - 96.

14. Kaminski, E., S. Stawicki, E. Wasowicz and M. Kasperek: Volatile odour substances produced by microflora. Nahrung 24 (1980) 103 -113.

15. Hanssen, H.-P.: Volatile fragrance compounds from the fungus Gloeophyllum odoratum (Basidiomycotina). Z. Nat. forsch., Sect. C: Biosci. 41 (1986) 825 - 829.

16. Drawert, F., R. G. Berger und K. Neuhäuser: Über die Biosynthese von Aromastoffen durch Mikroorganismen. 5. Mitt.: Lactone in Kulturen von Polyporus durus. Chem. Mikrobiol. Technol. Lebensm. 8 (1983a) 91 - 92.

17. Drawert, F., R. G. Berger and K. Neuhäuser: Biosynthesis of flavor compounds by microorganisms. 4. Characterization of the major principles of the odor of Pleurotus euosmus. Eur. J. Appl. Microbiol. Biotechnol. 18 (1983b) 124 - 127.

18. Hattori, S. Y., Y. Yamaguchi and T. Kanisawa: Preliminary study on the microbiological formation of fruit flavors. Technol. Aliment. 16 (1976) 143 - 151.

19. Collins, R. P.: Terpenes and odoriferous materials from microorganisms. Lloydia 39 (1976) 20 - 24.

20. Collins, R. P. and A. F. Halim: Production of monoterpenes by the filamentous fungus Ceratocystis variospora. Lloydia 33 (1971) 481 - 482.

21. Lanza, E. and J. K. Palmer: Biosynthesis of monoterpenes by Ceratocystis moniliformis. Phytochemistry 16 (1977) 1555 - 1560.

22. Tahara, S. and J. Mizutani: δ-Lactones produced by Sporobolomyces odorus. Agric. Biol. Chem. 39 (1975) 281 - 282.

23. Fischer, K. H., F. Senser und W. Grosch: Aromabildung durch Mikroorganismen. 1. Mitt. Bildung von Fruchtestern in Kulturen der Hefe Dipodascus magnusii. Z. Lebensm.-Unters.-Forsch. 177 (1983) 336 - 338.

24. Kieslich, K., W. R. Abraham, B. Stumpf and P. Washausen: Microbial transformation of terpenoids. In: Berger, R. G., S. Nitz and P. Schreier (Eds): Topics in flavour research. H. Eichhorn, Marzling-Hangenham 1985, pp. 405 - 425.

25. Ruttloff, H., A. Quehl und A. Leuchtenberger: Fermentationstechnische Gewinnung von Käsearoma unter besonderer Berücksichtigung der Richtung Roquefort. 1. Mitt. Literaturüberblick. Nahrung 28 (1984) 873 - 881.

26. Quehl, A.: Untersuchungen zur fermentationstechnischen Gewinnung von Edelpilzkäse-Aromakonzentrat mittels Penicillium roqueforti. Dissertationsschrift A, Akademie der Wissenschaften der DDR, 1985.

27. Rothe, M., H. Ruttloff, H. Herrmann and W. Engst: Problems of technical production and characterization of Blue cheese aroma concentrate. Nahrung 30 (1986) 791 - 797.

28. Hewitt, E. J.: Enzymatic enhancement of flavor. J. Agric. Food Chem. 11 (1963) 14 - 19.

29. Guadagni, D. G., J. L. Bomben and J. S. Hudson: Factors influencing the development of aroma in apple peels. J. Sci. Food Agric. 22 (1971) 110 - 115.

30. Guadagni, D. G., J. L. Bomben and J. G. Harris: Recovery and evaluation of aroma development in apple peels: J. Sci. Food Agric. 22 (1971) 115 - 119.

31. Vamós-Vigyázo, L., N. Kiss-Kutz and A. Hersiczky: Poster session, 3rd Internat. Flavour Symposium, Abstract 2.4, Kecskemet/Hungary, 11. - 14. 9. 1984.

32, Teutonico, R. and D. Knorr: Plant tissue culture: Food applications and the potential reduction of nutritional stress factors. Food Technol. 38 (1984) 120 - 127.

33. Stafford, A. , P. Morris and M. W. Fowler: Plant cell biotechnology: A perspective. Enzyme Microb. Technol. 8 (1986) 578 - 587.

34. Böhm, H,: Pflanzliche Zellkulturen als Produzenten von Naturstoffen. Wiss. Fortschr. 35 (1985) 302 - 305.

35. Berlin, J.: Secondary products from plant cell cultures. In: Rehm, H. J. and G. Reed (Eds.): Biotechnology, Vol. 4, VCH Verlagsgesellschaft, Weinheim 1986, pp. 630 - 658.

36. Lee, J. M. and G. An: Industrial application and genetic engineering of plant cell cultures. Enzyme Microb. Technol. 8 (1986) 260 - 265.

NATURAL AND ARTIFICIAL CHEESE FLAVOUR

J. ADDA[1]

I. Introduction.

Originally cheese making appears as a mean of preserving the milk components. To day, at least in affluent societies, cheese appears as a product which is valued mainly for its flavour.
Unfortunately it is not so easy to obtain a full balanced flavour in cheese on an industrial basis and furthermore this implies an extra cost as flavour develops only after a long maturing period.
It is not therefore surprising that so many attempts have been made to get a better knowledge of cheese flavour and its formation with the hope to control it, but also to see the attempts made by the flavour industry to provide the food industry with cheap substitutes, obtained by chemical synthesis, or to day by biotechnology, which could be used instead of cheese in many foods. More recently the cheese industry itself became aware of the possibility of saving on the cost of storage by trying to accelerate the ripening process which immediatly brought a new interest for the mechanisms through which flavour is formed, as it soon became obvious that our knowledge is not yet sufficient to overcome that problem.
The problem is indeed a difficult one, as one could say, that when we are talking about cheese flavour we do not know what we are talking about : Flavour varies obviously between cheese varieties, but even within one given type, the flavour can be very different even from one batch to another. We nevertheless describe all these differents products using the same word "cheese" which may be justified as they appear to have something in common, that we cannot describe, but we call cheesy, above which is a more specific character, often still difficult to describe, explain or imitate, as cheese aroma is rather the result of the perception of a large number of molecules belonging to very different chemical classes than the perception of some character impact compounds[1].
We will not try to review the existing knowledge, but will rather try to show through the available patents which part of this knowledge has been used by the flavour industry for nature identical or artificial flavour, and see which trends are being followed in the use of biotechnology either to accelerate the apparition of flavour in cheese technology or in order to produce cheese imitation.

[1]) Institut National de la Recherche Agronomique, Dijon 21034 FRANCE.

2. Nature and nature identical cheese aroma.

2.1. Components arising from the lactic fermentation.

A common feature to all cheese varieties is the lactic fermentation which appears as a critical step, more for the physico chemical conditions it creates than, for the flavour molecules which are formed from the well known lactose and citrate metabolism, but that point will not be discussed here as the practical applications are not important.

2.2. Components arising from milk fat.

Another common feature to all cheeses is the existence of milk fat characterized by short chain fatty acids. Its role is recognized as important for different sensory attributes of cheeses, but its main role is probably as a source of free fatty acids (FFA) a part of which is further metabolized by various type of microorganisms or involved in a series of reactions. High level of FFA are to be found in most types of cheese, where this level may vary according to the length of the ripening period. Up to 10 p 100 of total triglycerides fatty acids may be found under free form in a well ripened Camembert but it is in blue, in italian type of cheese and also perhaps in Cheddar that a correlation seems to exist between FFA and flavour intensity.
The addition of kid lipase, which preferentially hydrolyses short chain fatty acids is traditional for certain italian type of cheese such as Romano, Provolone and Fontina. In more traditional cheese, positive results have been claimed for Cheddar after the action of added fungal lipase[2], while other similar experiments have led to poorer results[3]. The existence of many conflicting results makes that the importance of FFA is not that clear even for italian type[4]. Most patents however make use of FFA for cheese flavor imitation, whatever type considered, but it is only for some type of the italian cheeses that a simple mixture of acids (butyric, hexanoic and octanoic 600 - 10,000 ppm) is claimed to impart a typical flavour to a bland curd or to process cheese[5].
It must be reminded that some of the most important fatty acids present in cheese result from the oxidative desamination of amino-acids. This explains for the presence of iso and aromatic acids which are to be found in surface ripened cheeses but also in Roquefort and other varieties[6]. Some patents make use of these amino acids derived acids and for exemple 1-phenylproprionic and isovaleric acids have been claimed to produce a Fontina type flavour at the level of 100 and 20 - 300 ppm respectively[7], while the use of calcium-3-methyl-butanoate is advocated

for improving Gouda cheese flavour in a cheese flavouring composition[8]. Besides alkanoic acids, milk fat also contains small amounts of oxo-acids, which can also results in cheese from microbial activity. With a few exceptions[9,10], oxoacids have not been given much importance in the papers devoted to flavour. However oxoacids such as 2-oxo-3-methyl butanoic, 2-oxo-4-methyl pentanoic, 2-oxobutanedioic and 2-oxopentanedioic acids are being incorporated together with pyruvic acid in blue cheese and Cheddar cheese flavour composition[11] and its has been claimed[12] that α-oxoisocaproic and α-oxomethylmercaptobutyric acids derived from leucine and methionine, upon incubation with Aspergillus flavus, give a flavour similar to that of Cheddar cheese.

Delta-hydroxyacids, normally present in small proportion, in milk fat, lead to the corresponding delta-lactones by cyclisation.

Hydroxyacids themselves have rather been ignored as important contributors to natural cheese flavour, but have been used in artificial preparations : alpha-hydroxy-isobutyric and alpha-hydroxy-isocaproic acids have been included at a 150-1000 ppm level, together with alpha-hydroxyphenyl-alkyl acids, in formulation for flavouring fresh lactic cheese, made for the manufacture of processed cheese or in Cheddar cheese formulation[13]. However it seems reasonable to think that, in such cases, the 4,5,6-hydroxyalkanoic acids were merely added as lactones precursors.

Lactones themselves, have received much more attention and have been identified in most cheeses. Their importance for natural Cheddar cheese flavour has been stressed[14] and lactones have been used in manv formulation to imitate Cheddar and blue cheese flavours[15] where they have been claimed to significantly improve the formulation[16].

Methylketones (MeK) are common constituents of cheeses. They may be formed either by decarboxylation of oxoacids, or by enzymic oxidative decarboxylation of alkanoic acids. The second mechanism is of importance to the formation of flavour in mould ripened cheeses where the growth of Penicillium species is permitted, and, above all, in blue cheeses where MeK appear as character impact compounds.

In cheese MeK are further metabolized by Penicillium sp., into the corresponding secondary alcohol, the importance of which is not to be ignored as secondary alcohols possess flavour thresholds similar to that of the MeK.

Insaturated ketones are also to be found in blue and mould surface ripened cheeses. The more common is 8-nonen-2-one[17], which seems to result from oxidative decarboxylation of 9-decenoic acid, an enzymic oxidative cleavage product of polyunsaturated fatty acids[18].

The various ketones have been used, at a great extent, in flavour composition.

8-nonen-2-one, which has a threshold value of about 0.2 ppm, has been claimed to impart by itself a flavour described as common to all kind of cheese, except that of blue cheese, and to give in combination with secondary alcohols and others volatiles a blue or Cheddar like flavour[19]. Some formulation have also been made using MeK and the corresponding secondary alcohols as main constituents but it has been claimed that significant improvment was obtained when using another fungal metabolite, 1-octen-3-ol consistently found in any cheese of the Camembert type[18,20,21]. 1-octen-3-ol was found to mask or to counteract the chemical note possessed by previous blue cheeses flavour composition[21], and included in a composition together with cinnamic acid. At the same period methylcinnamate was for the first time identified in Camembert cheese[23], by those who, later, described its use in a patent[24] together with 1-octen-3-ol. It must be emphazised that methylcinnamate has never been identified again in any type of cheese by any other group.
According to the above mentionned patent, the use of 1-octen-3-ol with methylcinnamate gives a specific blue cheese or Camembert cheese flavour to a basic mixture, made from MeK, secondary alcohols and acids, which otherwise has an unspecific cheesy note only.
With a high 1-octen-3-ol/methylcinnamate ratio (9/1) a Camembert type flavour would be obtained while with a lower ratio (2/1) a blue cheese flavour is perceived. Such formulations are intended to be added to a neutral process cheese.
MeK have also been used, associated to methylcaproate, ethylcaproate, and short chain fatty acids to give a composition which makes possible to confer foodstuffs with a goat milk cheese flavour[25].
Thus, even if sometime described as character impact compound, MeK mainly appear as able to impart a more cheesy note than a really specific one, which is in accordance with the fact that they have been identified in most type of cheese.

2.3. Components arising from proteins.

Perhaps more important that the lipids derived flavour compounds, are those arising from the proteins degradation.
It is clear nowadays that not enough attention has been given to the non volatile, water soluble components even if the role of the free amino-acids has been aknowledge since long but perhaps forgotten later. The enzymic degradation of caseins first leads to numerous peptides, the structure of which has not been elucidated, but which seem responsible for the flavour intensity of matured cheese[25] and for various specific notes[27]. The lack of knowledge on peptides activity in relation with their structure, with the exception of the bitter character which is well under-

stood, explains that peptides are not often cited in flavour application. One patent, however, describes the use of a cyclic dipeptide (asparagine-phenylalanine) as the active ingredient of a cheese flavour composition[28]. In normally ripened cheese, a part of the peptides are split to free aminoacids, which may represent up to 13 p 100 of total amino nitrogen, and each type of cheese has its own characteristic pattern.
It appears from several patents that aminoacids are either essential to, or improve remarkably cheese taste : for instance, the blue cheese taste is said to be intensified in an artificially flavoured foodstuff if a mixture of glutamic acid, lysine and methionine is added to a level high enough to obtain that the concentration of each added aminoacid is at least twice as large as that of any other free aminoacid present[23]. This recommandation is an agreement with the fact that glutamic acid and methionine are always present at a high level in blue cheese[29]. Another example would show that the same aminoacids mixture is used in a Gouda cheese imitation preparation[9].
Probably more important than the aminoacids themselves are the volatile components derived from them.
Microorganisms are able to decarboxylate the aminoacids to give volatile, as well as non volatile amines, the role of which has been mostly ignored. A few authors, however, made some studies in the amine content of cheeses and gave them some credit[30,31,32,34,35]. Their importance was finally fully recognized when the incorporation of a series of alkyl-amines in cheese flavour ingredient was suggested in a patent in order to impart a ripe Cheddar or Emmental cheese flavour to food[35]. Amino-acids will also serve as substrate for the formation of aldehydes which are then reduced to the corresponding alcohols. This explains why aldehydes appear only as trace component in normal cheese, or are even often indetectable. However aldehydes have been included in various formulation : 3-methyl butanal, phenylethanal, phenylpropanal and phenylpropenal have been added to Emmental cheese flavour together with benzaldehyde[11]. 2-phenylethanol, 3-methylbutanol and 3-methylthiopropanol are the more common alcohols found in cheese, originating from phenylalanine, valine and methionine respectively. The importance of 2-phenylethanol and 3-methylbutanol is underlined in a patent according to which these compounds added to unripened cheese at a ration of 1:3 to 1:4 impart an italian cheese type flavour[36].
Among aminoacids derived components, volatile sulfur compounds have been recognized to play a major role in the flavour of various type of cheeses. Methionine is an important precursor which leads to methanethiol as well by an enzymic pathway as by a chemical one[37]. Methanethiol, a highly reactive product, condenses itself into dimethyldisulfide, often identified in cheese and which has been used in artificial flavour as a source of

volatile sulfur[8] ; 3-mercaptopropionic acid, has also been used for the same purpose[16] although never identified in any cheese.
In some type of cheese such as, surface ripened cheeses, methanethiol is esterified by short chain alkanoic acids[38,39] and these esters are nowadays used in cheese flavour preparation and for instance methylthiobutyrate is found in the catalog of some flavour firms, but to my knowledge is not described as an useful ingredient in any patent.
Ethyl-4-(methylthio) butyrate, which possesses a similar structure, but has never been identified as a natural constituent of cheese, has been claimed to be useful to increase cheese flavour intensity[40], before the identification of the methylthioesters as natural molecule in cheese.
2,4-dithiapentane which has been isolated from Gouda[41] and Camembert[42], has been claimed[44] to impart or improve a Gouda type cheese flavour in products such as processed cheese, cheese snacks or cheese sauces when used in combination with 2,5-diethyl-3-methylpyrazine and/or trans-anethol[43] both constituents being natural cheese aroma components as proved later[41].
Aromatic aminoacids are the source of interesting aroma compounds such as phenol, 4-methylphenol, 4-ethylphenol, acetophenone and indole, some of which have been suggested as essential to Cheddar flavour[45,46], while others were found important by several authors[38,39] in surface ripened cheeses, and blue cheese.
Their importance in cheese flavour is emphasized by their inclusion in various patented flavour formulation. Phenol was added to Camembert and blue cheeses flavours[46] while phenol, 4-methylphenol and guaiacol were used in a composition claimed to reproduce Cheddar cheese flavour[47].

3. Non nature identical components.

A whole series of others phenols such as 2-methylphenol, 3-methylphenol, 3-ethylphenol, 3,5-dimethylphenol, 2,6-dimethoxyphenol, 4-methylguaiacol, 4-ethylguaiacol and alpha-naphtol which have never been identified in any cheese were also indicated, as possibly used, in the same patent.
This is not the only example were non nature identical molecules are included in cheese flavour formulations : a substituted bicyclic pyrazine, (5,6,7,8-tetrahydroquinoxaline) is included[48] in a preparation given to be incorporated in bland cream cheese, while 5-methyl-3,4,6,7-tetrahydro-2(H)-cyclopentapyrazine is used to provide a cheese flavour[49].
Another additive included in a Cheddar cheese flavour is a thienopyrimidine[50].
Another example of artificial flavoring is the one we could observe recently on a powdered preparation sold to the dairy industry as "old Gouda cheese flavour" in which the main volatiles were a series of

1,2-propanediol esters[51].
In summary it appears that the flavour industry has in the past made some profit of the available results, which for a good part have been obtained by the flavour industry itself, and has marketed rather simple compositions to match natural complex flavours. All these efforts could be ruined if new flavour regulations, or new labelling requirements make that synthetic formulations are not anymore accepted by the consumer. This explains a renewal of interest for biotechnology, which would allow to produce more sophisticated cheese flavour either by industrial fermentation or by modification of the traditional cheesemaking procedures.

4. Cheese flavour obtained by a biotechnological process.

4.1. Fermentation.

Cheese flavour can be obtained by solid or submerged fermentation processes. The best example is that of blue cheese flavour which is easily produced by the action of P. roqueforti mycelium or spore on various substrates (such as whey powder, ultrafiltrated skim milk, homogeneized milk etc...) containing milk fat or vegetable fat or oil, with or without a lipolytic pretreatment[52,53,54]. The process was first patented nearly 25 years ago[55,56], but has raised new interest in the recent years[57,58]. A similar work using P. caseicolum is under developpment[59].
The fermented medium can be distilled, to obtain MeK, which can be used in flavour composition or can also be incorporated readily in dairy or other food products. Recent results[60] show that such a blue cheese flavour can be used with advantages for cottage type or butter spreads, provided sensory differences are accepted. However, we are aware of industrial experiments which failed, as the flavour mixture proved itself unstable during the storage of the dairy preparation in which it was incorporated.
Other microorganisms have also been cultivated on dairy substrates to make products with a cheese flavour. For instance, concentrated milk inoculated with two different starters successively (Streptococcus thermophilus, Lactobacillus helveticus and S. lactis, S. diacetylactis S. acetaldehyde lactis respectively) leads to a flavoured base for processed type of cheese[61].

4.2. Enzymes preparation.

Another approach is the use of enzymes preparation. For instance a cheese like flavour was obtained by action of a crude enzyme mixture isolated from Candida rugosa on a butter fat emulsion[65]. Similarly

lipases[66,67] and fungal proteinases[67] were added to lactose hydrolysed whey or whey fractions to give products claimed as having a flavour profile characteristic of ripened cheese.
This reminds of the attempts made to accelerate cheese ripening through the use of enzyme added to the curd, with the hope that a more pronounced lypolysis and/or proteolysis would lead, more rapidly, to a highly flavoured product, thus saving on costs. Most attempts have led to flavour imbalance or even to flavour defects such as bitterness[68]. Some success have however been claimed as Cheddar ripening time could be halved without off-flavour formation[69].
Other successful results have also been obtained when using an enzyme mix made from Neutrase, a food grade B. subtilis proteinase, and an exopeptidase extracted from S. lactis[68].
In these examples, the role of the enzymes is mainly to accelerate the early steps of the ripening process and it may happen that even with a more pronounced proteolysis the significant flavour components fail to occur.
It explains why some groups are trying nowadays to more specifically accelerate the final steps of the series of reaction leading to recognized aroma compounds.

4.3. Microencapsulation.

This can be obtained by microencapsulation of bacterial cell free extracts, with substrates and cofactors. The milk fat coated microcapsules are then incorporated into cheese.
It was possible, with such a technique to obtain an eightfold increase of diacetyl level in Cheddar cheese[70] and other experiments[37,71] have led to similar results for 3-methylbutanal, methional and methanethiol.
A similar approach is being used with P. roqueforti spores encapsulated with a lipase and a low melting fat fraction as a substrate, in capsules consisting of high melting butterfat fraction[72]. Cheeses containing encapsulated flavour system developed significantly higher blue cheese flavour intensity in shorter time, than cheeses to which, lipase and spores were added directly.
Indeed the techniques are of interest even if it is doubtful that the increase of the level of one specific flavour component would be enough to really duplicate cheese flavour. The question which then arise is again, what do we call cheese flavour, and what will be in the future the consumers' acceptance for products which for most of them will not match the traditionnal one.

Literature.

(1) Badings, H.T. and R. Neeter : Neth. Milk Dairy J. 34 (1980) 9.

(2) Sood, V.K. and F.V. Kosikowski : J. Dairy Sci. 62 (1979) 1713.

(3) Law, B.A. and A.S. Wigmore : J. Dairy Sci. 67 (1984) 234.

(4) Woo, A.H. and R.C. Lindsay : J. Dairy Sci. 67 (1984) 960.

(5) Freytag, W.G., K.H. Ney and I.P.G. Wirotama : U.K. Patent, 1, 470, 256 (1977).

(6) Simmonart, P. and J. Mayaudon : 14th Intern. Dairy Congr. Rome 2 (1956) 541.

(7) Indian Patent 136, 717 (1976).

(8) Hofman, H.S. and D. Sloot : U.S. Patent 3, 883, 667 (1975).

(9) Wirotama, I.P.G. and K.H. Ney : Z. Lebensm. Unters. Forsch. 154 (1974) 67.

(10) Ney, K.H. and I.P.G. Wirotama : Fette Seifen Anstrichmittel 80 (1978) 249.

(11) Ney, K.H., I.P.G. Wirotama and W.G. Freytag : U.S. Patent 4, 020, 190 (1977).

(12) Tanaka, H. and Y. Obata : Agric. Biol. Chem. 33 (1969) 147.

(13) Begemann, W.J. and P.D. Harkes : N.L. Patent 7, 305, 105 (1973).

(14) Wong, N.P., R. Ellis, D.E. La Croix and J.A. Alford : J. Dairy Sci. 56 (1973) 636.

(15) Henning , G.J. : U.S. Patent 3, 520, 699 (1970).

(16) O'Keefe, P.W. : Diss. Abstracts Intern. B. 33 (1972) 1.

(17) Svensen, A. and E. Ottestad : Meieriposten 58 (1969) 50.

(18) Karahadian, C., D.B. Josephson and R.C. Lindsay : J. Dairy Sci. 68 (1985) 1865.

(19) Yamamoto, R., I. Yamamoto and T. Yoshizawa : U.S. Patent 3, 904, 780 (1975).

(20) Moinas, M., M. Groux and I. Horman : Lait 53 (1973) 529.

(21) Dumont, J.P., S. Roger, P. Cerf and J. Adda : Lait 54 (1974) 501.

(22) Ney, K.H., I.P.G. Wirotama and W.G. Freytag : U.S. Patent 3, 865, 952 (1975).

(23) Moinas, M., M. Groux and I. Horman : Lait 55 (1975) 414.

(24) Moinas, M., M. Groux and I. Horman : C.H. Patent 552, 949 (1975).

(25) Groux, M. and M. Moinas : C.H. Patent 578, 842 (1976).

(26) Mc Gugan, W.A., D.B. Emmons and E. Larmond : J. Dairy Sci. 62 (1979) 398.

(27) Aston, J.W. and L.K. Creamer : N.Z. J. Dairy Sci. Technol. 21 (1986) 229.

(28) Dietrich, P., W. Pickenhagen and A.Y. Smith : C.H. Patent 608, 344 (1979).

(29) Kosikowski, F.V. and A.C. Duhlberg : J. Dairy Sci. 37 (1954) 167.

(30) Golovnya, R.V., I.L. Zhuravleva, G.A. Mironov and R.M. Abdullina : Moloch. Prom. 31 (1970) 8.
(31) Spettoli, P. : Ind. Agr. 9 (1971) 42.
(32) Ney, K.H. and I.P.G. Wirotama : Z. Lebensm. Unters. Forsch. 146 (1971) 343.
(33) Ney, K.H. and I.P.G. Wirotama : Z. Lebensm. Unters. Forsch. 149 (1972) 275.
(34) Wirotama, I.P.G. and K.H. Nev : Z. Lebensm. Unters. Forsch. 154 (1974) 67.
(35) Ney, K.H., I.P.G. Wirotama and G. Freytag : N.L. Patent 7, 213, 536 (1977).
(36) Freytag, W.G., K.H. Ney and I.P.G. Wirotama : N.L. Patent 7, 315, 975 (1974).
(37) Lindsay, R.C. and J.K. Rippe : In Biogeneration of Aromas. Parliment T.H. and R. Croteau Eds. ACS Symp. Series (1986).
(38) Dumont, J.P., C. Degas and J. Adda : Lait 54 (1976) 553.
(39) Parliment, T.H., M.G. Kolor and D.J. Rizzo : J. Agric. Food Chem. 30 (1982) 1006.
(40) Pittet, A.O., J.V. Pascale, S. Patton and M.H. Brodnitz : U.S. Patent 3, 870, 800 (1975).
(41) Sloot, D. and P.D. Harkes : J. Agric. Food Chem. 23 (1975) 356.
(42) Dumont, J.P., S. Roger and J. Adda : Lait 56 (1976) 595.
(43) Sloot, D. : N.L. Patent 7, 308, 548 (1973).
(44) Mc. Gugan, W.A. : J. Agric. Food Chem. 23 (1975) 1047.
(45) Ramshaw, E.H. : CSIRO Food. Res. Q. 45 (1985) 20.
(46) Moinas, M., M.J.A. Groux and I. Horman : C.H. Patent 552, 949 (1975).
(47) Henning, G.S. : U.S. Patent 3, 520, 699 (1970).
(48) Pittet, A.O., M.E. Mason, E.T. Theimer and M.S. Tibbetts : U.S. Patent 3, 686, 177 (1975).
(49) Pittet, A.O., R. Muralidhara and E.T. Theimer : U.S. Patent 3, 705, 158 (1973).
(50) Katz, I., R.A. Wilson, W.S. Evers, M.J. Vock and G.W. Verhoeve : U.S. Patent 3, 726, 692 (1973).
(51) Adda, J. : Unpublished results.
(52) Dwidedi, B.K. and J.E. Kinsella : J. Food Sci. 39 (1974) 83.
(53) Jolly, R. and F.V. Kosikowski : J. Food Sci. 40 (1975) 285.
(54) Jolly, R. and F.V. Kosikowski : J. Dairy Sci. 58 (1975) 1272.
(55) Watts, J.C. and Nelson : U.S. Patent 3, 072, 488 (1963).
(56) Knight, S.G. : U.S. Patent 3, 100, 153 (1963).
(57) Ruttloff, H., A. Quehl and A. Leuchtenberger : Nahrung 28 (1984) 873.
(58) Kanisawa, T. and H. Itoh : Nippon Shokulin Kogyo Gakkaishi 31 (1984) 477.
(59) Kanisawa, T. : Nippon Shokulin Kogyo Gakkaishi 32 (1985) 787.

(60) Rothe, M., H. Ruttloff, H. Herrmann and W. Engst : Nahrung 30 (1986) 791.

(61) Paillier, J.C. : European Patent 144, 274 Al (1985).

(62) Kasik, R.L. and A.S. Luksas : U.S. Patent 3, 667, 968 (1972).

(63) Kasik, R.L. and A.S. Luksas : U.S. Patent 3, 765, 905 (1973).

(64) Boudreaux, D.P. : European Patent 137, 536 Al (1985).

(65) Lee, K.C.M., H. Shi, A.S. Huang, J.T. Carlin, C.T. Ho and S.S. Chang in Biogeneration of Aromas. Parliment, T.H. and R. Croteau Eds ACS Symp. Series (1986).

(66) Crossman, T.L. : U.S. Patent 4, 500, 549 (1985).

(67) Johnson, J.D. and D.L. Southworth : U.S. Patent 3, 780, 182 (1973).

(68) Law, B.A. : in Advances in the Microbiology and Biochemistry of Cheese and Fermented Milk Daries B.S. and Law B.A. Eds Elsevier Appl. Sci. Publ. (1984).

(69) Sood, V.K. and F.V. Kosikowski : J. Dairy Sci. 62 (1979) 1865.

(70) Magee, E.L., N.F. Olson and R.C. Lindsay : J. Dairy Sci. 64 (1981) 616.

(71) Braun, S.D. : Thesis The university of Wisconsin-Madison (1984).

(72) Arnadottir, A.T. : Thesis The university of Wisconsin-Madison (1986).

COCOA FLAVOUR EVALUATION USING GUIDE SUBSTANCES
CORRELATED WITH COCOA PROCESSING

G. Ziegleder[1)]

Cocoa flavour is very valuable and, therefore, provides an interesting field of research. Flavour industries are looking for compounds of "cocoa impact character", which should solely produce the aroma impression of "cocoa" or "chocolate" and thus work as substitutes. There is only little chance to find such compounds within the natural flavour complex, since every roasted aroma consists of some hundreds of components which interact and enhance each other. The question is whether or not we really need substitutes, since cocoa probably will remain on the market in the future.

Our object is to optimize cocoa technology for producing the natural cocoa flavour with the greatest yield and of the highest quality. Therefore, we are looking for analytical methods to estimate aroma intensity and for those that enable an objective quality control. Objective analytical data are necessary to relieve the industrial quality control of time-consuming organoleptical tests and to give advice to the producers of semimanufactured cocoa goods. The routine analytical check of aroma in food industries employs time-saving methods that can easily be performed.

For that purpose suitable "indicators" are of need to evaluate the aroma. The concentration of typical indicative compounds may be correlated with the cocoa flavour character. This relationship can be completely accidental; the substances are indicators without necessarily contributing to the quality under consideration. Of course, only those compounds that can be estimated rapidly and exactly are useful key substances.

Cocoa and chocolate flavour are influenced by fermentation, roasting and conching. As roasting is the dominant aroma process, it is most important to be able to measure the roasting intensity. Some examples will be given how roasting effects on cocoa flavour could be measured by means of gaschromatography (GC) or highpressure-liquid-chromatography (HPLC). In all cases the cocoa flavour was isolated by steam-distillation.

1) Fraunhofer-Institut für Lebensmitteltechnologie und Verpackung,
Schragenhofstr. 35, D-8000 München 50, FRG

Figure 1 compares two gaschromatograms of steam distillates of unroasted and roasted cocoas. The chromatograms have specific deviations at least within two regions. Those compounds, that change during the roasting process could, in principle, be useful as roasting indices. Such special regions of interest within the chromatograms will be considered more thoroughly in the next figures. The first region contains mainly pyrazines, the second one phenylic compounds and pyrrols.

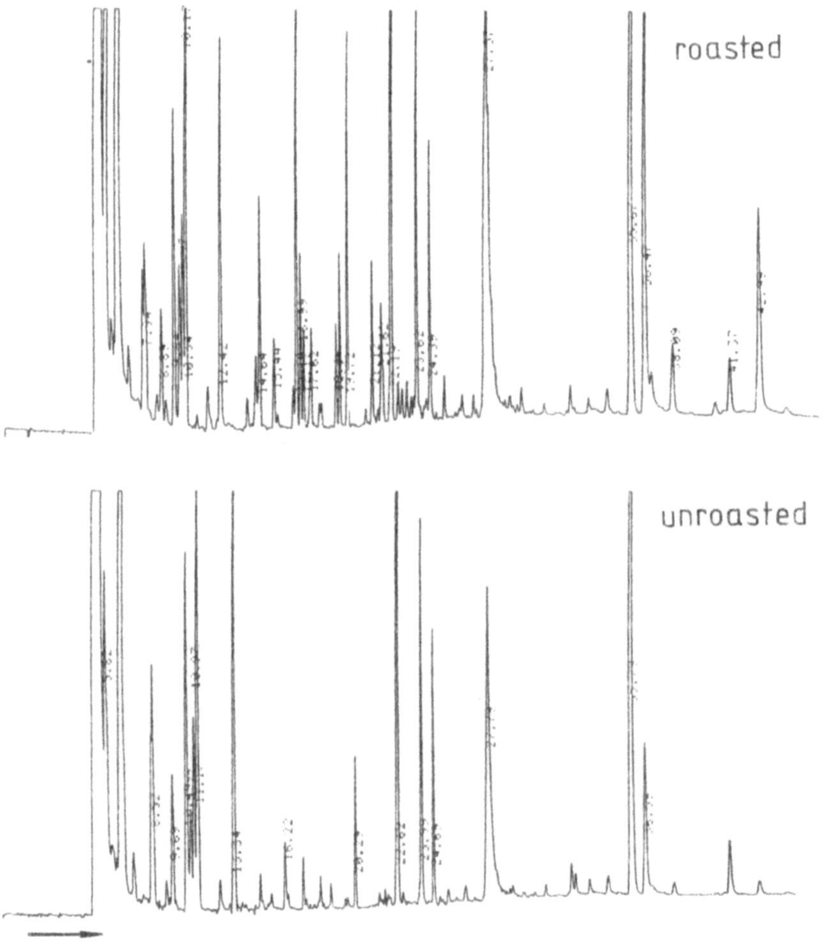

Fig. 1: capillary gaschromatograms of steam distillates of unroasted and roasted cocoas. GC-conditions: 50 m Carbowax capillary column, temperature programme 60 ° to 180 °C, 5°/min, FID and injector 200 °C (Ziegleder 1984).

Figure 2 demonstrates the changes within the complex of pyrazines. The concentration of every pyrazine derivative increases in a special way during the roasting process (Ziegleder 1982). Tetramethylpyrazine with its high content in unroasted cocoa soon reaches a maximum level at medium roasting intensities. Whereas trimethylpyrazine increases steadily, the different isomers dimethylpyrazines don't do so unless strong roasting conditions occur. As a result not only the total amount of pyrazine increases but also the relative ratios of the different derivatives change ("fingerprint"). For example the concentration ratios between trimethylpyrazine and tetramethylpyrazine or between 2.5-dimethylpyrazine and tetramethylpyrazine are useful for indicating the degree of roasting. In this respect these compounds may be called key compounds. Figure 3 shows the correlation of the ratios with organoleptical assessment. At normal degrees of roasting the cocoa aroma is developping its whole richness while overroasted samples reveal a specific burnt taste and coffee-like bitterness. Relativeley high levels of dimethylpyrazines indicate such overroasting. It is very important to note that pyrazine fingerprints are influenced exclusively by roasting. It has been proven that the origin of the cocoa used and its further treatment may influence the total concentration of pyrazine but not derange the ratios mentioned above. Barel et al. 1985 studied the change of tetramethylpyrazine content over the fermentation time of cocoa.

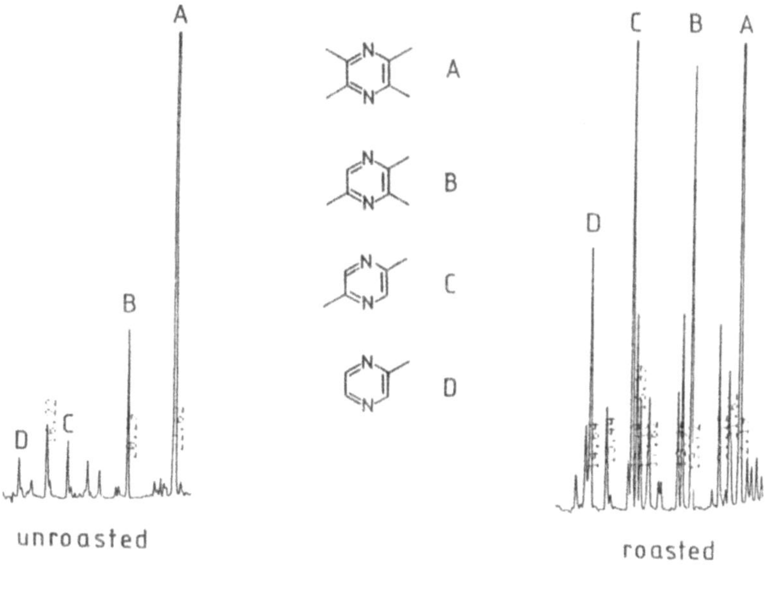

Fig. 2: region of pyrazines within the gaschromatograms of figure 1: Change of pyrazines with roasting. A = tetramethyl-, B = trimethyl-, C = 2,5-dimethyl-, D = methylpyrazine

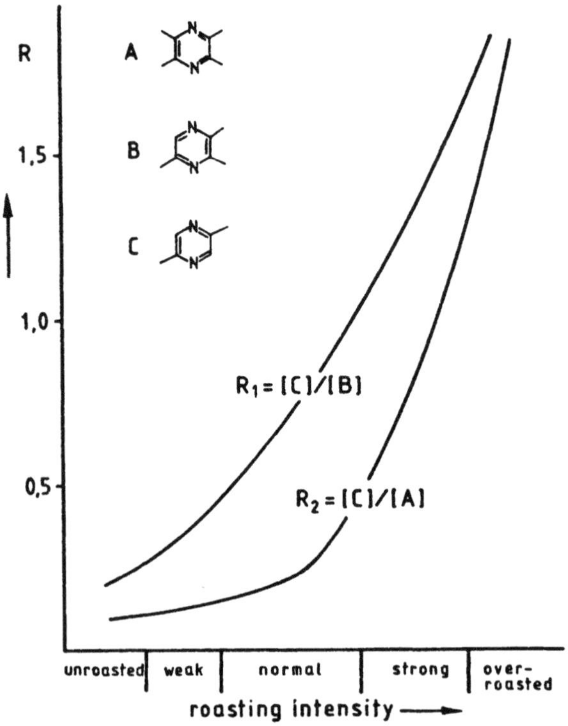

Fig. 3: Correlation of the concentration ratios of different pyrazines with the roasting intensity (Ziegleder, 1982). The degree of roasting was evaluated according to the "quality specification for cocoa liquor" (Arbeitskreis Kakaomasse, 1979)

Another method of application for the analytical evaluation of the degree of roasting was described by Meili (1985). In this case the second region of the flavour chromatogram in Figure 1 is used. Here 2-phenylethanol, that remains constant during the roasting process, works as a reference compound for other flavour constituents (Figure 4). Some of them, the phenyl-2-alken-2-ales are generated via aldol-condensation of Strecker aldehydes. The concentration of methyl-5-phenyl-2-hex-2-enal and phenylethanol works very effectively as indicator of roasting intensity.

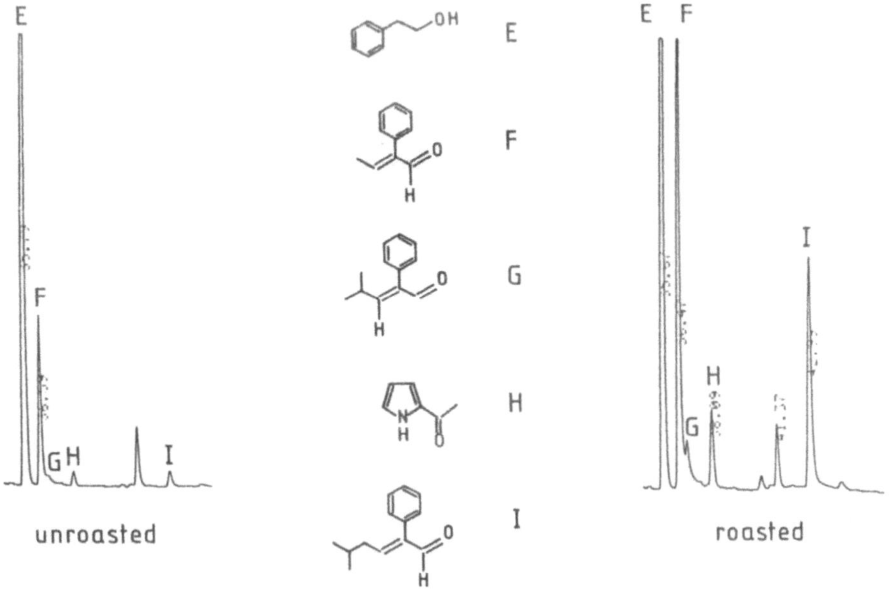

Fig. 4: region of some lower volatile compounds within the gaschromatograms of figure 1:
Changes of phenyl-enales and acetylpyrrole during roasting. E = 2-ethanol, F = 2-phenyl-but-2-enal, G = 2-phenyl-4-methyl-pent-2-enal, H = 2-acetylpyrrole, I = 2-phenyl-5-methyl-hex-2-enal.

The correlation of the ratio between methyl-5-phenyl-2-hex-2-enal and 2-phenylethanol with the roasting duration is demontrated in Figure 5. The concentrations of the other phenyl-enal-derivatives or of 2-acetylpyrrole may be used similarly as indices (Meili 1985). Such compounds, present in unroasted cocoa already, are drastically increased by the roasting process.

The phenyl-enales are of greater interest than just in their function as indicative compounds. They have been found in few flavours, as a predominant flavour only in cocoa, and carry an intense flowery and a heavy aromatic note. They seem to be of great value for the richness of cocoa flavour. Usually, they are main components of artificial chocolate flavours, as well.

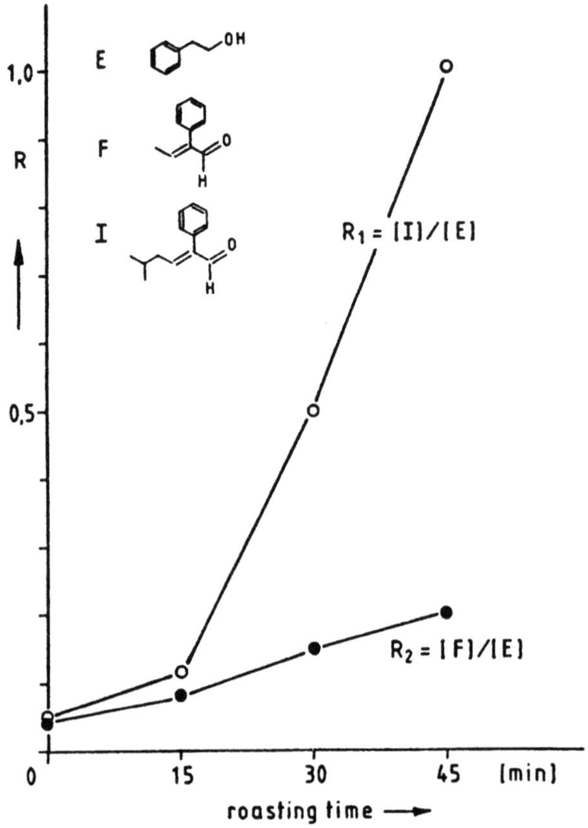

Fig. 5: Correlation of roasting duration with the concentration ratios of 2-phenyl-5-methyl-hex-2-enal and 2-phenylethanol respectively 2-phenyl-4-methyl-pent-2-enal and 2-phenylethanol (according to M. Meili 1985)

Due to its importance for flavour formation, the roasting process is a primary object of analysis: for instance the determination of the optimum temperature or the optimum duration of roasting. Of further interest is the influence of the water content during the heat treatment, which can be studied using furfuryl compounds. The furane content in cocoa is usually small. The concentration of furfural, furfurylic alcohol, 2-acetyl furane, 5-methyl furfural and 5-methyl furfurylic alcohol only reach high levels if the cocoa is steam- or water treated before or within the first part of the roasting process.

Such techniques sometimes are used to reduce the microbial load, in order to enhance flavour development or in the course of alkalization during cocoa powder manufacturing. Obviously furfurylic compounds result from a breakdown of reducing sugars. If cocoa is roasted with a high water content mainly furfuryl compounds will develop whereas the formation of other important flavour components (pyrazines, phenyl-enales) is suppressed (Figure 6) effecting a flat and poor taste. Therefore, the water-treated cocoa should be thoroughly dried before roasting. Pyrrol, 2-acetyl pyrrol and 2-formyl pyrrol, as typical by-products of furfurylic compounds, are influenced by the same processes. If unroasted cocoa is treated with water or sugar solutions, dried afterwards to the initial humidity and normally roasted an intense aroma appears that contains the different furanes, pyrrols, pyrazines and phenylenales at high levels. Whenever such special techniques are applied flavour control by gas

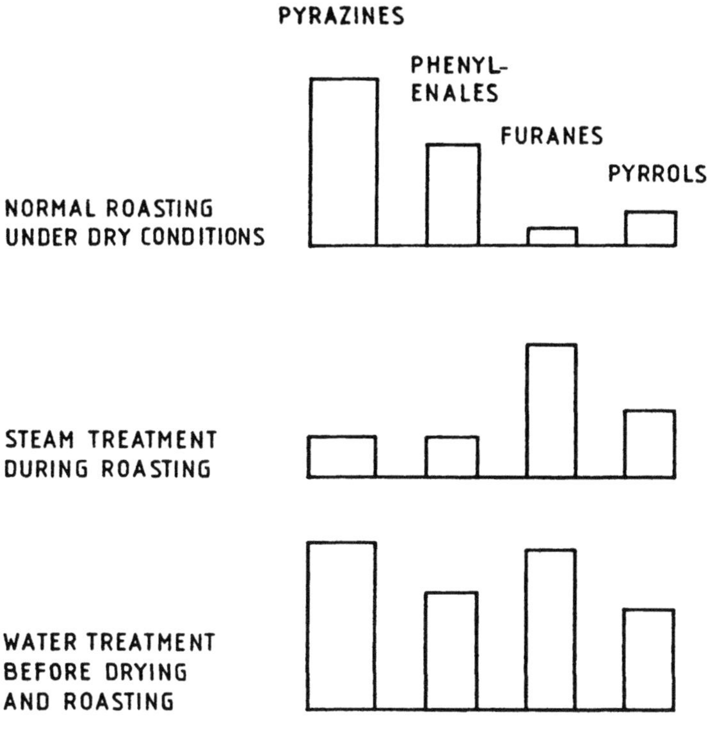

Fig. 6: Influence of water treatment on the development of some groups of flavour compounds (schematized).

The relative amount of highly volatile compounds of cocoa may be estimated by a simple gas chromatographic analysis of the headspace vapor (Ziegleder, 1981). This static headspace technique can easily be performed and if used under standardized conditions delivers results that allow a first evaluation of cocoa flavour. The concentration of organic compounds in the headspace is a function of the water activity of the cocoa. Therefore, it is necessary to bring all samples, which originally have different moisture contents, to almost the same level of water activtiy. Among the measured highly volatiles mainly i-pentanal as the Strecker-degradation product of leucine gives some information about cocoa flavour quality and intensity. The use of i-pentanal as an index during fermentation, roasting and the further cocoa technology has proven valuable. A limit for the use of this indicator is set by the fact that its concentration is influenced by many different factors. For that reason it does not provide an absolutely characteristic value and can only be applied to check certain processing stages.

These few examples demonstrate that gas chromatography is very effective for measuring the influence of technology on the quality of cocoa flavour. As we know, the utilisation of capillary gas chromatography may be rather time-consuming and, therefore, not in any case be suitable for industrial application. Sometimes HPLC (high performance liquid chromatography) can be more useful. Several sample preparation steps can be avoided using HPLC (since flavour distillates in water can immediately be injected without complications); and even some well-known problems with HPLC, especially concerning detection, may turn into benefits. Thus, reverse phase chromatography with ultra-violet detection as well as with electro-chemical detection have proven to be very useful in analyzing cocoa flavour. As a UV-measurement works without destroying the indicated compounds, it can readily be combined with a second detector, as well.

It is advantageous, especially for measuring the degree of roasting that pyrazines are selectively detected at 280 nm. Pyrazines represent the main part of the cocoa aroma detectable at this wave length (Figure 7). Further detectable compounds (acetyl-pyrrol, furfural) are separated well enough and most of the other aromatic compounds (as quantitatively predominant aldehydes, ketones and esters) are only indicated very weakly (Ziegleder and Sandmeier, 1983). The information to be gained from these HPLC-spectra may therefore be comparable with that from gas chromatography (Figure 7). The HPLC-spectra should be used as a fingerprint technique; with some experience in application the visual interpretation may help to distinguish several roasting parameters from one and another.

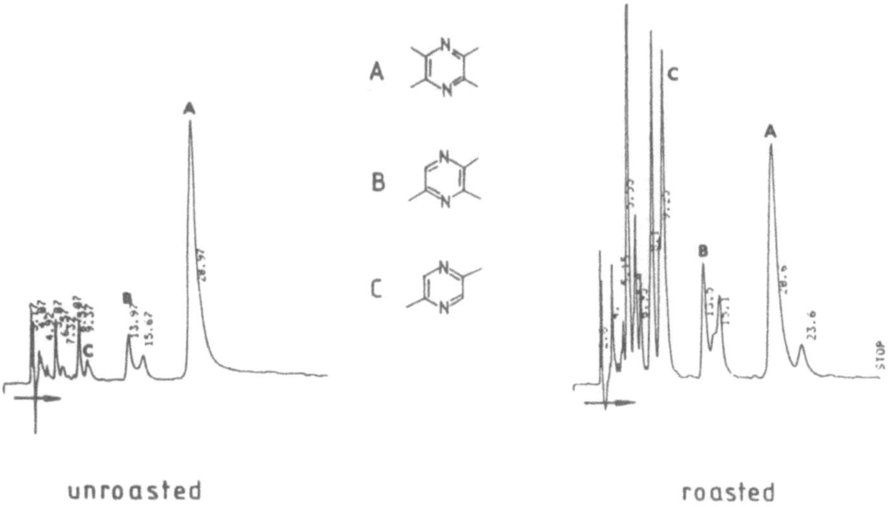

Fig. 7: HPLC-chromatograms of steam distillates of unroasted and roasted cocoas. HPLC-conditions: column Nucleosil 5 C 18, eluent: methanol/water 40:60, UV-detector 280 nm, 100 µl sample loop. Peaks A, B, C mainly consisting of tetramethyl-, trimethyl- and 2.5-dimethylpyrazines (Ziegleder and Sandmeier, 1983)

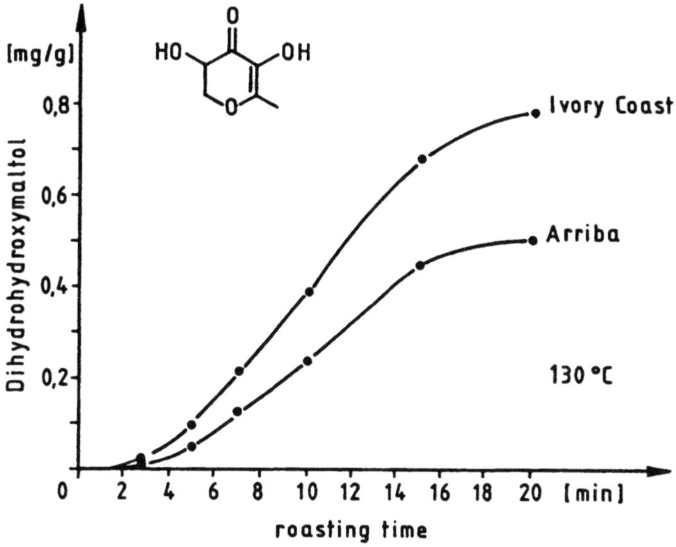

Fig. 8: Correlation of the concentration of dihydrohydroxymaltol in cocoa steam distillates with the roasting time at 130 °C. HPLC-conditions: column RP 18, mobile phase: methanol/water 10/90 (type A), 80/20 (type B), 0,09 m acetic acid, 0,005 m KH_2PO_4; linear gradient from A/B 86/14 to A/B 28/72; detectors UV 280 nm and /or ELDC + 0.7 V, range 8 µA (Schnee and Eichner, 1985)

Two methods for the application of cocoa flavour analysis using electrochemical detection have been reported in recent years. Schnee and Eichner 1985 studied the applicability of using the concentration of dihydrohydroxymaltol (2,3-dihydro-6-methyl-4H-pyran-4-on) as an index of roasting intensity (Figure 8). The compound was estimated within steam distillates and extracts of cocoa and analyzed by HPLC in combination with UV-detection and/or ELDC (oxidation potential at + 0,7 V). Obviously, this pyron derivative is formed at moderate temperatures and relatively high humidities. For that reason the method indicates the early stages of roasting and drying.

A serious flavour defect of raw or roasted cocoas is known as "smokiness". It is caused by volatile phenols which are absorbed from wood smoke during the drying process. These volatile phenols may be estimated selectively if steam distillates of cocoa are analyzed by means of HPLC with electrochemical detection at + 1.2 V oxidation potential. Cocoas of good quality are to a great extent free of such phenols, while smoky cocoa beans contain several phenolic derivatives. Comparable gas chromatographic results have been published recently (Guyot 1986). Smokiness of cocoa liquors or cocoa butter sometimes is reduced by deodorization or thin-layer treatment. The effectivity of those techniques can analytically be judged using the level of volatile phenols as an index.

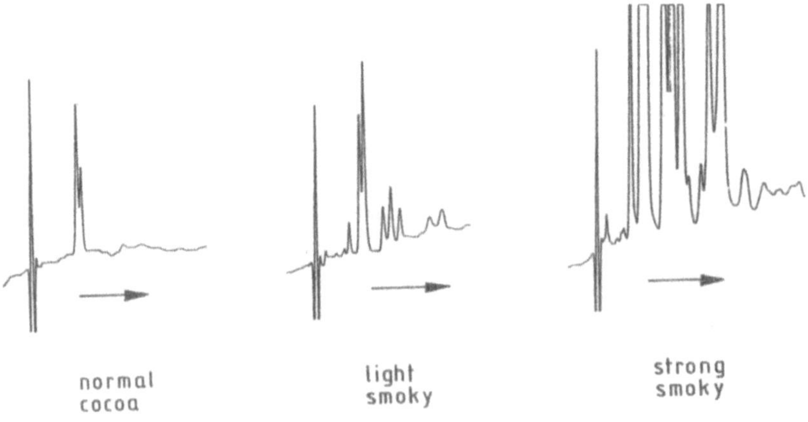

Fig 9: HPLC-spectra of steam distillates of cocoas with smokiness of different intensity. HPLC-conditions:
column 5 C 18, eluent: THF/acetonitril/water (15/15/70), 1 g/l acetic acid and 1 g/l $LiClO_4$; electrochemical detector + 1.2 V, range 100 nA (Sandmeier, 1987).

The cocoa aroma complex eludes a strict definition, since it represents different types and during cocoa processing it is changed from an inital rough, roasted and bitter aroma to the finer chocolate aroma. It is improved like a raw diamond which reaches its true value by polishing. Considering this intimate connection of flavour quality with processing technology, it becomes obvious that flavour studies must always be put into relation with technological parameters as well as with sensorial evaluations. Rapid methods for flavour analyses are useful to control this close connection.

References

Arbeitskreis Kakaomasse: Qualitätsanforderungen an geröstete Kakaomasse. Süßwaren 23 (1979) 20

Barel M, Leon D, Vincent JC: Influence du temps de fermentation du cacao sur la production des pyrazines du chocolat. Café Cacao Thé 29 (1985) 277-285

Guyot B, Gueule D, Morcrette I, Vincent JC: Characterisation chemiques des fèves de cacao a gout de fume. Café Cacao Thé 30 (1986) 113-119

Meili M: Etudes des correlations entre le compartement technologique des substances aromatiques et leurs propriétés physio-chimiques, These (1985) Univerité de Technologie de Compiègne

Sandmeier D: Rauchgeschmack bei Kakao und Kakaoprodukten; Zucker- u. Süßwarenwirtsch. 40 (1987), 40-43

Schnee R, Eichner K: Beiträge zur Kakao- und Kaffeeanalytik mittels HPLC und electrochemischer Detektion. Lebensmittelchem. Ger. Chemie 39 (1985) 95

Ziegleder G: Leichtflüchtige Kakaoaromastoffe als Leitsubstanzen bei der Kakaoverarbeitung. Zucker- u. Süßwarenwirtsch. 34 (1981) 105-109

Ziegleder G: Gaschromatogrpaphische Röstgradbestimmung von Kakao über methylierte Pyrazine, Deutsche Lebensm.-Rundsch. 78 (1982) 77-81

Ziegleder G: Möglichkeiten zur analytischen Beurteilung des Kakaoaromas, Süßwaren 28 (1984) 422-426

Ziegleder G, Sandmeier D: Röstgradbestimmung von Kakao mittels Hochdruckflüssigchromatographie. Deutsche Lebensm.-Rundsch. 69 (1983) 343-3477

FORMATION OF COCOA FLAVOUR BY MAILLARD REACTION

W. Rödel[1)], D. Habisch[2)] and H. Ruttloff[1)]

The Maillard reaction, the reaction of reducing sugars with compounds containing amino groups, is an important step in the formation of natural cocoa flavour. In contrast to other roasting flavours as meat e. g. the presence of peptides in addition to free amino acids is essential. This has been demonstrated by Mohr et al. [1, 2] using precursor fractions isolated from cocoa. From this it is easy to see why there are insufficient results to produce cocoa flavour by heating sugars with single amino acids or peptides [2, 3]. More successful seems to be an appropriate mixture of amino acids and peptides from partially hydrolyzed proteins as patented in USA [4]. Based on this fact we proved proteins of their aptitude to produce cocoa flavour. We investigated the conditions of the reaction and analyzed the volatile compounds formed by thermal reaction of partially hydrolyzed gelatine with glucose using gas chromatography/mass spectrometry. The thermal reaction was achieved by thin layer roasting. Tab. 1 and 2 survey the parameters under investigation for preparation and methods of evaluation of the thermal reaction products.

Table 1: Parameters investigated for Maillard reaction

a) Parameters of material

- source of protein
- rate of hydrolysation
- source of enzyme
- amount of sugar
- amount of water

b) Parameters of procedure

- temperature of roasting
- time of roasting

1) Zentralinstitut für Ernährung Potsdam-Rehbrücke, Akademie der Wissenschaften der DDR, Arthur-Scheunert-Allee 114/116, Bergholz-Rehbrücke, DDR-1505

2) Zentralinstitut für Organische Chemie, Akademie der Wissenschaften der DDR, Rudower Chaussee 5, Berlin, DDR-1199

Table 2: Methods for evaluation of the products

a) Sensory analysis

- . difference tests
- . ranking tests
- . scoring tests
- . profile method

b) Instrumental analysis

- . volatile compounds (GC/MS)
- . α-amino-N (Spier et al.)
- . glucose (enzymatical)
- . browning (colorimetrical)

A major problem was the selection of a suitable source of protein. Most of the proved proteins (Tab. 3) failed to produce a pleasant cocoa flavour. When using yeasts the final products showed a yeasty off-flavour, and faba bean protein resulted in a beany off-flavour. First of all bitterness of partial hydrolysates from different sources restricted their use. This especially concerned casein but also proteins from whey, egg, wheat, maize, and vicia faba. From this aspect the best source of protein seems to be gelatine. This agrees with calculations from Ney [5] on the bitterness of peptides from the composition of amino acids. Thus the Q-value of gelatine is close to that of cocoa protein and below the borderline for the appearance of bitter peptides. Therefore we used gelatine for further research.

As mentioned peptides are essential for the formation of cocoa flavour Therefore the hydrolysis rate of the protein is an increasing factor. Fig. 1 shows the improvement of the sensory quality with increased hydrolysis of gelatine. This is indicated by a decrease of the ranking sum for the whole quality and an increase for the partial quality 'cocoa-like'. Less than about 20 % of hydrolysis of gelatine gives lightly coloured, little flavoured products. The source of the enzyme is without consequence. That means, gelatine partial hydrolysates with the same rate of hydrolysis prepared by thermitase, alkaline protease or pancreatin result in products comparable in their sensory quality.

Table 3: Source of proteins tested in Maillard reaction

	off-flavour	bitter	Q[1] (Ney 1986)
gelatine			1280
bone meal			1280
wheat gluten		+	1420
maize gluten		+	1480
egg protein		+	
whey protein		+	
casein		+	1600
faba bean protein	+	+	
baker's yeast	+	+	
brewer's yeast	+	+	
(cocoa protein)			1124

1) $Q = \frac{f}{n}$

f = free energy of amino acids
n = number of amino acids

Q 1350 no bitter peptides
Q 1350 bitter peptides

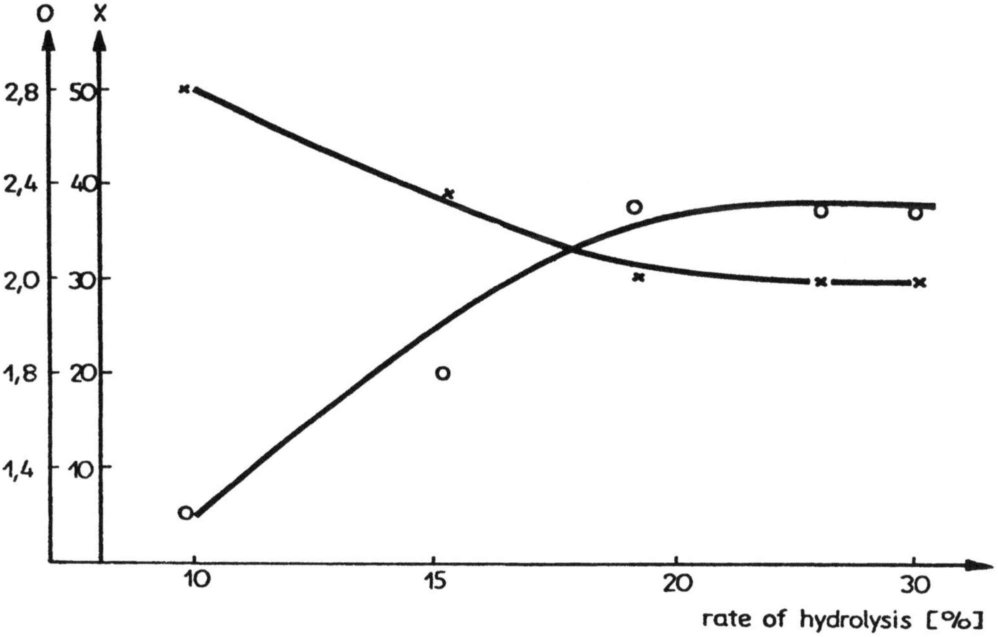

Figure 1: Effect of hydrolysis of gelatine [%] on sensory data of the Maillard product

O = 'cocoa-like' (scoring)
x = sensory quality (ranking sum)

As is shown from literature [5, 7] moisture in the mixture influences the Maillard reaction. Browning doesn't take place either without water or with more than 90 % water. Normally there is a maximum in the range of low water content depending on the system. In our system less than 6 % water results in only lightly coloured products with little flavour (Fig. 2). The turnover of α-amino nitrogen is low. It reaches the borderline of about 40 % if 6 % water or more are in the mixture. Nevertheless, the turnover of glucose is almost complete in all experiments. Water seems to be essential for diffusion of the reactants of the Maillard reaction.

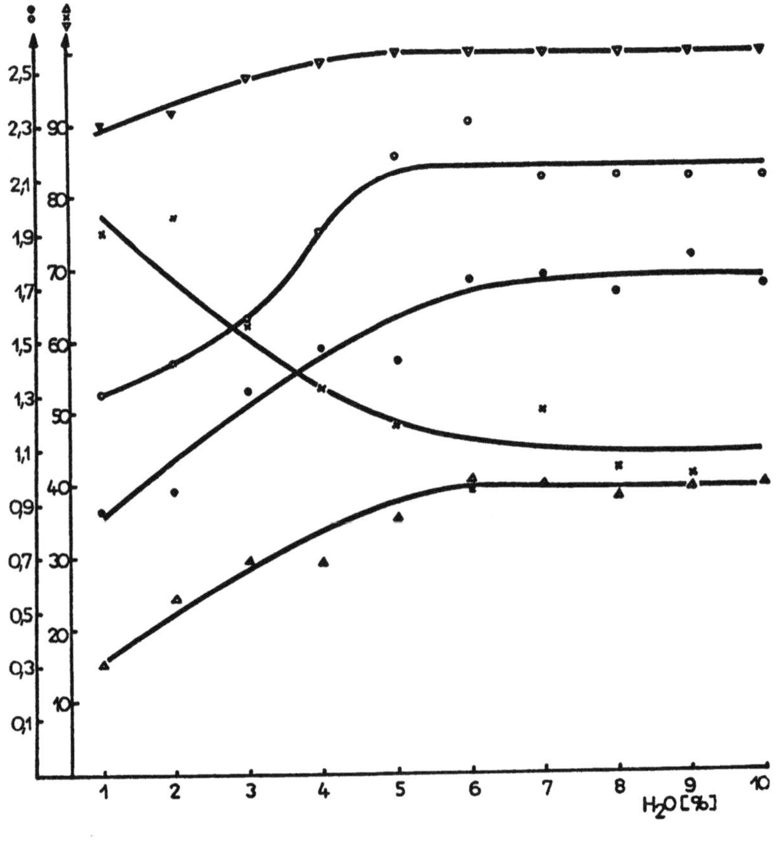

Figure 2: Effect of water on analytical and sensory data

x = sensory quality (ranking sum)
o = 'cocoa-like' (scoring)
o = browning
Δ = turnover of α-amino nitrogen [%]
v = turnover of glucose [%]

Using peptides and amino acids from natural cocoa Mohr et al. [2] pointed out: only glucose, fructose or sorbose yield cocoa flavour. In our experiments glucose, fructose and inverted sugar produce a similar flavour. An interesting point is the quantity of the sugar component. Increasing the quantity of glucose from 20 to 100 g per 100 g protein partial hydrolysate the cocoa flavour decreases and a caramelic/malty note occurs. Independent from the quantity of glucose the turnover of α-amino nitrogen stops at 40 %. The turnover of glucose is almost complete even if a high quantity is in the mixture. This was confirmed by an additional experiment with a successive addition of glucose (Fig. 3). The reaction starts at 20 g glucose per 100 g gelatine partial hydrolysate. The sugar turns over completely in a short time, the α-amino nitrogen diminishes more slowly to about half the amount. After a new addition of glucose the sugar disappears quickly again, but the α-amino nitrogen remains almost constant. This means that a part of the α-amino nitrogen isn't accessible for the Maillard reaction. The same fact was observed when roasting cocoa beans [8] or isolated flavour precursors of cocoa [6].

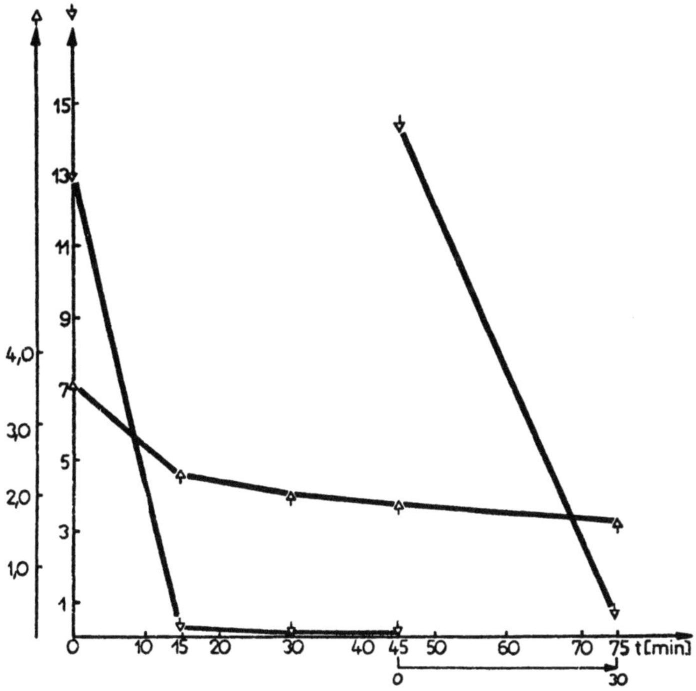

Figure 3: Effect of glucose addition and roasting time on the content (g/100 g) of α-amino nitrogen (♤) and glucose (♡)

The most sensitive parameters of the reaction are the heating conditions. To study the influence of temperature and time of heating we used mathematical-statistical methods. After finding the range of the two parameters with first order designs the data were fixed with second order plans. The so-named BN-plan was particularly favoured in our laboratory, because the research of two parameters demands only eight samples. In comparison the Box-Hunter plan requires 13 samples. With this method we investigated the effect of temperature and time of heating on different sensory and analytical criteria. It can be demonstrated by the whole sensory quality. The crest-line analysis according to Hoerl shows the sensory quality in relation to temperature and time of roasting (Fig. 4). Computation of the data results in maxima of 144 °C and 21 minutes. The functions at the maxima (Fig. 5) show good sensory quality only in a narrow range of temperature. On the other hand the time of heating doesn't influence the quality as sensitively. These findings were confirmed by a Box-Hunter plan.

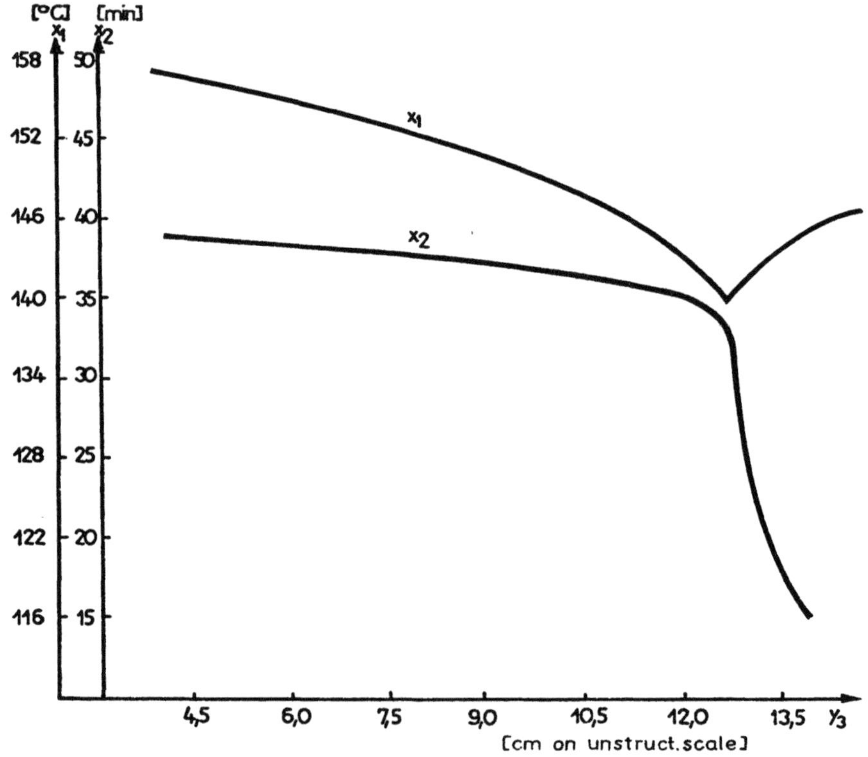

Figure 4: Crest-line for sensory quality (y_3) in dependence of temperature (x_1) and time (x_2) of roasting

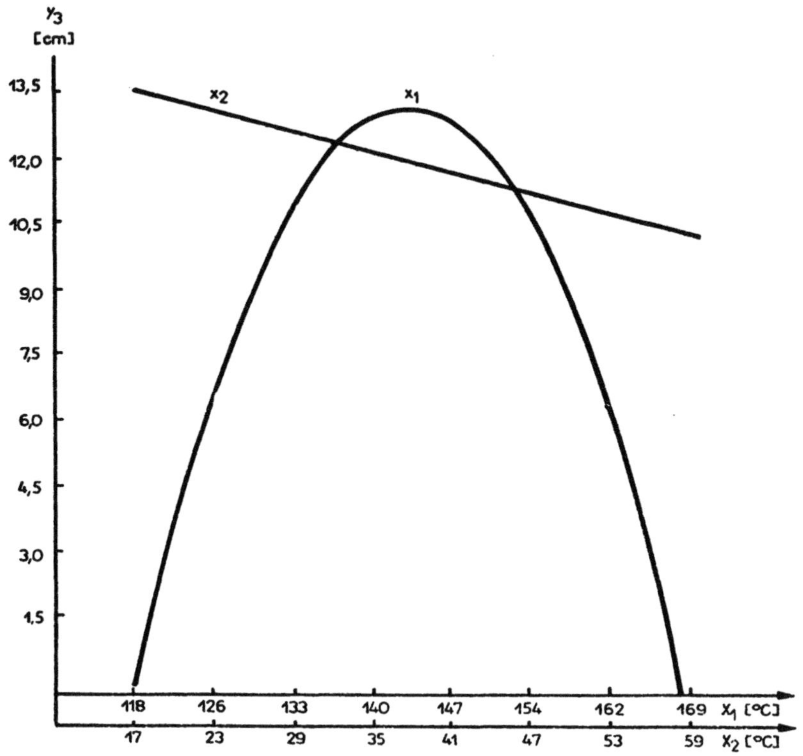

Figure 5: Functions of the maxima for sensory quality (y_3) in dependence of temperature (x_1) and time (x_2) of roasting

How does such a product on the base of gelatine partial hydrolysate prepared by optimal conditions look? It is a brown powder similar to natural cocoa. The sensory quality in comparison to cocoa is demonstrated by the profile analysis of products prepared from milk mass and 5 % cocoa and respectively 5 % of the Maillard product. Fig. 6 shows the profile of the product with cocoa. In this profile are involved the whole sensory quality, the off-flavour, and the following sensory qualities: cocoa-aromatic, milk chocolate-like, caramel-like, vanilla-like, sweet, fatty, nougat-like, malty, astringent and bitter The mass prepared with the Maillard product shows a profile rather similar.[1)]

[1)] I would like to thank Dr. Rothe and his group for carrying out the profile analysis.

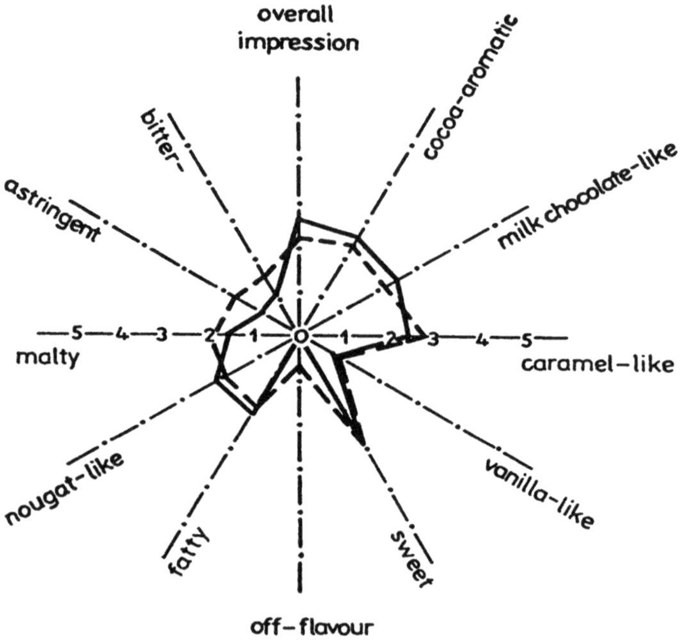

Figure 6: Profile analysis of milk mass with 5 % cocoa respectively Maillard product (5-point scale)
——— = 5 % cocoa standard, ----- = 5 % substitute

The analysis of the volatile compounds in the Maillard product by capillary gas chromatography/mass spectrometry resulted in 49 compounds [9]. Many of them are described as cocoa components in literature. Amongst them particularly pyrazine and furan compounds are written as characteristic for cocoa and other roasted flavour notes.

Summary:

- gelatine partially hydrolyzed by enzymes is an appropriate source for producing cocoa flavour by Maillard reaction
- flavour formation is influenced by temperature and time of roasting, furthermore by the content of water and sugar
- optimization of the reaction conditions by mathematical-statistical methods yields a product comparable with cocoa
- 49 volatile compounds are identified (partially tentative) by GC/MS. Pyrazines, furans and other compounds typical for cocoa and Maillard products are involved.

References

[1] Mohr, W., M. Röhrle and T. Severin: Fette, Seifen, Anstrichm. 73 (1971), 515

[2] Mohr, W., E. Landschreiber and T. Severin: Fette, Seifen, Anstrichm. 78 (1976), 88

[3] Lopez, A. and V. C. Quesnel: Rev. Int. Choc. 26 (1971), 19

[4] Rusoff, I. I.: US-PS 2 835 592 (1958)

[5] Ney, K. H.: Gordian 86 (1986), 84

[6] Eriksson, C. (Edit.): Maillard Reaction in Food. Pergamon Press, Oxford, New York, Toronto, Sydney, Paris, Frankfurt 1981

[7] Rockland, L. B. and G. F. Stewart (Edit.): Water activity: Influences on Food Quality. Academic Press, New York, London, Toronto, Sydney, San Francisco 1981

[8] Reineccius, G. A., P. G. Keeney and W. Weissberger: J. agric. Food Chem. 20 (1972), 202

[9] Rödel, W. and D. Habisch: Nahrung (in press)

CALCULATION OF COCOA SUBSTITUTION EFFECTS BASED ON SENSORY PROFILE DATA

M. Rothe, I. Voigt and H.-P. Kruse[1)]

Cocoa is one of those aroma-rich foods where a partial substitution may have an economic interest. The practical realization of this target, however, is a rather difficult task. One side of the problem is the fact that in many countries the use of flavouring or flavour extending additives in cocoa products is limited by national food laws and international agreements.

On the other hand cocoa flavour is unique and cannot be exchanged or simulated by flavourings easily. In patent literature we can find a lot of proposals for cocoa substitution. Heated yeast, wheat by-products or beets, e. g. are in discussion as well as compositions based on pyrazines and other heterocyclics, maltol or vanillin, often in combination with bitter and astringent agents.

The practical application of such substitutes, however, seems to be low and the promised effects doubtful. As in the case of other Maillard products character impact components are lacking in cocoa flavour. Considering the several hundred of aroma components identified in cocoa and chocolate flavour the composition from single components seems to be impossible.

Absolute as well as partial flavour substitution generally leads to the problem of comparability of the traditional with the substituted food or beverage. In most cases each deviation from the traditional quality is registered by the consumer as negative and as a quality decrease. For aroma-rich foods like cocoa this fact is valid, especially. Each sensory testing of substitution effects has to consider this point.

As to Fig. 1 there are three different possibilities which can cause a quality deviation in products with a partial substitution of cocoa. First an off-flavour can be present in the substitute. Another possibility is a changed ratio of the sensory attributes which leads to deviating quality impressions. Finally substitutes with diminished

---

[1)]Zentralinstitut für Ernährung Potsdam-Rehbrücke, Akademie der Wissenschaften der DDR, Arthur-Scheunert-Allee 114/116, Bergholz-Rehbrücke, DDR-1505

cocoa-aromatic notes must reduce the hedonic value of pure cocoa by means of a dilution effect.

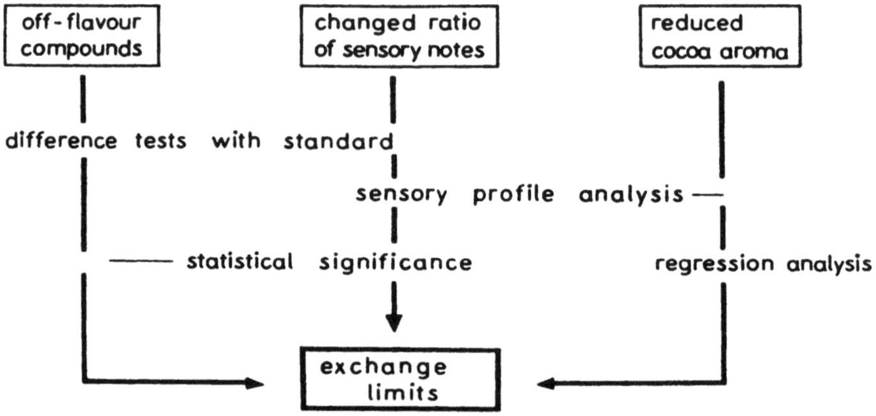

Fig. 1. Flavour decrease in cocoa substitution products (origin and checking)

Two principle ways of sensory analysis can be used to check such quality changes as shown in Fig. 1:

1) substitution of different parts of cocoa within cocoa-containing foods or beverages and comparison of the quality with the standard quality by means of difference tests. The maximum exchange level can be calculated as the least concentration without significant difference to the standard,

2) measuring the intensity of special sensory notes or of the overall sensory impression in partially substituted cocoa products in dependence on the cocoa content. Calculation of regression lines for concentration-intensity relationship, e. g., leads to the determination of maximum exchange limits. An example for this second way will be discussed.

Sensory profile analysis has some advantages for such studies. Here several sensory attributes as well as the overall quality can be quantified parallely within one test session. This makes various calculations from one test serie possible.

Test panel	: 8–10 trained people
Method	: intensity testing of 11 attributes and evaluation of overall impression against standard (5% cocoa)
Evaluation	: unstructured scale, 10 cm, with mark for standard
Samples	: 2–15% cocoa powder in milk paste (nonfat dry milk / sucrose / speciality fat)
Graphical demonstration	: Circle profilogram

Fig. 2. Sensory profile analysis of cocoa powder in milk pastes (test conditions)

The conditions used in our test are tabulated in Fig. 2. Pastes made from nonfat milk powder, sucrose and a speciality fat have been used as a model system. Before realization of the test program the test panel was trained with the medium for six weeks. The average value of two sensory sessions with each between 8 and 10 people was used as a base for the intensity values of the different attributes indicated as vectors in the following graphical profilograms.

Some results out of these series are shown as examples in Fig. 3 and 4. Different amounts of cocoa powder are included into the paste madium and tested. The evaluation was carried out in comparison with a reference standard containing 5 % cocoa powder in the medium. The attributes of this standard are also indicated in the figures. The profilograms indicate average results of the panel on a 5-point-scale.

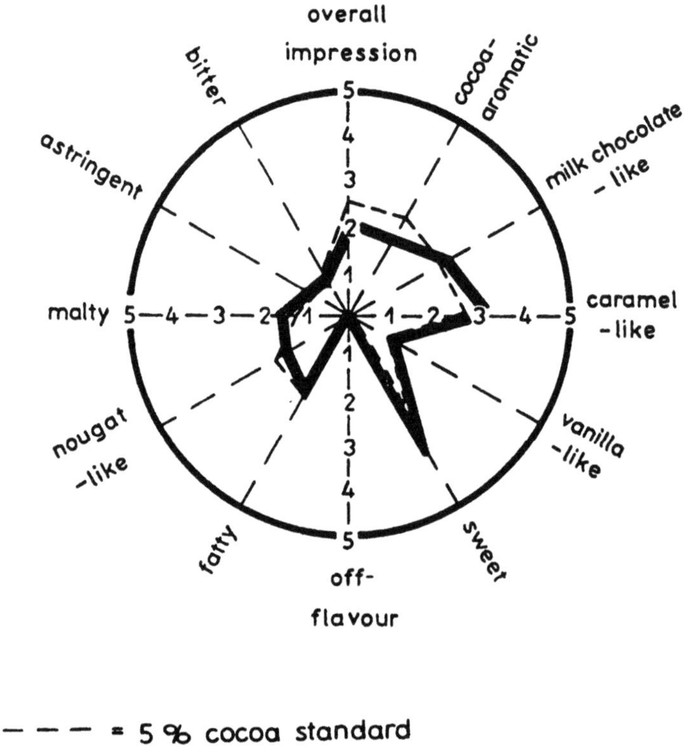

Fig. 3. Sensory profilogram of milk paste with 3 % cocoa (5-point scale; average of 3 cocoa powders)

The test principle as well as the influence of different cocoa amounts are demonstrated in Fig. 3 and 4. Cocoa concentrations of 3 and 10 % are compared with the 5 % standard showing decreases or increases in most of the tested sensory attributes.

When the single sensory attributes are checked as to their stimulus - response effect caused by different cocoa concentrations in the medium - the results indicate the relationship as linear in a semi-logarithmic diagram. This linearity is valid within the medium part of the concentration/intensity curve shown in Fig. 5. In the area between 2.5 and 12 % cocoa powder in the chocolate-type medium high correlation coefficients are found. The linear regression lines calculated from data between 2.5 and 12 % cocoa powder in the medium show

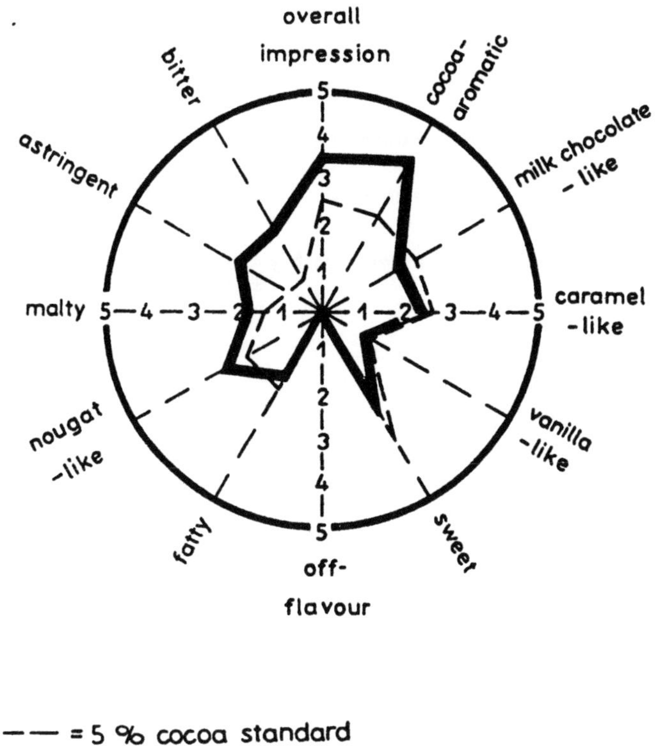

Fig. 4. Sensory profilogram of milk paste with 10 % cocoa (5-point scale; average of 3 cocoa powders)

some quality differences between the 3 cocoa powders as demonstrated by

- the absolute distance of the lines and
- the slope of the lines which is found, however, rather similar here.

By use of the regression functions comparison and weighting of the 3 cocoa powders under test is possible as shown in the right part of Fig. 5.

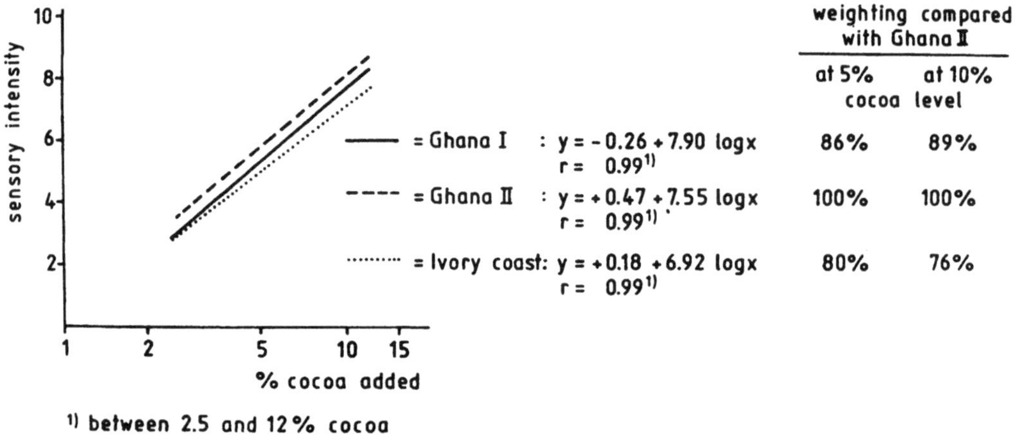

Fig. 5. Relationship between cocoa-powder content and sensory intensity of the 'cocoa-aromatic' note in milk pastes (3 cocoa varieties)

When the stimulus/response curve is extended over the limits of 2.5 and 12 % cocoa additive within the mass the curve approximates to a minimum and maximum level, respectively. The lower level indicated in Fig. 6 is identical with the sensory recognition threshold of the sensory attribute tested. The maximum level indicates a saturation effect. In our example above an amount of 12 % cocoa powder in the medium increasing concentrations of coca enhance the sensory response only weakly.

In Fig. 6 three sensory attributes are demonstrated besides of the overall impression. Two of these notes ('cocoa-aromatic' and 'bitter') are found highly weighted within cocoa and chocolate flavour. The inverse effect of the 'caramel-like' sensory note is caused by the fact that the nonfat milk powder included in the paste medium has its own caramel-like aroma which is depressed by increasing concentrations of the cocoa additive.

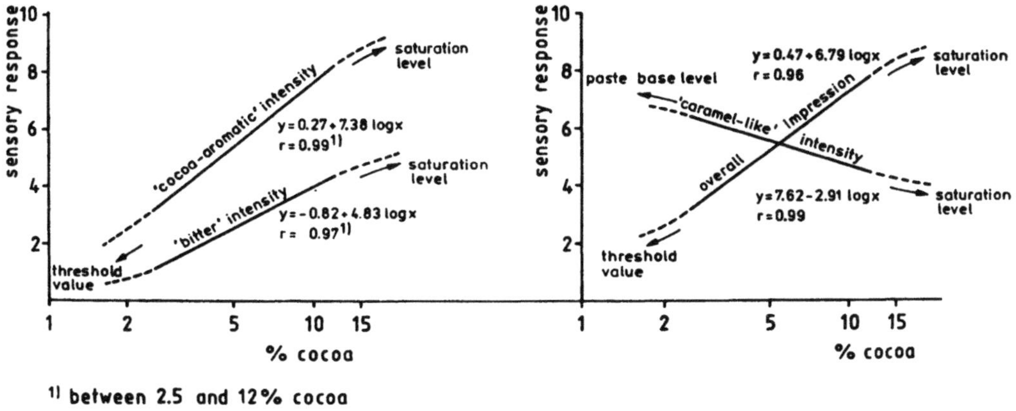

Fig. 6. Relationship between cocoa powder content and sensory response of 4 sensory notes (average regression lines of 3 cocoa powders)

In the following phase of the program the calculated regression lines of 'overall impression' and 'cocoa-aromatic' note were used as a base for a systematic sensory comparison of natural cocoa powder with different cocoa substitutes. For this purpose the substitutes were combined with cocoa powder in a 1 : 1 ratio. This combination is necessary because the character of all substitutes depends somewhat on the presence of natural cocoa and its special properties. On the other side such a high substitution ratio as 50 % has been found essential to get a significant and reproducible value for most of the sensory quality notes under test.

The quality of 3 cocoa substitutes tested in this way is shown in the figures 7 to 9 by their profilograms. The sensory test procedure again included the comparison with a 5 % cocoa standard. In the case of the profilograms demonstrated the pastes contained 5 % cocoa powder + 5 % of each substitute. The substitutes can be characterized as follows:

1) cocoa by-product produced and applied with nonfat dry milk as carrier
2) product according to 1), but without carrier
3) substitute based on roasted cereals.

The profilograms in Fig. 7 to 9 show typical differences. With respect to the 'cocoa-aromatic', 'bitter' and 'astringent' sensory notes which are highly weighted in cocoa-containing foods substitute 2 seems to be best appropriated. This advantage is, however, cancelled by a rather intensive 'off-flavour' accompanied by an increased 'malty' note. Obviously these changes reduce the value for 'overall quality'. Substitute 1 shows this off-flavour only in reduced intensity and therefore should have better chances in cocoa substitution. Substitute 3 is free of off-flavour and shows a profilogram very similar to cocoa. Its 'cocoa-aromatic' intensity, however, is very low thus increasing the standard value of 5 % cocoa only insufficiently. This third substitute has the character of a cocoa extender substance but can substitute cocoa powder only in rather limited amounts.

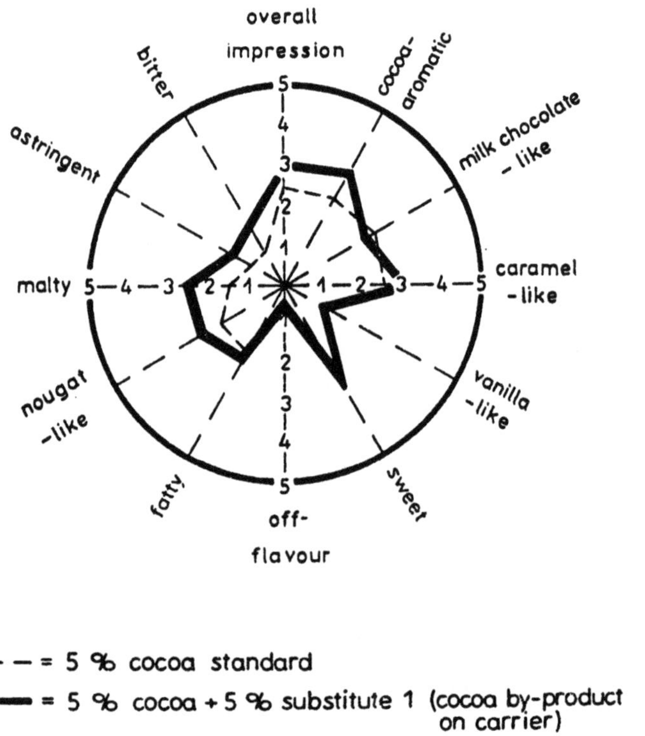

Fig. 7. Sensory profilogram of substitute 1 in 1 : 1 combination with cocoa powder

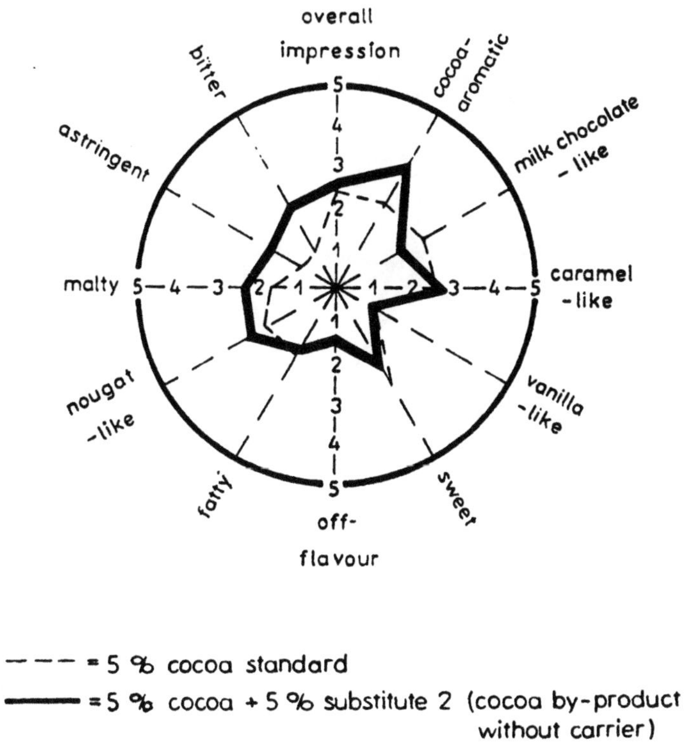

Fig. 8. Sensory profilogram of substitute 2 in 1 : 1 combination with cocoa powder

Again calculations are possible by means of the regression functions mentioned. They make such suppositions calculable and enlarge the significance. For the comparison with cocoa powder in the area between 2.5 and 12 % cocoa the regression lines calculated from the 'overall impression and the intensity of the 'cocoa-aromatic' note were used.

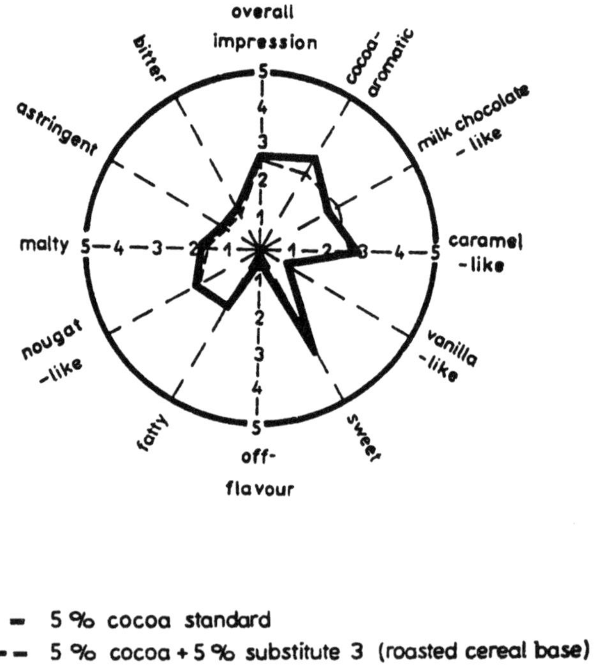

Fig. 9. Sensory profilogram of substitute 3 in 1 : 1 combination with cocoa powder

Fig. 10 combines such regression lines for all three substitutes under test. It can easily be observed that in the case of substitute 1 the regression line 'cocoa-aromatic' for the combined cocoa/substitute product is lower compared with the line for pure cocoa powder. The difference indicates the percent cocoa value of the substitute. The slope of the line is rather similar to that of cocoa.

When testing the 'overall impression' in this way we find, however, much higher differences. This is true for the slope of the regression line, especially. As a practical consequence the application of this substitute should be limited to products with a low cocoa content, e. g. of 5 % or less.

In the case of substitute 2 this negative behaviour of the 'overall impression' in dependence on the amount of cocoa in the final product is still more obvious. Here increasing amounts of the additive enhance the 'cocoa-aromatic' note while the 'overall impression' increases only very slowly. This is a consequence of the off-flavour in this substitute mentioned earlier.

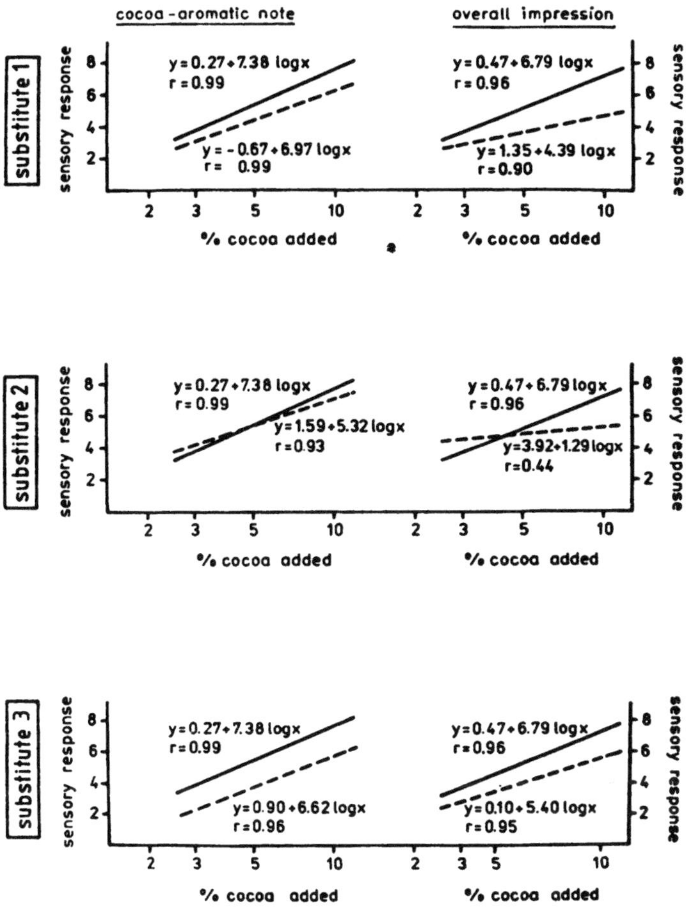

Fig. 10. Sensory characteristic of 3 cocoa substitutes by regression lines of 2 sensory notes

Substitute 3 shows a better adaptation to the sensory profilogram of pure cocoa products. On the other hand the differences in the intensity of several sensory notes are rather high indicating a low contribution to the typical attributes of cocoa and the overall impression. Here again the exchange value of the substitute must remain low in order to avoid quality changes. Substitutes of this type will dilute the typical cocoa aroma and thus have to be applied in small percentages of the cocoa content only.

The calculation of such dilution effects is also possible via the regression functions mentioned. By equalizing the two functions of pure cocoa powder and of the 1 : 1 substitute the percent value of each substitute can be calculated. Fig. 11 summarizes data obtained in this way.

cocoa content of the medium %*	substitute 1		substitute 2		substitute 3	
	cocoa-aromatic note	overall impression	cocoa-aromatic note	overall impression	cocoa-aromatic note	overall impression
3	70	91	111	132	62	70
5	68	76	96	87	59	63
7,5	67	66	86	63	56	58
10	66	60	79	50	55	55

*All values calculated for 1:1 cocoa: substitute ratio

Fig. 11. Calculated percent value of 3 cocoa substitutes for two sensory notes at different concentrations (% of pure cocoa powder)

According to Fig. 11 only one case can be found (3 % cocoa, substitute 2) where an equal exchange of the substitute against cocoa is possible. In all other cases tested the cocoa exchange value was found below 1, i, e, compared with pure cocoa more substitute has to be applied if an equal quality is expected.

For all 3 substitutes tested the conclusion must be drawn that the cocoa content of the food to be partially substituted should not exceed 5 %. In products containing higher cocoa levels a quality deviation is easier possible.

Cocoa reduction in cocoa-containing foods and beverages is a quality problem which can be discussed critically. If such an exchange should be considered we have to make us sure that no change of the traditional quality occurs at all. The calculations discussed here are a useful tool to check this possibility.

FOOD FLAVOR: THE PSYCHOLOGICAL CATEGORY

H. Haenel[1)]

Nutrition comprises 3 fields,

chemical:	the food, the composition
biochemical:	the nutrition metabolism, the physiological effect
psychological:	the arousal, the choice, the pleasure, incited by flavor.

1. Definitions

Flavor is the oronasal sensation connected with the intake of a mouthful, including aroma, taste, temperature, physical structure, kinesthetics and, here included, color. Flavor is physiologically perceived by the 5 senses.

Flavor is psychologically perceived in the mind and here confronted with the individual state of the mind; variables are appetite, expectation, experience, habit, cognition, emotions, craze, abhorrence a. o.

The psychological steps include registration, classification and recognition of the flavor, and comparison with former experience. This evaluation leads to preference or aversion, to the motivation to eat or not to eat.

A preferred flavor is valued according to its hedonic value as pleasing, rewarding, delightful or neutral. A flavor is rejected because of distaste, danger, disgust or inappropriateness as food (A.E. Fallon, P. Rozin; 1983).

2. Flavor quality: Measurement and judgment

A mystique exists about flavor: sophisticated techniques of the analytical flavor chemist and of the computer programmer cannot evaluate the hedonic dimension of a flavor that results from individual impression of its qualities (A.C. Noble; 1976). It is impossible to measure by any means, why a food is preferred by some, neutral to others and rejected by the rest. Unknown are the reasons for the individual differences in this hedonic evaluation, or even

---

[1)] Zentralinstitut für Ernährung Potsdam-Rehbrücke, Akademie der Wissenschaften der DDR, Arthur-Scheunert-Allee 114/116, Bergholz-Rehbrücke, DDR-1505

such differences in the same individual under different conditions.

Many methods have been developed to measure sensory space and flavor hedonics as likes or dislikes, to receive information about the relative contribution of sensory dimensions to the total volume of evaluating a food preference (A. Drenowsky, H.R. Moskowitz; 1985), but with limited success.

The sensoric quality and intensity of a flavor and of flavor components can be expressed as the average judgment of a group of panelists, for experimental or marketing purpose.

But panelists are not going to eat what they try, they recommend for other people.

Panelists are trained and thus manipulated, as they had to adapt their original, individual opinion to the average result of the group.

The individual reaction to a flavor stimulus is quite different: the decision to prefer or reject, to eat and drink, to start and stop, or not to eat and drink. This depends on the individually perceived hedonic value (Table 1).

Table 1 Food Stimulus

1. Sensoric perception by panelists;
   "Objective" psychophysical quantification of characteristics.
   Average judgment; recommendation

2. Individual sensoric perception and psychical classification of a food: decision to eat or not to eat according to (preexisting or provoked) appetite.

Panelists flavor evaluation is cognitive and uses questionnaires, ratings and scales. Individual flavor evaluation is an emotional process, generating desire or refusal.

3. Flavor and appetite

In the generation of appetite flavor plays the prominent role. But this is not a two-sided, but a many-sided intricate process with plenty of internal and external pressures (Table 2): Hedonic expectation and food stimulus, sociocultural and local environment, temporal and individual factors. A flavor is either related to actual appetite motivating in a given situation. Or it is connected with permanent liking or disliking, the generell disposition for a special food:
"I love peanuts, I hate sauerkraut".

Table 2 Generating Appetite

- Hedonic (health...) expectation:
  - Experience, knowledge, attitude
  - Preference vs aversion
  - Hunger vs satiety
- Food stimulus:
  - Flavor (aroma, taste, tactile sensation)
  - Color, size, shape, package, hygiene, choice, image, reward, placebo effect/illusion
- Environment:
  - Culture
  - Locality, company, chronological situation, special occasion, temperature
- Additional individual factors:
  Genetic/metabolic pattern; biological, social, emotional situation; neophobia vs curiosity; self control vs excess; habits; disease; stress-reaction and adaptability; acuity of taste

In affluence the pleasure connected with flavor is the overwhelming motivation to eat and to drink, in humans as - for instance - in rats. (Rats preferred delicacies under extreme cold, when they had the choice to lab chow under convenient temperature) (Table 3) (M. Cabanac; 1985). With a cafeteria diet rats became obese, but not with a lab chow (A. Sclafani, D. Springer; 1976).

Table 3 Hedonic Motivation of Rats

Lab chow		Delicacies
Warm room	vs	Cold room (- 15 °C)
	- Same Nutritional Value -	

All rats ran into the cold and devoured the delicacies as quick and as much as they could, the more tasteful the quicker.

4. Individual hedonic perception of flavor

Each individual prefers his flavor pattern, individually acquired, different according to flavor quality and intensity, for instance a special grade of preference for sweetness (D.A. Booth; 1986). The overriding general preference for familiar flavors shows individual variations. Flavor sensations are changing, during lifetime, time of day, during eating and biological changes.

There are genetic roots. Newborn infants, just starting metabolical independence, had a distinct preference for sweet taste and rejected sour, bitter and salty (Dresor et al.; 1977). A sugar solution produced more attractiveness than milk (Fig. 1). Individuality is

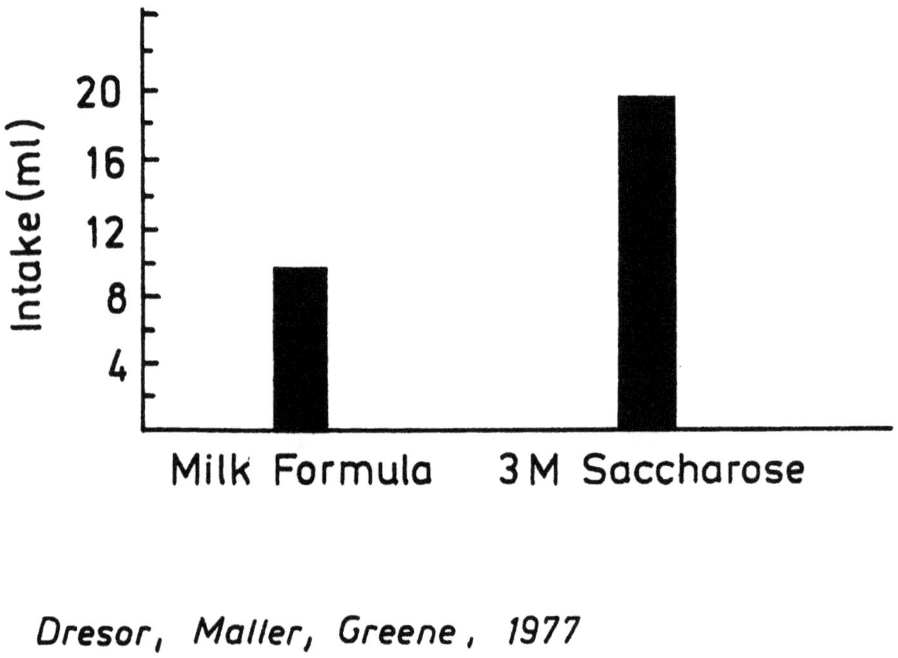

Fig. 1 Intake of Milk Formula or Isocaloric Saccharose Solution by 1 - 3 Day Old Infants (N = 15) within 3 Minutes

revealed when smelling different aromas (Table 4) (J.E. Steiner; 1977). The genetic pattern is - investigated in young children in the absence of any social inputs - sufficient, to secure homeostasis and health with free choice nutrition, when given unprocessed foods (C.M. Davis; 1939).

But the innate disposition is overwhelmed by social learning processes, leading to flavor preferences and aversions, typical for a family, a group, a social class, a region. Small children prefer mild flavors, not too salty, too sour, too spicy, even not too sweet,

Table 4 Judgment (in %) of Facial Expression of 25 Newborns, Smelling Different Aromas

6 Observers

	Acceptance	Indifference	Rejection
Butter	62	32	6
Banana	55	25	20
Vanilla	46	46	8
Shrimp (fishy)	22	25	53
Bad Egg	0	0	100

J.E. Steiner, 1977

as long as they are not habituated to stronger sensations: Salty snacks, chillipepper and other strong spices, coffee, alcohol, cigarettes, though originally rejected, turn into preferences. An unknown flavor will be observed with reserve or rejected. But mere exposure, and repeated offer will enhance preference, will cause familiarity, promoted by biological or social well-being. A familiar food, experienced as enjoyable ranks higher in preference than a familiar food without special hedonic characteristics. But this ranks higher than an unknown food. Neophobia has been accompanying man in evolution and history and still acts today. But the individual experience that a preferred food has resulted in acute disease, may change preference into aversion (Fig. 2) (D.M. Zahorik; 1977). (Unfortunately, this is not the case in chronic disease.) The same may happen, when opening an embryonated egg at the breakfast table, or after extreme overfeeding with a desired food or a special alcoholic beverage.

The individual preference of a flavor follows a narrow ridge between to much and to little, compared with expectation and habit. Liking turns into disliking if a flavor is deviating from expectation: the taste too sweet, salty, bitter, sour or not intensive enough, the aroma not pure, revealing an off-flavor, too weak or too strong, the texture too grainy or too smooth, too dry or too moist, too crispy or too soggy, the color too dark or too light (H.R. Moskowitz, B.E. Jacobs; 1986).

The phenomenon of sensory specific satiety is characteristic for the individual inconstancy in flavor preference. Eating a pudding reduces gradually its first palatability, turns preference temporarily into aversion. But this satiety is pudding-specific, other flavors cause new sensory-specific appetite, if the stimulus is

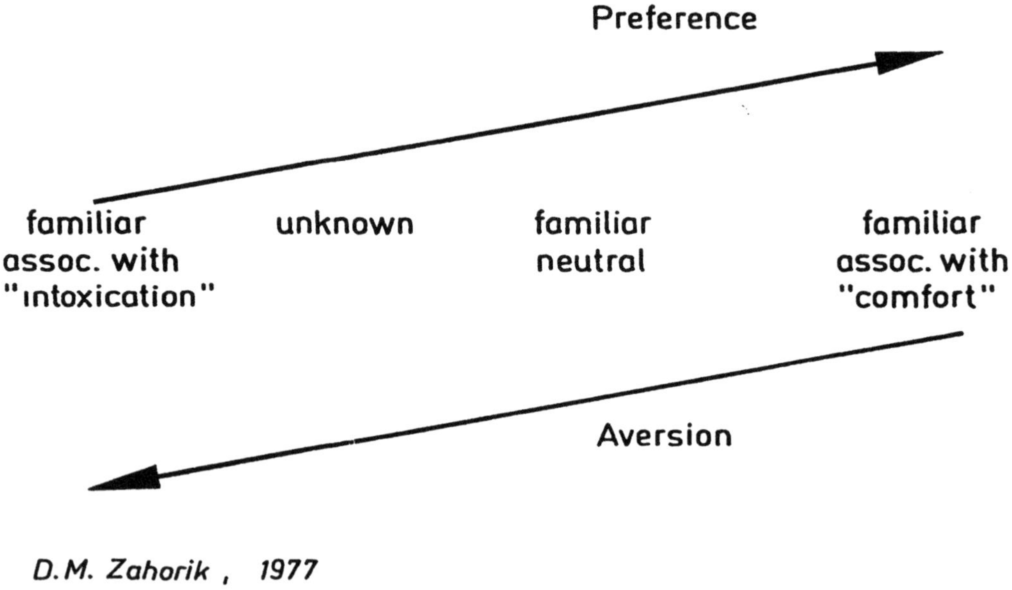

Fig. 2 Preference and Aversion, Depending on Different Experiences with Food

attractive and strong enough. The consumption of sandwiches with 3 different additions was by 15 % higher, as if only one was offered. The many flavor stimuli at a banquet from hors-d'oeuvres, soup, fish, main dish to ice-cream and cheese increased energy consumption by 60 % and more: variatio delectat (B.J. Rolls et al., 1982; B.J. Rolls, 1986).

Obesity, prevalent in affluence, is largely a psychological problem. Interference is possible with the avoidance of repeated sensory specific appetite: Offer of simple meals with few constituents. The drive for pleasure should be more directed towards quality, small bits, spices, slowly eating consciously, eating with stops, not gorging down big portions of food rich in meat, fat, beer, sugar, and salt.

A survival advantage in human evolution was the capacity to eat above need, and to store surplus energy as body fat for poorer times. The need or desire to switch between foods was helpful in this regard, lead to a balance of intake of more and of less preferred food. Diverse foods eaten during a day or a meal contribute to the balanced nutritive value of a diet.

Other individually changing hedonic perceptions are the picae of pregnant women, the desire for unusual flavors. Pregnant women preferred stronger salt concentrations (had also higher thresholds for sour, bitter and - the least - sweet). The preference for more salt disappeared after delivery (J.E. Brown, R.B. Thoma; 1980).

5. Physiological reality and psychological appearance

Any real, touchable thing can reproducibly be described by chemical and physical measuring. The situation with describing flavor is different, because psychological aspects are indispensable for description. Internal and external pressures operate in an incalculable, obscure, inconstant network, which reveals itself only in the output: the decision to prefer or to reject a special flavor quality and intensity.

Color can be important in the perception of flavor (C.M. Christensen; 1983; J.A. Maga, 1974; H. Moir, 1936). Foods colored contrary to expectation may be refused: green eggs, blue milk, purple asparagus are rejected; people may feel violently sick upon blue foods (A.S. Kotyla, F.M. Clydesdale; 1978). Even minimal deviations from an expected color may impair acceptability.

Lack of the brown color mislead experienced panelists in the identification of colorless coca cola (U. Kynast; 1979) (Table 5).

Table 5 Identification of Colourless Coca-Cola
30 Pertinent Panelists

Identified as Coca-Cola	2
Described as "something similar"	4
Unable to describe flavor	16
Wrong description	6

U. Kynast, 1979

But blind testing gave correct recognition. Increasing color intensity of strawberry juice lead to a 12 % increase in sweetness perception of same sugar concentration. Higher and lower concentrations of redness were judged as unacceptable or unpleasant, either perceived as unripe or unnatural (Johnson et al.; 1983). Similar experiences were made with butter and margarine.

Apparently flavor is connected with a traditional color conception, conditioning flavor by color. Sweetness perception, sugar concentration, color acceptability and pleasurable taste sensations operate in context. That is the reason we have included color in the flavor definition.

The image of a brand, skilled advertising, inciting package, a simple food in a first class restaurant, can increase flavor hedonics. A cheap brandy filled into a Hennessy bottle may have a higher hedonic value as the Hennessy brandy offered in a coffee cup. Illusions about the magic properties of a food as enhancing beauty, youth, and power can act in the same direction. This belief in extraordinary efficiency of a food ("natural" food, yoghurt, garlic etc.) may cause the impression of a result, and may even cause the result. Flavor may act as placebo (O. Fennema; 1985). Foods which children have been compelled to eat for health and performance reasons, decrease in preference, lower flavor hedonics (B. Rolls; 1986).

Connections between moods and flavor preferences confirm the inconstancy of flavor preferences. Healthful foods were more often preferred in positive emotions, junk foods tended to be more often preferred in negative emotions. Presumably moods partly determine food selection, the foods taken affect the moods (Table 6) (B. Lyman; 1982).

Eating leads to changes in plasma composition, causing a selective modulation of the rates at which neurons convert nutrients to neurotransmitters: A carbohydrate-rich food raised brain tryptophan

Table 6 Decreasing Frequencies of Food Preferences with Moods (Extract from 22 common emotions)

Healthy Food	Junk Food
Self-confidence	Joy
Friendliness	Boredom
Happiness	Frustration
Solemness	Amusement
Relaxation	Depression
Excitement	Guilt
Joy	Anxiety
Love	Worry
Anxiety	Loneliness
Frustration	Jealousy
.	.
.	.

(B. Lyman, 1982)

level stimulating production and release of the neurotransmitter serotonin, leading to changes in mood, sleepiness, pain sensitivity and food preferences (R.J. Wurtmann, J.J. Wurtmann; 1984).

A special flavor memory is the base of life-long lasting preferences. Remembering mothers favourite dishes, her kirschtorte, her sülze and her brathering produces life-long childhood-specific preferences. But her intention to press for the consumption of "healthy, ironrich" spinach led to an aversion, still alive.

Direct comparison of the quality of several brands of orange juice resulted in the perception of distinct differences. This was not the case with a time difference of 24 hours from brand to brand (H. Glaeser, U. Greiner; 1986).

6. Summary and Conclusions

The perception of flavor is mediated by the 5 senses. The psychological internalization of a percepted flavor is the chain of registration, classification, recognition and comparison with former experience, leading to acceptance or rejection. This is an individual process, according to the individual state of the mind. Flavor can be described by sophisticated techniques. But the individual hedonic dimensions - enjoyment, delight, pleasure - resist measurement.

Flavor quality and intensity can be expressed as average cognitive judgment of panelists for experimental or marketing purpose. The individual flavor evaluation is an emotional process, leading to preference and aversion, depending on flavor hedonics. This is a manysided, intricate process under internal and external pressure. Life history, genetic traits and social influences, familiarity and neophobia, good and bad experiences, short term stimuli and long term habits, moods, image, small deviations from expected flavor quality or intensity, hunger, appetite and sensory specific satiety are examples.

The psychological appearance of a flavor may be different from physiological reality or panelist judgment. Image, illusion, placebo effects, nostalgic experiences, mystic expectations, compulsion may influence the perceived flavor hedonics.

Preference turns into aversion after acute infections or toxic experiences, connected or thought to be connected with the intake of a special food. This does not happen with harm caused by chronic consumption: The instinct for avoidance is lacking in chronic pathogenity. As the individual is helpless, society has to intervene.

Literature

1. Brown, J.E., and R.B. Toma: Taste changes during pregnancy. Am. J. Clin. Nutrit. 43 (1986) 414.
2. Cabanac, M.: Preferring for pleasure. Am. J. Clin. Nutrit. Suppl. 42 (1985) 5.
3. Christensen, C.M.: Effects of color on aroma, flavor and texture judgements of foods. J. Food Sci. 48 (1983) 787.
4. Davis, C.M.: Results of the self-selection of diets by young children. Canad. Med. Ass. J. 41 (1939) 257.
5. Drenowsky, A., and H.R. Moskowitz: Sensory characteristics of food - new evaluation techniques. Am. J. Clin. Nutrit. 42 (1985) 924.
6. Dresor, J.A., O. Maller and L.S. Greene: Preference for sweet in humans infants, children, and adults, S. 161. In: Weiffenbach, J.M. (ed.): Taste and development: The genesis of sweet preference. DHEW Publ. (NIH) 77 - 1068 US Dept. Health, Educ. Welfare, Bethesda, USA, 1977
7. Fallon, A.E., and P. Rozin: The psychological bases of food rejection by humans. Ecol. Food Nutrit. 13 (1983) 15.
8. Fennema, O.: The placebo effect of foods. Food Technol. Austral. 37 (1985) 516.
9. Glaeser, H., and U. Greiner: Untersuchungen zum sensorischen Gedächtnis des Konsumenten - Unterscheidbarkeit verschiedener Orangensaftmarken bei gleichzeitiger und zeitversetzter Verkostung. Flüssiges Obst 53 (1986) 2.

10. Johnson, J.L., E. Dzendolet and F.M. Clydesdale: Psychophysical relationship between sweetness and redness in strawberry-flavored drinks. J. Food Protect. 46 (1983) 21.

11. Kotyla, A.S., and F.M. Clydesdale: The psychophysical relationships between color and flavor. CRC Crit. Rev. Food Sci. Nutrit. 10 (1978) 303.

12. Kynast, U.: Beziehungen zwischen Farbe und Geschmack bei Lebensmitteln und Getränken. Ernährungsindustrie 10 (1979) 15.

13. Lyman, B.: The nutritional values and food group characteristics of foods preferred during various emotions. J. Psychol. 112 (1982) 121.

14. Moir, H.: Some observations on the appreciation of flavor in food stuffs. Chem. and Industry 55 (1936) 145.

15. Moskowitz, H.R., and B.E. Jacobs: The relative importance of sensory attributes for food acceptance. Acta Aliment. 15 (1986) 1.

16. Noble, A.C.: A chemical approach to aroma. Cereal Foods World 21 (1976) 59.

17. Rolls, B.J.: Sensory specific satiety. Nutrit. Rev. 44 (1986) 3.

18. Rolls, B.J., E.A. Rowe and E.T. Rolls: How sensory properties of foods affect human feeding behavior. Physiol. and Behavior 29 (1982) 409.

19. Sclafani, A., and D. Springer: Dietary obesity in adult rats: Similarities to hypothalamic and human obesity syndromes. Physiol. Behavior 17 (1976) 461.

20. Steiner J.E.: Facial expressions of the neonate infant indicating the hedonics of food-related chemical stimuli, S. 173. In: Weiffenbach, J.M. (ed.): Taste and development: The genesis of sweet preference. DHEW Publ. (NIH) 77 - 1068 US Dept. Health, Educ. Welfare, Bethesda, USA, 1977.

21. Wurtman, R.J., and J.J. Wurtman: Nutrients, neurotransmitter synthesis, and the control of food intake, S. 77. In: Stunkard, A.J., and E. Stellar: Eating and its Disorders. Raven Press, New York 1984.

22. Zahorik, D.M.: Associative and non-associative factors in learned food preferences, S. 181. In: Barker, L.M., M.R. Best and M. Domjan (eds.): Learning mechanism in Food selection. Baylor University Press, 1977.

TIME-INTENSITY STUDIES ON TASTE AND FLAVOUR SUBSTANCES - AN ACTUAL REVIEW

Nina Baryłko-Pikielna[1/]

Flavour is a multidimensional phenomenon. Usually three "dimensions" are used for its description: qualitative identity, intensity and hedonic character. However, another important dimension is here missing: it is the time. All sensory phenomena occur in time.

In conventional sensory techniques the time as a parameter or "dimension" is not taken into consideration. The results are "time-free", reflecting only the maximum of perceived intensity and not its dynamics in time. In such approach all temporal informations of the sensation are completely lost.

Data concerning changes of sensory response due to time may be of a special interest in flavour research. Final flavouring effect depends not only on close similarity or identity of flavouring agent with a natural one in terms of chemical and sensory properties but also on similarity in time-release from food in the oral cavity. Various events occuring there after introducing a portion of food or beverage greatly influence above characteristics.

Events in oral cavity and their temporal relation to flavour intensity

The portion of food after entering the mouth is changed and modified in many ways (Fig.1.). Depending on the initial temperature of foodstuff heat transfer takes place, increasing or decreasing temporaly the temperature of the oral cavity. In foods containing solid fat part or whole fat is melted, depending on its physical characteristics. Increasing volume of saliva, stimulated by food (especially when sour or hot) may change in products of initial lipid-continuus phase into water-continuus one. This change is accompanied by dissolving all water-soluble solids, among them sensory-active ones. Mixing of food with saliva is facilitated by paralell mechanical work of mastication. Particles of food became

[1/] National Food and Nutrition Institute, Powsińska 61/63, 02-903 Warszawa, Poland

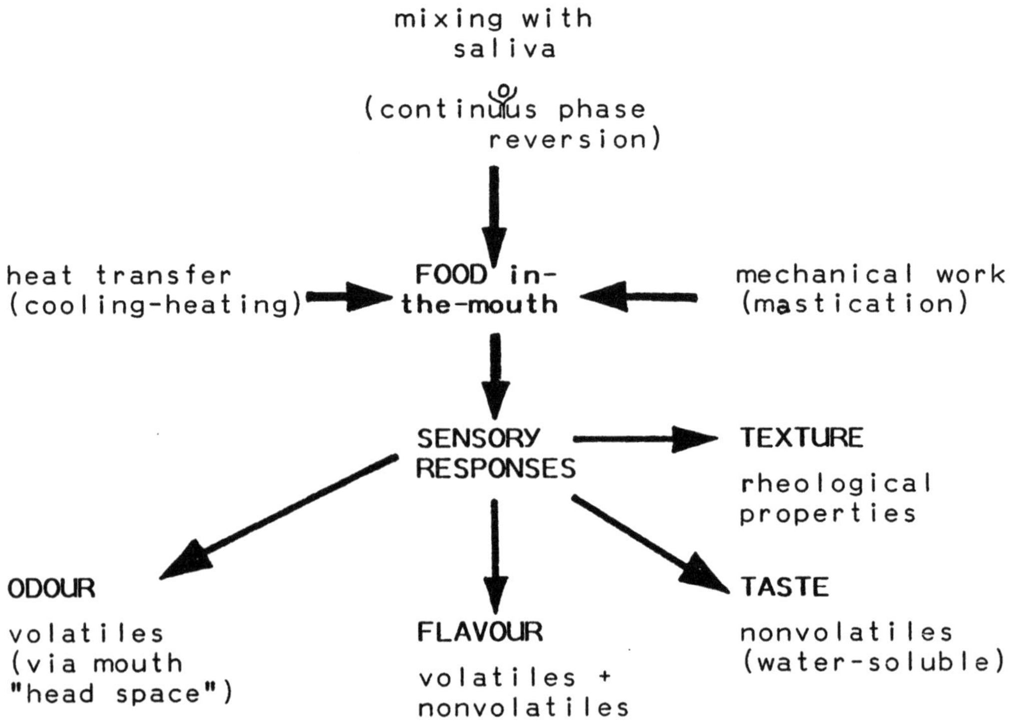

Fig. 1. Modification of food in the oral cavity related to development of sensory sensations

gradually smaller and come easier in contact with saliva. Warming-up the food to mouth temperature causes partial release of volatiles from oil and water phases into the air phase and consecutive stimulation of olfaction via "head space" of oral cavity.

All above phenomena have an obvious relation to temporal shape of intensity perception.

The idea of time-intensity measurement

Although temporal parameter is not directly taken in account in conventional sensory procedures, it is well known that it does influence the results of sensory tests, mainly by adaptation and fatigue, both time-dependent phenomena (O'Mahony, 1986). In constant stimulus method depending on type and concentration of stimulating substance, different time is required for gustatory sites to respond (Yamamoto et al., 1982). Birch and his coworkers noted, that perception of maximum intensity is related not only to the concentration of a stimulus but also to its duration (Birch et al., 1980).

Time-intensity procedure allows quantitative measurement of changes in the intensity due to time what offers an extra portion of informations, contributing to elucidation of some questions in sensory perception as well as having an important impact to solve practical problems in flavouring application.

The idea of T-I method is to follow and to record stimulus intensity in certain predetermined time intervals (rather very short ones) during the whole duration of perceived sensation - instead of taking only its maximum value, as it is in conventional sensory methods. The T-I curve offers many useful informations.

Typical T-I response curve is shown in fig.2. Time (in seconds) is a horizontal axis, stimulus intensity a vertical one. Initial time ("zero time") is the time of introduction of the sample into the mouth. From the curve following data can be read: time to appearing measurable sensation - "lag time" (1), time to reach maximum intensity (2), rate of appearance (3), maximum intensity (4), rate of extinction (5), special events occuring at certain time - as swallowing or expectoration (6), total duration of the sensation (7) and the area under the curve, informing about the "body" or "volume" of the sensation. If the same procedure is used for measuring the results of multiple consecutive introductions of the sample in to the mouth, some additional informations are available: time between introductions

(9), decrease (or increase) in maximum intensity by second or further introductions in relation to the first one (10), baseline "shift" (11) and ratio of the area under each consecutive curve relative to the first introduction area(12).

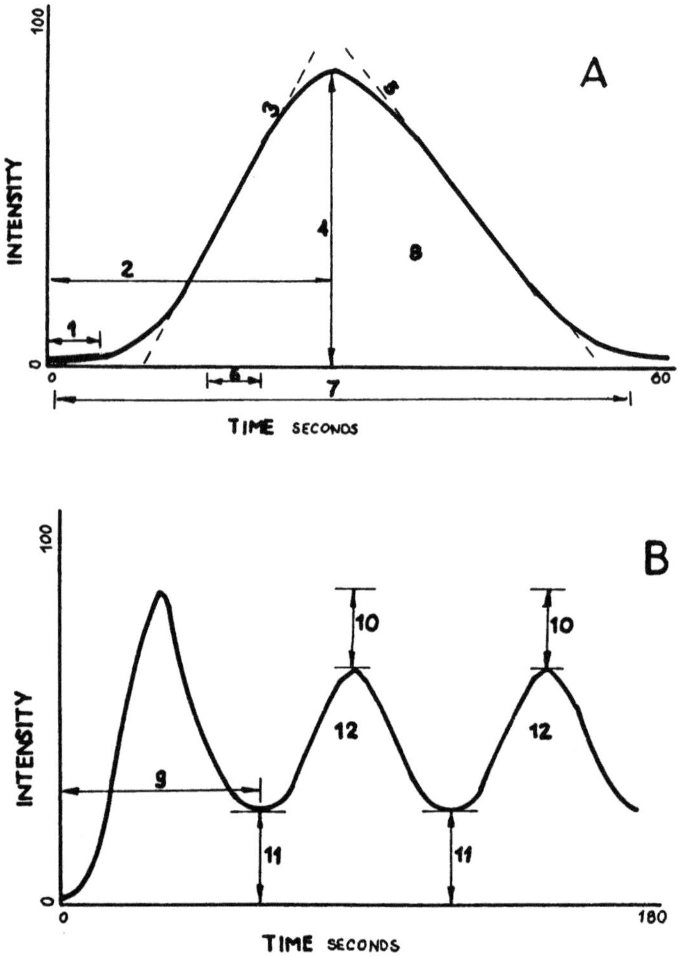

Fig. 2. Typical Time-Intensity response curve for single (A) mouth entries (A) and multiple mouth entries (B). The definable include: 1 - "lag time", 2 - time to maximum intensity, 3 - rate of appearance, 4 - maximum intensity, 5 - rate of extinction, 6 - events related to expectoration or swallowing, 7 - total duration, 8 - area under the curve, 9 - time between introductions, 10 - drop in maximum intensity relative to first introduction, 11 - baseline "shift", 12 - ratio of area under the curve relative to the first introduction area. From Lee and Pangborn (1986)

Of course above informations are to be extracted after completing the evaluation; during evaluation judges shall focus their attention only to follow stimulus intensity at the moment and on its recording, what is enough difficult task. Special training of judges is usually needed to develop their instant reflex because the duration of the whole sensation in most cases is no more then 5 - 45 seconds (Lee and Pangborn, 1986).

Development of time-intensity procedures

Conciousness that flavour or taste sensations may differ not only in intensity but also in duration goes back to ninehundred fifties and can be marked as informal start-point of the development of T-I procedures. That time, when collecting first experiences in flavour profiling several reaserchers, among them also the author, noticed and recorded that various flavour notes differ remarcably not only in intensity but also in their duration (Baryłko-Pikielna, 1957).

Formally, first measurement of T-I response to bitter compounds in water and beer and peppermint flavour in chewing gum has been conducted in the same year by Neilson (Neilson, 1957). He used the chart with intensity scores and given time intervals in which judges should record the perceived intensity of the stimulus. The time of recording intensity notes was controlled by stop-watch . T-I curve was constructed by investigator after completing the evaluations.

Similar procedure was used by Jellinek (1964) to measure T-I responses to bitterness and total flavour intensity in beer simultaneusly. One may wonder how experienced must be her judges to made this complicated task with the reasonable accuracy.

To release their judges from the necessity of watching time, Meiselman (1968) and few years later Mc Nulty and Moskowitz (1974) used sound signal pointing time intervals for intensity recording using Magnitude Estimation method

In all above procedures T-I curves were constructed by the investigator after the evaluation on the base of rather limited number of discrete data.

The next important step toward the procedure improvement was generation of continuus T-I curve directly by each judge using moving chart recorder. This technique was first introduced by Larson-Powers and Pangborn (Larson-Powers and Pangborn, 1978). In the procedure horizontal intensity scale was attached to the chart recorder; the recorder was running with the known speed and a judge was moving pen along the scale according to at moment perceived intensity, generating the curve. The "start" and "stop" of the recorder was operated by the judge via footpedal. The procedure was used to study T-I taste responses (sweet and bitter) to various natural and artificial sweeteners in gelatin along with recording T-I changes in gel hardness. Similar procedure was used in the labolatory of the author in preliminary studies of various bitter substances, as caffeine, chinine, sucrose octaacetate (Baryłko-Pikielna, unpublished data 1984).The above procedure offered advantage of direct generation of T-I curve;however all transfering the data to further statistical treatment reminded to be done manually. Some progress in the last part of procedure was introduction mechanical digitizer for reading individual points on the curve generated via external potentiometer.

The big step forward to improve T-I response measurement was complete computerization of measuring technique, independently developed in few sensory laboratories (Tahagi and Ashura 1984, Lee 1984, Guinard et al., 1985), utilizing microprocessors to run data collection and instructing judges via computer terminal. Fully programmed handling of individual data allows to present individual or averaged curves and to compute and analyse all characteristics shown in Fig.2. By this technique the accuracy of the data increase greatly because they are collected in very short time intervals (0.1 sec or shorter), practically continuusly. Also the storage of the data on floppy or hard discs is much more convenient for their further comparison, including comparison across various experiments run at different time .

The scheme of computerized T-I measurement equipment is shown in Fig.3.

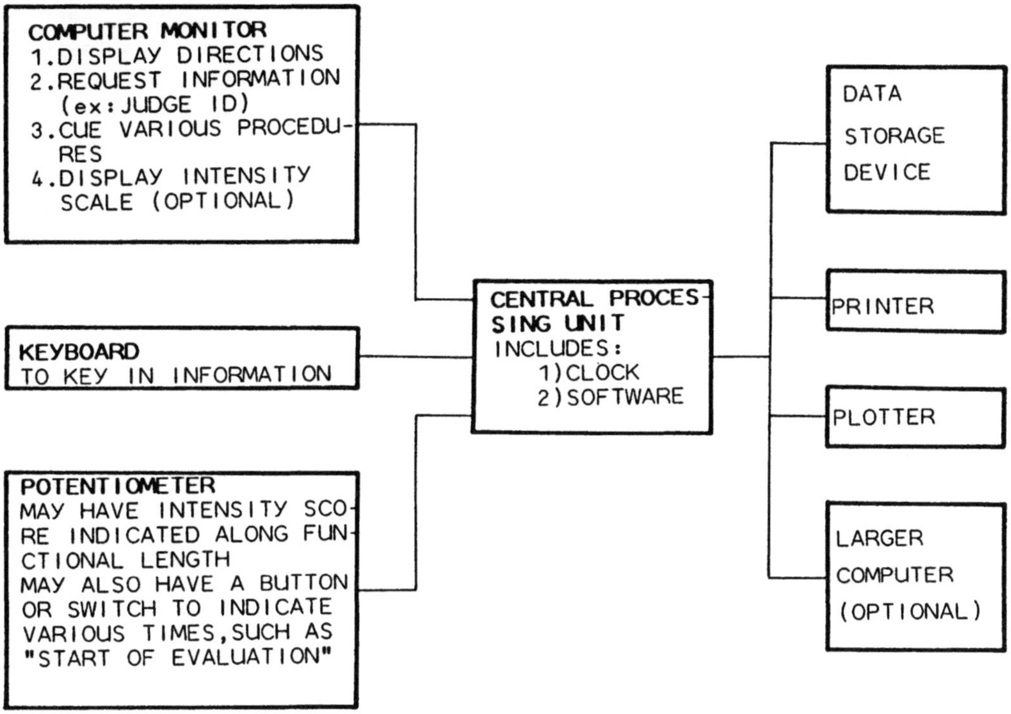

Fig.3. Basic elements of a microcomputer - based T-I procedure

Main elements of computer -based T-I procedure are: potentiometer operated by judge, computer monitor displaying directions concerning the particular evaluation technique, keybord to key-in instructions - all connected to central processing unit, combined on the other side with data-storage unit, printer, plotter and (optionally) with larger computer. The potentiometer can be operated by finger-touch or by button-switch moving along intensity scale according to in-the-moment perceived intensity of evaluated attribute.

Above technique is adaptable to most of available microcomputers. As reported by Lee (1985) computerized technique of T-I response measurement has been found by judges as convenient, interesting and even entertaining. Of course, they should first familiarize themselves with the technique what needs a preliminary training period.

Application of time-intensity procedures

T-I procedure hase found an application in basic sensory research as well as in applied studies on various aspects of sensory quality of food and beverages.

In basic research it was used for studying the temporal course of sensory adaptation (Meiselman, 1975), gustatory responses to sucrose and thaumatin solutions (Birch et al., 1980) and gustatory responses to four basic tastes (Takagi and Asakura, 1984), as well as to various sweeteners and their interrelation (Harrison and Bernard, 1984).

In applied studies T-I measurements help to learn more about the real behaviour of taste and flavour substances in natural environment of food. Therefore they may contribute greatly in flavour research and in successful application of flavourings by new products development. Some examples can illustrate the potential application of the method and kind of information provided.

Very often we have to do with the problem of aromatization of fatty-predominated, emulsion-type products. Fig.4 shows the influence of chemical and physical characteristics of lipids in two butter-like products on the intensity and temporal course of flavour perception. Both products were identically formulated and have the same amount of flavouring agent added except that one product contained predominantly olein-, whereas another product - stearin palm oil fraction.

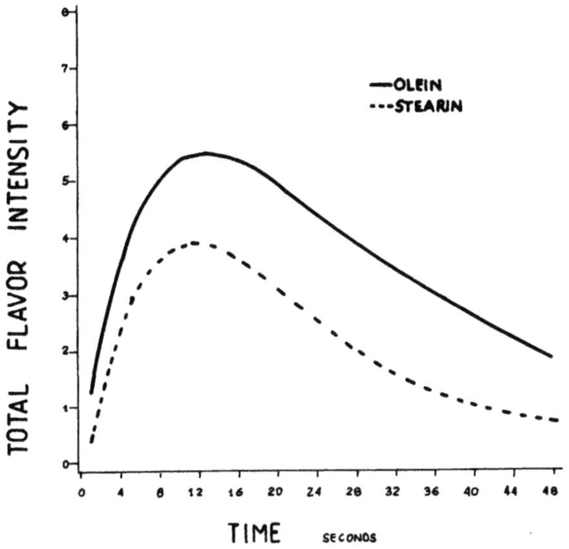

Fig. 4. T-I curves for total flavour of a butter-like product formulated with either an olein palm oil fraction or a stearin palm oil fraction; the compositions otherwise identical. From Lee (1986)

In the product with rapid-melting low-solids fraction maximum intensity was much higher and the duration of whole response much longer comparing with the product containing stearin fraction. Quicker melting released more flavour substances to saliva leading to taste sensation and to the mouth "head - space" leading to olfactory one. In the opposite, product with stearin palm oil fraction demonstrated slow melting with slow and uncomplete release of the flavour; additionaly some unmelted fat diminish flavour intensity by coating mouth surface resulting in definitely lower and shorter stimulation.

Another example shows the relation between oral viscosity and perception of sweetness and flavour in the "real" food, namely in canned desserts (Fig.5).

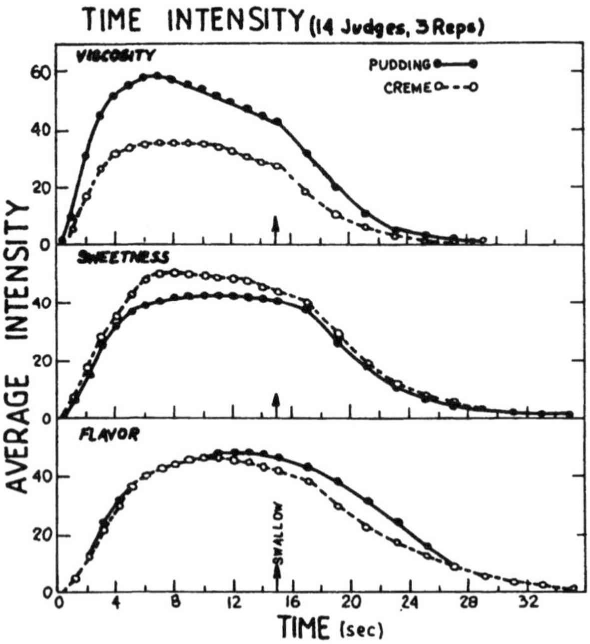

Fig.5. T-I curves for oral viscosity, sweetness and chocolate flavour of canned pudding and creme. From Pangborn and Koyasako (1981)

As one can see, sweetness is oppositely related to the viscosity of the product. Less viscous creme is easier mixed with saliva, releasing sucrose to get in contact with gustatory receptors, observed as higher intensity of a sweet taste. Decreasing viscosity of the pudding due to manipulation in the mouth is accompanied by 12 sec lasting plateau of maximum sweetness before its extinction. It is interesting that the time course of the flavour perception is the same for both products on the onset and in maximum value but slightly different in the extinction part.

Quite many foods display gel structure with various texture or hardness depending on concentration of geling agent and temperature. The relationship between the time course of breaking down gel structure and intensity of sweetness and flavour may be observed in Fig.6. Maximum intensity of flavour and sweetness occured after about 10-12 sec paralelly with dramatic decrease of hardness due to break-down of gel structure. The picture illustrates very clearly that the same taste or flavour substance may give quite different sensory effect depending of the product-context.

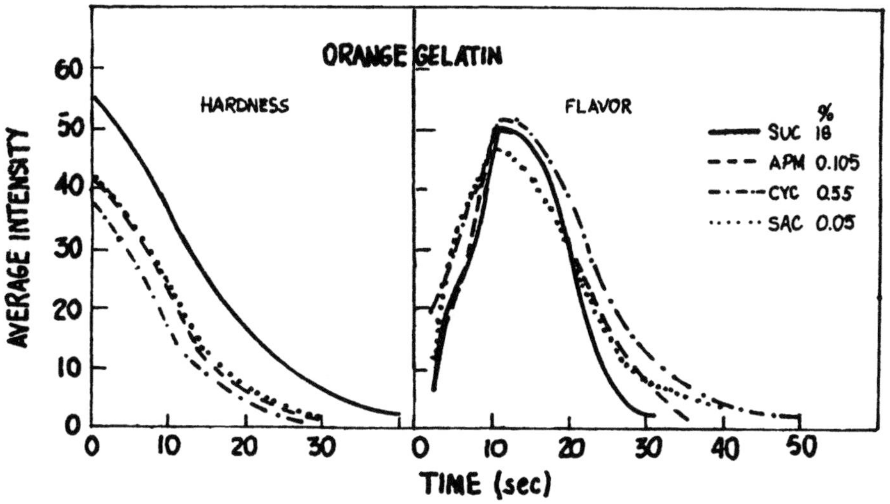

Fig.6. T-I curves for hardness and flavour characteristics of orange gelatin made with four sweeteners. From Larson-Powers and Pangborn (1978)

In the routine practice of sensory testing we are concerned on the sensory response (usually as maximum value) to single entry of food sample into the mouth. However in the normal eating practice we do many consecutive entries of food; over-all sensory impression is a synthetic result of them all. T-I technique in its to-day computerized edition offers a possibility to follow sensory responses to the model or natural stimulus in the situation of several consecutive entries of a sample into the oral cavity. Above procedure has been first introduced by Guinard for studying bitterness of iso-alpha-acids in beer (Guinard et al., 1986a) and than to study astringency in wine (Guinard et al., 1986b). Fig.7 illustrates the last experiment.

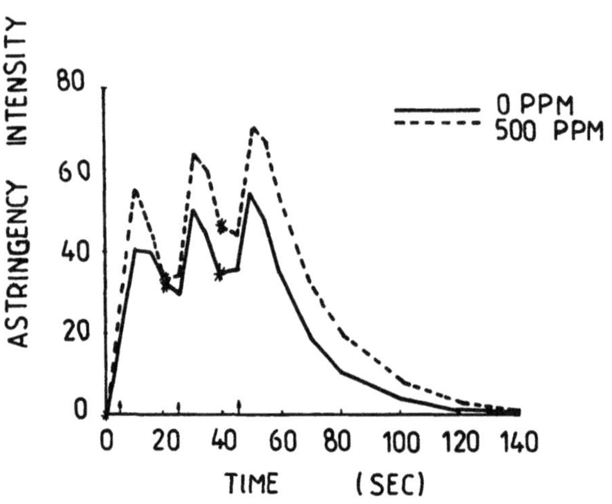

Fig.7. Average T-I curves of astringency of 0 and 500 ppm of tannic acid in wine upod three succesive ingestions of 8-ml volumes at 20 sec time intervals. Sample uptake and swallowing are indicated by a star and an arrow, respectively. From Guinard et al. (1986b)

The astringency of wine with two levels of added tannic acid (0 and 500 ppm) was studied by three consecutive ingestions of 8 ml sample of wine every 20 sec. Higher amount of tannic acid resulted in higher maximum intensity in all three introductions of the sample; the shape of time-intensity response remained the same. By rather high frequency of sample ingestions (20 sec) the cumulative effect may be observed as an increasing maximum intensity in consecutive entries and very flat "valleys" between them. The last effect did not appear when the frequency of entries was lower (40 sec) (Guinard et al., 1986b).

The above examples show the informative potential and variety of applications of time-intensity procedure both in basic and applied sensory science. To-day computerized technique facilitates its convenient use for judges as well as quick and effective data collection and consecutive handling for an investigator.

In the flavour research above method may provide valuable informations concerning the behaviour of flavouring when incorporated to the food. One can predict that along with the computerization of sensory laboratories in near future, the T-I method will be more broadly used in applied flavour research and new product development.

As pointed by Lee and Pangborn (1986), it offers better approximation to the real phenomena associated with in-mouth flavour and texture perception, since it incorporates the role of time to a degree not possible in "time-free"sensory procedures. Using this method we can get better approximation of global sensations related tothe real world of eating and drinking beverages.

Literature

[1] Baryłko-Pikielna, N.: O składnikach smakowitości mięsa. Przem. Spoż. 11 (1957) 1, 26

[2] Guinard, J.X., Pangborn, R.M., Shoemaker, C.F.: Computerized procedure for time-intensity measurements. J.Food Sci. 50 (1985) 543

[3] Guinard, J.X., Pangborn, R.M., Lewis, M.J.: Effects of repeated ingestion on temporal perception of bitterness in beer. Am.Soc.Brew.Chem.J. 44 (1986a) 28

[4] Guinard, J.X., Pangborn, R.M., Lewis, M.J.: The time-course of astringency in wine upon repeated ingestion. A.J.Enol.Vitic. 37 (1986b) 184

[5] Harrison, S.K., Bernard R.A.: Time-intensity sensory characteristics of saccharin, xylitol and galactose and their effect on the sweetness of lactose. J.Food Sci. 49 (1984) 780

[6] Jellinek, G.: Introduction to a critical review of modern methods of sensory analysis (odour, taste and flavour evaluation) with special emphasis on descriptive analysis (Flavour Profile Method). J.Nutr.Diet.1 (1964) 3, 219

[7] Larson-Powers, N., Pangborn, R.M.: Paired comparison and time-intensity measurements of the sensory properties of beverages and gelatins containing sucrose or synthetic sweetners. J.Food Sci. 43 (1978) 41

[8] Lee, W.E.III: Evaluation of time-intensity sensory responses using personal computer. J.Food Sci. 50 (1985) 1750

[9] Lee, W.E.III: A suggested instrumental technique for studying dynamic flavour release form food products. J.Food Sci. 51 (1986) 249

[10] Lee, W.E.III, Pangborn, R.M.: Time-intensity: the temporal aspects of sensory perception. Food Technol.40 (1986) 11, 71

[11] McNulty, P.B., Moskovitz, H.R.: Intensity-time curves for flavoured oil-in water emulsions. J.Food Sci. 39 (1974) 55

[12] Meiselman, H.L.: Magnitude estimation of the coarse of gustatory adaptation. Percept. Psychophys. 4 (1968) 193

[13] Meiselman, H.L.: Effect of response task on taste adaptation. Percept. Psychophys. 17 (1975) 591

[14] Neilson, A.J.: Time-intensity studies. Drug Cosmetic Ind. 80 (1957) 452

[15] O'Mahony, M.: Fatigue and adaptation . Food Technol. 40 (1986) 11, 67

[16] Takagi, M., Asakura, Y.: Taste time-intensity measurements using a new computerized system Proc.18th Japanese Symposium on taste and smell. Japanese Assoc. for the study of taste and smell. Osaka Univ.,Osaka

[17] Yamamoto, T., Kato, T., Matsuo, R., Araie, N., Azuma, S., Kawamura, Y.: Gustatory reaction taim under variable stimulus parameters in human adults. Physiol. Behav. 29 (1982) 79

METHODS OF SELECTION OF DESCRIPTORS AND GRADING SCALES IN THE SENSORY PROFILE ANALYSIS OF FOODS

J. Pokorný and J. Davídek[1)]

ABSTRACT

The preparation of a sensory profile analysis includes the selection of a descriptor list and of grading scales. The following methods were found useful for the selection of descriptors for sensory profiles: (1) presenting samples to a group of inexperienced persons for the description of flavour character; (2) selection from a general list of descriptors; (3) ranking of a list of descriptors according to their importance; (4) selection of typical or objectionable flavour notes; (5) suggestions of a group of experts. The selected list is reduced to 8 - 16 notes, tested, adapted, and tested again. The finally included descriptors are tested whether their meaning is clear and unambiguous. Intensities of individual flavour notes are evaluated with use of various scales. The disadvantage of generally used ordinal category scales is that nonparametric methods only can be used for statistical analysis. Magnitude estimation and use of unstructured oriented graphical scales are better for descriptors of weak to strong intensities as the results may be evaluated by parametric methods, however, deviations occur in case of very weak or very strong flavour notes. The Procrustes analysis is usually possible with both the above scales but produces some bias. The evaluation of sensory profiles by discussion was found unreliable, very sensitive to various factors, and is not recommended. When sensory profiles of several samples are compared, the determination of statistical significance of their differences is very difficult because nonparametric methods available for this purpose are relatively insensitive.

INTRODUCTION

The sensory profiling has a psychophysical basis [12, 13]; various sensory profile techniques [11] have increased in importance in the last ten years both for the development of new products and for their evaluation from the standpoint of sensory quality, and several methods have been internationally standardized [5]. Applications of the profiling techniques are growing so rapidly that there has been no time to investigate the principles accordingly, and to test the influence of various factors. Mistakes are often made in the selection of descriptors, in the selection of rating scales, and in the statistical evaluation of sensory data. In this paper we present our experience in the selection of flavour descriptors and of rating scales.

---

1) Department of Food Chemistry, Prague Institute of Chemical Technology, Suchbátarova 5, CS - 166 28 Prague 6, Czechoslovakia

SELECTION OF FLAVOUR DESCRIPTORS

The list of flavour descriptors should be prepared with great care because the success of an analysis will largely depend on it. Some methods used in our laboratory are shown in Tab. 1. The first procedure mentioned in Tab. 1 consists in presenting a sample of the product tested to a group of instructed assessors but not to experts. We prefer to use a panel of college students (mostly graduated students) with some basic knowledge in the sensory analysis and fair knowledge in food science but without any great experience in sensory profiling. Their responses reflect the attitudes and descriptions of an average consumer. They are told to taste the sample, and to give a list of flavour notes they perceive (the minimum of 4 - 5) with description of their intensities. The usual number od descriptors given varies between 4 - 10 with the average of 6. An example (evaporated unsweetened milk Tatra Grand, Czechoslovakia) is given in Tab. 2, indicating intensities of the flavour notes as well.

Table 1. Methods of the descriptor selection

(1) Frequency of use in flavour description.
(2) Frequency of selection from a large general list of descriptors.
(3) Ranking of descriptors according to their importance.
(4) Selecting descriptors of typical flavour notes.
(5) Selecting descriptors of main variables in the product to be tested.
(6) Including in the list of descriptors characterizing serious flavour defects.
(7) Listing suggestions of a group of experts.

It is evident from Tab. 2 (next page) that most respondents used very simple descriptors which are generally used in everyday life (such as the four basic tastes) but are not typical for evaporated milk. The salty taste was detected by 65 % of the respondents even when its intensities were rated as very weak. On the contrary when selecting from a list of 116 descriptors (based on suggestions of a group of experts), respondents used the descriptors specific for evaporated milk with much higher frequency while the frequency of the basic tastes (salty and bitter) was lower than in the case of free description; specific descriptors ommitted in the list (e. g. boiled flavour) were lower when selecting from a list (the assessors could add them to the list) than in case of free description. Persons with small experience are evidently not able to find (and perhaps, to look for) suitable specific descriptors, especially when they are not in common daily use.

What is then the usefulness of a free description? It may lead to discovery of other descriptors than those suggested by experts, and they will supply various homonyms and synonyms used in current speech. The selection from a general list of descriptors is a very useful basis for developing a list for particular products. We are presenting lists of 100 - 140 descriptors, mostly including terms known to all respondents, and avoiding special vocabulary known only to experts. The list developed by Harper [3] and by Dravniek et al. [1, 2] served as a basis for our list, but our lists were more specialized for particular food products

Table 2. Selection of descriptors for sensory profiling of evaporated milk by free description or by selection from a list of descriptors

Flavour descriptor	Method of free description		Selection from a general large list of descriptors
	frequency (%)	average intensity	
Sweet	97	medium	100
Sour	70	medium	80
Salty	65	very weak	46
Bitter	43	rather weak	19
Buttery	27	weak	68
Caramel-like	27	weak	21 a
Boiled, boiled milk	16	weak	8 a
Creamy	16	weak	100
Fresh milk	14	weak	100
Nutty	8	weak	8
Rancid, rancid butter	8	weak	20
Astringent	5	weak	31
Molten butter, butter fat	5	very weak	6 a
Alkaline	5	very weak	12
Dried milk	3	weak	2 a
Egg-like (raw egg)	3	weak	0 a
Tart	3	weak	13
Cheese-like	3	very weak	37
Fatty	0	-	56
Soapy	0	-	47
Vanilla-like	0	-	44
Metallic	0	-	32
Gingerbread-like	0	-	20
Oily	0	-	19
Roasted	0	-	17
Floral	0	-	15

Note: a = not included in the list of descriptors.

by introducing special descriptors, e. g. 12 descriptors of fruit flavours were inserted into the list used for evaluating canned fruit samples, or 13 descriptors of floral notes were inserted into the list for evaluating candy or tooth paste.
The list of descriptors obtained in the above way still contains 40 - 50 items which is too complicated for the characterization of food flavours if not by experienced experts. Therefore, we have attempted to eliminate the descriptors unimportant for the overall flavour. An example, concerning characterization of the gingerbread flavour, is given in Tab. 3. A list of 23 descriptors was presented to a group of 196 graduate students who had to rank them according to their importance in the overall flavour of gingerbread. Average rankings were obtained from the results (sum of ranks divided by the sum of respondents), and these average rankings were again ranked. It is evident that some flavour notes are generally considered as important, and another group as unimportant. There was no great difference between the ranking of male and female respondents, only that men considered the "spicy" flavour note as more important than the "ginger" flavour (the Czech name for gingerbread has nothing common with ginger) while the reverse was true for lady respondents. Similar difference was found between "soapy" and "salty". High ranking of cocoa flavour is due to high consumption of chocolate-coated gingerbread. The methods are particularly suitable for selection of a list for grading fresh prod-

Table 3. Ranking of descriptors for characterizing the gingerbread flavour according to their importance

Descriptor examined	Average ranking			Final ranking		
	men	women	total	men	women	total
Sweet	2.6	1.8	2.1	1	1	1
Caramel-like	4.2	5.5	5.1	4	5	4
Roasted	4.8	5.9	5.5	5	6	6
Cocoa-like	3.3	3.6	3.5	2	3	3
Coffee-like	5.1	6.3	5.8	6	7	7
Burnt	7.6	8.4	8.0	9	11	9
Spicy	6.1	5.1	5.4	7	4	5
Honey-like	3.6	3.3	3.4	3	2	2
Fatty	9.1	8.7	9.2	10	12	12
Floral	7.4	7.8	7.7	8	9	8
Bitter	9.4	10.3	9.9	12	15	15
Astringent	12.6	12.9	12.8	17	17	17
Bread crust	9.2	9.9	9.6	11	14	14
Malty	9.4	7.7	8.3	13	8	10
Ginger-like	10.3	7.9	8.7	16	10	11
Sour	9.7	10.3	10.0	15	16	16
Soapy	9.5	23.0	9.3	14	23	18
Salty	13.3	9.2	9.3	18	13	13
Stored, old flour	13.7	15.2	14.6	19	18	19
Rancid	16.3	17.8	16.0	20	20	20
Mouldy	16.5	16.8	16.7	21	19	21
Sulphuric	22.0	20.0	20.8	23	21	23
Ammoniac	18.7	20.5	19.7	22	22	22

ucts of good quality.

For the differentiation between products of top quality and of the good quality it is not so useful to rate the most important flavour notes (they are present in both types of products in nearly the same intensities) but to rate intensities of the most typical, characteristic flavour notes. For the selection of the typical flavour notes of gingerbread we presented a list of 18 descriptors to 230 respondents (graduate students with fair knowledge of food technology, and basic training in the sensory analysis). They had to select 2 – 5 descriptors typical for gingerbread, and 2 – 5 ones considered as objectionable, if detected in gingerbread. The results are summarized in Tab. 4. While the sweet flavour was rated as the most important flavour note, the most typical flavour notes were roasted, honey-like, and caramel-like notes which were also extremely rarely judged as objectionable. Burnt and sour flavours were the only ones judged as objectionable by the majority of respondents.

Important flavour notes are often (but not always) the most intensive flavour notes, therefore, the importance may be judged on the basis of an intensity profile as well. From the data shown in Tab. 5, the caramel, roasty, spicy, sour, and cheese flavour notes may be found as the most important descriptors for the evaluation of soy sauce (based on 278 responses; Soy sauce, Vitana, Byšice, CS) while putrid, rancid, or floral flavour notes seem less important from this standpoint. In the application to the analysis of products produced by different manufacturers these descriptors are particularly important which are more variable (e. g. in

Table 4. Selection of descriptors of typical and objectionable flavour notes in gingerbread

Descriptor examined	Percentages of responses judging as		
	typical	neutral	objectionable
Caramel-like	66.0	32.6	1.3
Malty	23.5	67.4	9.1
Bread-like	17.0	70.4	12.6
Burnt	4.3	42.2	53.5
Fruity	17.4	71.7	10.9
Floral	1.7	70.4	27.8
Spicy	43.5	44.3	12.2
Ginger-like	29.1	62.6	8.3
Honey-like	81.7	17.4	0.9
Nutty	24.8	71.3	3.9
Sour, acid	1.3	44.8	53.9
Sweet	56.7	43.3	0.0
Fried	0.0	55.7	44.3
Cocoa-like	20.9	73.4	5.7
Coffee substitutes, chickory	7.8	82.2	10.0
Roasted	85.7	13.7	0.6
Coffee-like	2.6	90.0	7.4
Sharp	0.4	70.8	28.7

Table 5. Selection of important flavour descriptors based on intensities of the flavour notes (analysis of gingerbread)

Descriptor examined	Percentages of ratings of the respective intensities								
	1	2	3	4	5	6	7	8	9
Sulphuric	39.6	23.7	20.5	9.7	3.6	2.5	0.4	0	0
Roasted	12.5	21.4	20.7	10.7	13.4	10.4	6.8	3.9	0
Caramel-like	34.6	18.2	21.1	6.4	8.9	6.1	2.9	1.8	0
Floral	74.7	17.1	5.1	2.7	0	0.3	0	0	0
Ethereal	52.7	23.7	8.2	7.5	4.3	3.2	0.4	0	0
Sour, acidic, vinegar	32.2	20.2	11.6	9.9	12.2	9.1	3.3	1.2	0.4
Buttery	32.2	29.7	22.8	8.0	4.3	2.9	0	0	0
Rancid	63.9	18.2	11.1	2.5	3.2	0.4	0.4	0.4	0
Cheese-like	27.2	15.2	20.8	6.4	6.7	11.3	7.8	3.2	0.4
Stored, old flour	62.7	16.9	10.6	6.7	0.7	2.1	0	0.4	0
Putrid	74.6	11.8	6.4	5.0	1.1	0.4	0.7	0	0

Scale used for the description of intensity:
1 = imperceptible, absent; 2 = present only in traces; 3 = very weak; 4 = rather weak; 5 = important minor flavour note; 6 = main flavour note; 7 = the most important, dominating flavour note; 8 = the flavour note absolutely governing the total impression; 9 = the only flavour note present.

the case of gingerbread, they were the intensities of caramel, cocoa, chocolate, flour, and honey flavour notes. The same is valid for the examination of changes of sensory quality during storage of food products. On the storage of cakes the intensity of sweet flavour remained constant, and was thus not indicative, but the intensities of rancid, oily, soapy, and flour, stored flavours substantially changed.

For the differentiation between products of a standard, consumer quality and of

Table 6. Selection of descriptors of flavour notes judged as unsuitable in the product (% total responses)

Descriptor examined in gingerbread	Judged as unsuitable	Descriptor examined in chicken soup	Judged as unsuitable
Sweet	0	Chicken-like	0
Caramel-like	3	Meaty	6
Roasted	14	Lardy	16
Cocoa-like	4	Salty	0
Coffee-like	17	Acidic, sour	39
Burnt	68	Sweet	40
Spicy	12	Spicy	1
Honey-like	0	Peppery	19
Fatty	39	Rancid	52
Flour, dough like	32	Sharp	40
Bitter	48	Liver-like	15
Astringent	69	Skinny	17
Bread-like	51	Gluey	58
Malty	50	Egg white	22
Ginger-like	48	Egg yolk	16
Sour, acidic	66	Fried	52
Soapy	88	Milky	58
Salty	90	Buttery	33
Stored	86	Vegetables (leaves)	9
Rancid	86	Vegetables (roots)	0
Mouldy	86	Astringent	59
Sulphuric	88	Muddy	60
Ammoniac	86	Rotten	61

unacceptable products it is necessary to insert some descriptors of serious quality defects (such as rotten, mouldy, rancid, burnt, stored, etc.).

When new products are being developed it is necessary to avoid flavour notes generally judged as unsuitable flavour notes in the particular product. For this purpose we have presented a list of descriptors to a group of graduate students who had to decide in every descriptor whether they judge the respective flavour note unsuitable as a part of the total flavour of the product; 93 answers were collected in case of gingerbread, and 111 in case of chicken soup; in Tab. 6 the percentages of responses are given. Flavour notes judged by 25 % or more respondents as unsuitable in the products should be absent or present only at low intensity levels in the products developed. Note that the presence of salty taste has been judged as unsuitable at the same degree as the percentages of rancid or mouldy tastes in gingerbread, and that only 40 % respondents considered the presence of sweet taste in chicken soup as unsuitable, and that only 61 % judged the rotten flavour as unsuitable in the same product.

When a list of descriptors has been finally developed, it usually contains 8 - 12 descriptors in our laboratory; higher number of descriptors is allowed when very experienced assessors participate at the rating. Before using the final list of descriptors we recommend to test whether meanings of the descriptors are clear to the assessors. Results are sometimes surprising as Tab. 7 shows (results collected from 203 questionnaires, mostly presented to graduate students or college educated food chemists). As evident considerable uncertainty was found even in case of terms often used in describing the flavour character, such as metallic, roasted, or pungent. Most respondents could not tell the flavour of malt (however, about 25 % partcipated at an excursion to a malt factory), do not know linseed varnish or the flavour of raw beans. It is possible to teach them by presenting them standards but when the term is not being used in the daily practice, and the assessor has not used it often and for a long time, he will not be able to use the term properly. Therefore, it is preferable to use only the terms generally known and commonly used in the population. The unknown terms are often avoided, or their intensity is underestimated.

Now, when a list has been prepared we present it to a group of experts for reviewing, and make last corrections when they find it advisable, which is particularly useful when rare flavour defects have to be detected.

Table 7. Determination, how clear are the meanings of flavour descriptors

Descriptor examined	Frequency of responses (% total)				
	a	b	c	d	e
Sweet	97.3	2.7	0	0	0
Caramel-like	84.6	15.4	0	0	0
Astringent	79.3	16.3	1.9	1.0	1.4
Sharp	63.9	26.9	5.8	3.4	0
Soapy	61.1	23.1	11.1	3.8	1.0
Mint-like	57.7	16.8	11.1	3.8	10.6
Cold, cooling	56.7	24.0	11.5	4.3	3.4
Rancid	52.4	28.4	12.0	5.6	1.4
Warm	51.4	27.9	12.0	5.8	2.9
Pungent	48.1	34.1	14.4	3.4	0.0
Metallic	37.5	26.4	18.8	9.6	7.7
Roasted	32.2	36.0	17.3	9.6	4.8
Tart	27.4	19.2	10.1	8.7	34.6
Tallowy	26.4	31.2	23.1	14.9	4.3
Malty	15.9	34.1	19.7	18.8	11.5
Alkaline	13.5	25.0	28.4	15.4	17.8
Varnish-like (linseed varnish)	11.1	10.1	16.8	32.2	29.8
Sulphuric	11.1	23.6	30.3	18.7	10.3
Terpenic	1.9	7.7	15.4	29.8	45.2
Beany	1.4	7.2	9.6	25.0	56.7

SELECTION OF GRADING SCALES

While relatively great attention is being paid to the selection of proper flavour descriptors, the choice of scales is often careless, without testing their usefulness for the respective analysis. Several grading scales have been standardized [4], and many others have been suggested in the literature. For sensory profiling of food materials we have used the scales given in Tab. 8.

The use of nominal scales is only rare but we have obtained good results particularly with less experienced assessors who are very useful because their feelings are close to those of average consumers, but who would not be able to use more complicated scales. They are offered the sample and a list of 100 - 150 descript-

Table 8. Scales used in the sensory profiling

Kind of scale	Types used
(1) Nominal scales	lists of descriptors
(2) Ordinal category scales	simple scales: numerical alphabetical descriptive combined scales: numerical with description alphabetical with description
(3) Ratio scales	magnitude estimation
(4) Graphical scales	oriented series of categories structured linear scales unstructured linear scales starlike diagrams

Table 9. Use of nominal scales for sensory profiling of sardines in oil

Descriptor used	Frequency of response (% total)		Significance of differences
	Isabela	Socra	
Sweet	11.4	12.9	-
Sour	51.4	66.1	+
Salty	97.1	91.9	-
Bitter	48.6	35.5	-
Astringent	17.1	17.7	-
Alum-like	11.4	24.2	+
Metallic	45.7	50.0	-
Meaty	68.6	64.5	-
Fishy	94.3	100.0	-
Spicy	88.6	75.8	+
Peppery	28.6	38.7	+
Allspice	11.4	17.7	+
Bay leaves	0	9.7	+
Smoked	20.0	0	+
Buttery	25.7	9.7	+
Oily	97.1	87.1	-
Fatty	2.3	2.2	-
Rancid	0	16.1	+
Woody	17.1	8.1	+
Bark-like	11.4	0	+
Muddy	17.1	22.6	-
Pond water	11.4	12.9	-
Grassy	0	6.5	+
Leaf-like	20.0	8.1	+
Mushrooms	11.4	9.7	-
Cheese-like	0	14.3	+
Wet dog	11.4	8.1	-
Soup flavouring	17.1	8.1	-
Light	22.9	4.8	+
Sharp	0	12.9	+
Pungent	0	35.5	+
Lemon-like	0	6.5	-
Hot, biting	0	6.5	-
Chemicals	8.6	0	+

Table 10. Distinguishing between cake flavourings using nominal scales

Descriptor examined	Percentages of responses in analyzing cakes flavoured with			
	nougat	milk caramel	milk cocoa	lemon
Sweet	100	100	100	100
Sour	6	11	8	46
Salty	44	16	31	31
Bitter	31	32	46	7
Spicy	19	11	23	32
Vanilla	100	89	46	92
Cinnamon	19	26	13	23
Buttery	44	58	38	77
Creamy	12	11	14	29
Milky	50	37	31	7
Lard-like	12	12	14	2
Fatty	11	16	13	3
Fried	12	1	14	0
Gingerbread	12	5	13	0
Burnt	6	6	31	0
Roasted	31	58	46	15
Coffee-like	11	53	22	0
Coffee substitute	44	21	23	0
Chickory (roasted)	0	0	16	0
Chocolate	38	74	69	0
Cocoa	31	63	77	0
Nutty (walnuts)	38	42	13	2
Metallic	19	16	15	2
Astringent	19	16	15	3
Alum-like	3	2	14	0
Alkaline	19	5	14	2
Bread-like	18	16	3	3
Almonds	11	0	0	0
Woody	12	5	15	8
Light	62	21	14	23
Plastics, chemicals	0	10	0	7
Floral	2	5	0	23
Lemon-like	0	0	0	85
Orange-like	0	0	0	38
Essential oils	0	0	0	15

ors. They select the descriptors, the flavour notes of which they are able to detect (usually 8 - 15 descriptors after short experience – 6 sessions, and 12 - 20 descriptors after longer experience – at least 1 month of sensory profiling). An example is given in Tab. 9 on the characterization of two samples of sardines in oil (Isabela, Spain - 53 responses; Socra, USSR - 48 responses). It is evident that big differences could be found in many descriptors. Flavours percepted by the majority of assessors (50 - 100 %) may be considered as the main flavour notes (salty, fishy, meaty, spicy, oily, and sour in both samples). Assessors with less experience have difficulties in perceiving minor flavour notes which are carriers of specific character of the product. Therefore, the percentage of responses is lower in the case of minor flavour notes, and may be considered as proportional to the intensity and/or importance. Significant differences were found in 16 minor flavour notes.

Another example is in Tab. 10; in this case four bakery products of the same composition (produced by Opavia, Opava, Czechoslovakia) were compared, differring only in their flavouring (the results were calculated from 34 - 39 responses each). The sweet and vanilla flavours were the only main flavours found common for all the four samples, but the composition of their flavouring considerably affects the flavour profile both in minor and in major flavours. The character of the product is clearly defined in this way, e. g. between milk-cocoa and milk-caramel flavoured products, 19 descriptors were found statistically different in the frequency of responses (P = 0.95).

The use of ordinal category scales is the most frequently used type of scales, and most often, combined numerical and descriptive scales are used. The disadvantage of numerical scales is that both assessors and operators evaluating the results are tempted to treat them as equidistant ratio scales, and use improper methods of statistical evaluation. Five to six point scales are popular in determining the sensory quality [10] but for the sensory profiling we have found 7 - 9 point scales as more suitable, more sensitive, and not difficult to handle even by less experienced assessors.

Great care should be given to the proper selection of descriptive terms. An example is given in Tab. 11. We have asked the respondents to give the most suitable description of intensity of the highest-intensity degree in a 5-point ordinal category scale. It is interesting people preferred the description "very intensive" to "extremely intensive" which may seem more proper at the first glance; the assessors gave 18 various suggestions for this description.

Table 11. Selection of the most suitable descriptor of the top intensity in a 5-pt. ordinal category scale (free description)

Descriptor proposed	Number of responses	Percentage of the total
Very intensive	34	19.1
Extremely intensive	22	12.4
Extraordinarily intensive	18	10.1
Very strong	17	9.6
Excessively strong, too strong	12	6.7
Unusually strong	11	6.2
Intensive	8	4.5
Strong	8	4.5
Extremely strong	7	3.9
Overwhelming	7	3.9
Other responses	34	19.1

In another experiment they were given a list of 5 descriptors with the task to select the most proper term for the highest intensity degree in a 5-point scale, and to rank all expressions according to their suitability. "Very intensive" was rated here as the least suitable one among the group of descriptors, contrary to free choice (Tab. 12). The rating "very strong" and the rating "the strongest possible" differ from the rest of the set (P = 0.95) after Kramer [9]. The description found the most appropriate was never mentioned in case of free description.

Table 12. Selection of the most suitable descriptor of the top intensity in a 5-pt. ordinal category scale (from a list of 5 descriptors)

Descriptor examined	Number of ratings as the most suitable term	Sum of rankings
Very strong	0	140
Extraordinarily strong	0	98
Extremely strong	4	70
The strongest possible	22	36
The strongest degree of intensity	2	76

The meaning of terms describing hedonic responses is particularly vague. We presented a group of 274 assessors a list of terms, and ask them to give (on an unstructured scale) the meaning of the term (the left side of the scale = the best flavour possible; the right side of the scale = the worst flavour possible). Some examples are given in Fig. 1. The terms of extreme hedonic qualities ("excellent" and "extremely bad" gave relatively consistent responses situated at either end of the

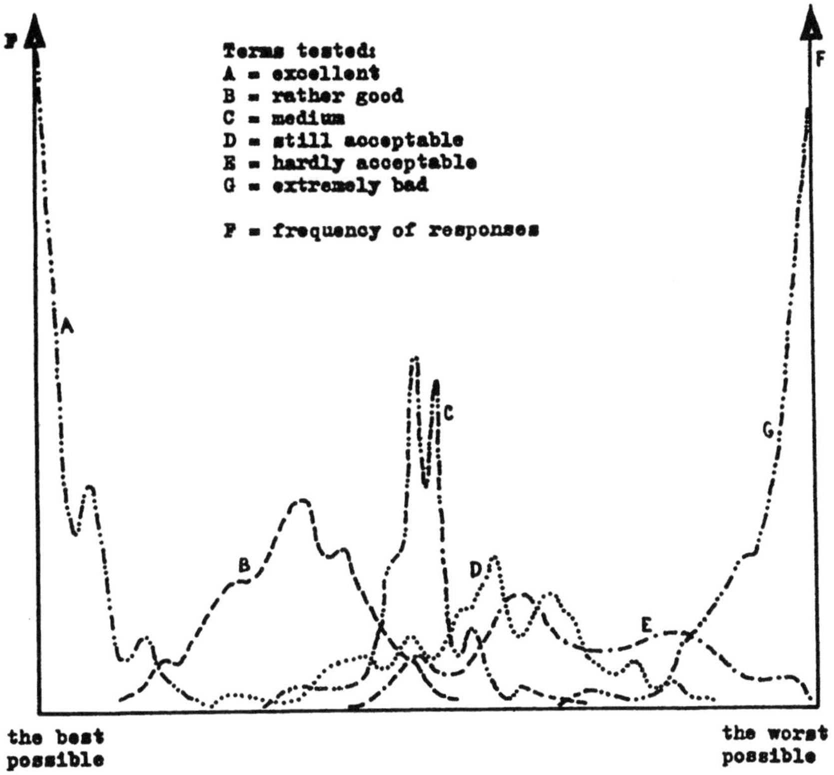

Fig. 1. Conversion of hedonic descriptive scales into graphical scales

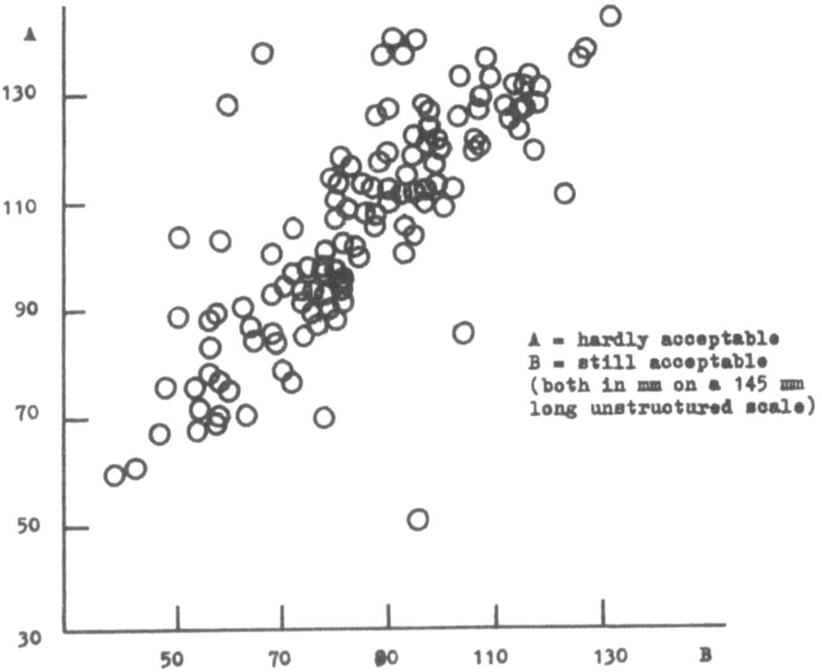

Fig. 2. Relation between ratings of hedonic descriptors on graphical unstructured scales

scale, and the term "average, medium" had also relatively exact meaning. The less defined terms like "rather good" were spread on a half of the scale, and terms like "hardly acceptable" and "still acceptable" had the distribution of responses scattered along the major part of the line. Such term should, obviously, be avoided in the descriptions of intensities or acceptabilities in category scales.

We assumed that the scatter was caused by a shift of responses to one or the other extreme of the line by individual assessors; in such a case the difference between the grading by one assessor would be nearly constant. Therefore, we have plotted the individual responses in a bidimensional graph (Fig. 2). Indeed, a linear relationship was found in this case between the two responses but the uncertainty is high so that the use of Procrustes analysis will increase the variance by about 10 - 20 %.

The above experiments showed that the descriptors defining the acceptability are not always clear. It may be assumed that the descriptors defining the intensities are also often not sufficiently clear. Therefore, they were tested using a group of 76 assessors. They were presented a list of 18 descriptors of intensity, and they had to evaluate how clear is their meaning (100 % = absolutely clear; 0 % = absolutely undefined or unknown). The average ratings are summarized in Tab. 13. It is evident that the degree of 20 - 50 % of uncertainty is involved in most ratings of intensity, only therefore, that the descriptors are not clearly defined

Table 13. How clear is the meaning of descriptors concerning intensities

Descriptor examined	The meaning is clear from %	Descriptor examined	The meaning is clear from %
Imperceptible	96	Relatively weak	70
Undetectable	90	Medium	64
Very strong	89	Average	60
Extremely strong	77	Rather weak	58
Rather strong	77	Negligible	57
Pretty strong	76	Extremely weak	53
Extraordinarily strong	75	Stronger than the average	47
Extraordinarily weak	75	Not too strong	45
Very weak	73	Not very strong	34

Table 14. Dependence of magnitude estimation from the numerical value of the standard sample

Numerical value of the standard sample	Interval of ratings	
	absolute	% of the standard
1	1 - 10	100 - 1000
10	10 - 28	100 - 280
100	100 - 200	100 - 200
1000	1000 - 1600	100 - 160

in human mind. Such descriptors as "not too strong" or "not very strong" should be avoided.

Ratio scale would be the most suitable scale because results of the analysis could be treated with the use of parametric methods, therefore, the magnitude estimation proposed by Stevens [15] becomes popular, and the magnitude estimation is increasingly used, especially in computerized sensory laboratories. McBride [9] has shown some inconveniencies of the method which are based on the lack of ratio scaling in human mind when handling with numbers larger than 10. Therefore, the results depend on the units used in the rating, and on the concentration [8]. In agreement with McBride we have found deviations from the Stevens' law in extreme concentration ranges, e. g. in evaluating the bitterness of tea infusions [7]. The choice of units is left free according to Stevens' suggestions but in our experiments the best results are obtained if the standard is equal to 10 (Tab. 14). It is necessary to fix the standard unit, and not to leave it to the assessor's decision. In case of medium stimulus intensities, magnitude scales may be considered as ratio scales.

We have obtained best results with use of unstructured graphical linear scales, which is in agreement with McBride's experiments [9], even when deviations from the linearity occur at extreme stimulus intensities [14]. At medium intensities (20 - 80 % of the scale) intensities of the response are proportional to intensit-

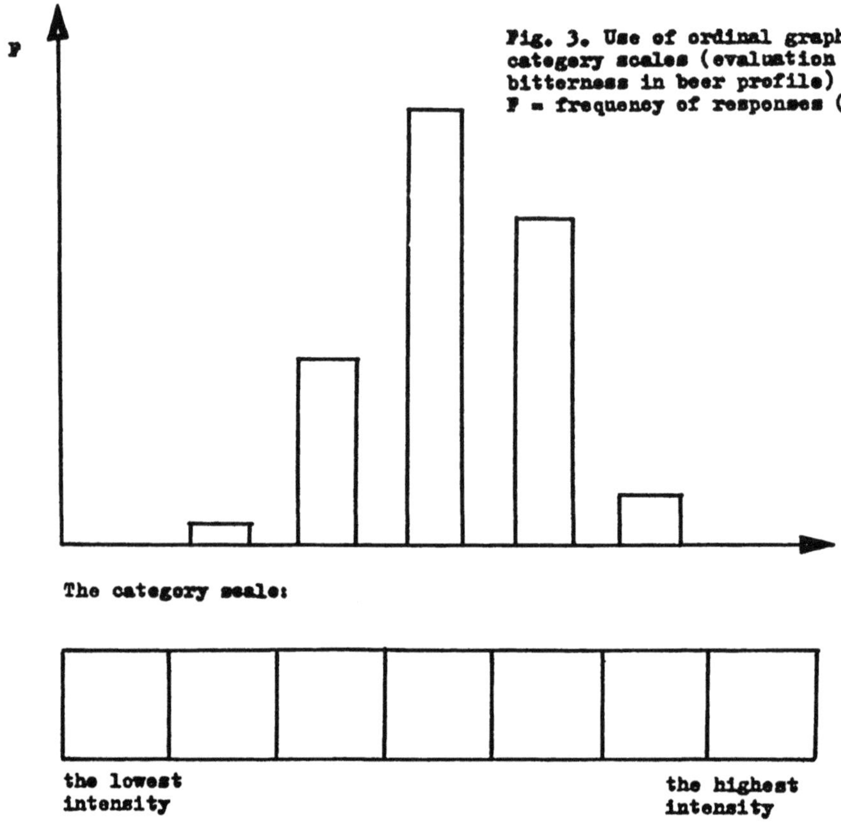

Fig. 3. Use of ordinal graphical category scales (evaluation of bitterness in beer profile)
F = frequency of responses (%)

ies of the ———————————————————— stimulus. The medium difference between two ratings by different assessors is about 10 - 15 % of the scale, depending on the stimulus intensity. The elimination of verbal description improves the reproducibility but the description has, nevertheless, an importance because most assessors evaluate first their percept verbally, and only then find a proper location on the scale.

The use of category scales containing no verbal descriptors (Fig. 3) is very similar to graphical scales but deviations from the linearity occur at the two extreme categories from each side.

Structured graphical scales are often recommended because the anchoring points facilitate the decision. We have compared structured and unstructured scales using a panel of 226 assessors who were presented an appetizer which was evaluated both by an unstructured and by a structured scale. The results shown in Fig. 4 make evident that the anchoring points attract the responses, and another accumulation of responses was observed in the middle between two neighbouring anchoring points. Therefore, the results are biased and difficult to evaluate, and we do not recommend the use of structured scales.

Starlike diagrams are often used for the presentation of sensory profiles, and sometimes, the ratings are directly made in the diagram. We have found, however, that the results are not the same when the starlike diagram are used directly or

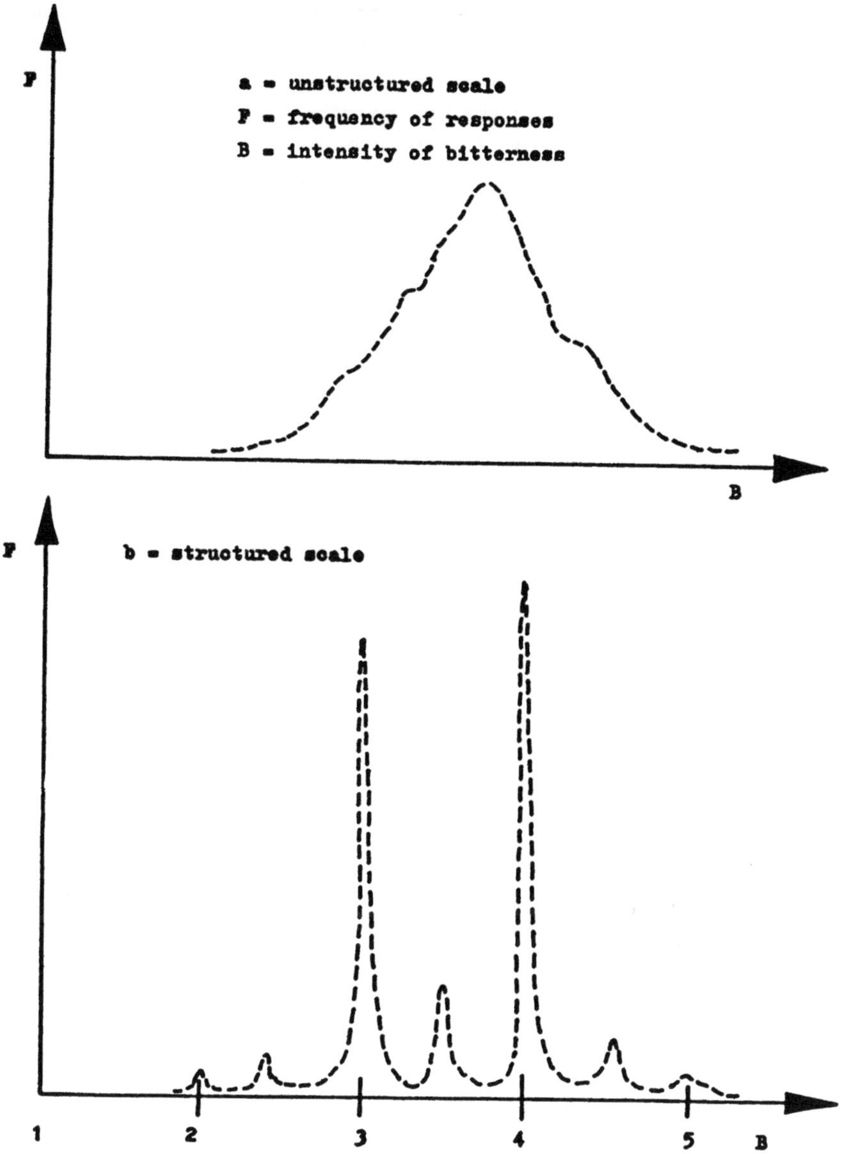

Fig. 4. Effect of structurating graphical scale (determination of bitterness of an appetizer)

unstructured linear graphical scales are used which are transferred into the diagram afterwards. In case of the linear scales assessors do not hesitate to use the extreme end corresponding to the zero intensity but hesitate to use the other extreme corresponding to the maximum intensity possible. On contrary on the starlike diagrams, people hesitate to put their rating marks near the centre (at the zero intensity) because the construction of the diagrams is then difficult but they less hesitate to use the extreme ends (the maximum intensity), probably because the ends of various descriptors are close to one another, and their marks do not look too far away even at the extreme rating. An example of an appetizer is shown in Fig. 5. The sample was served to 163 assessors to be graded either on a starlike diagram or on a linear scale. The average ratings show that the two profiles are slightly different. The difference is not great but nevertheless, statistically significant. Either one or the other method of grading should be used (it is difficult to decide which is better), and the results obtained by the alternative method cannot be compared without some transformation which has to be determined from case to case.

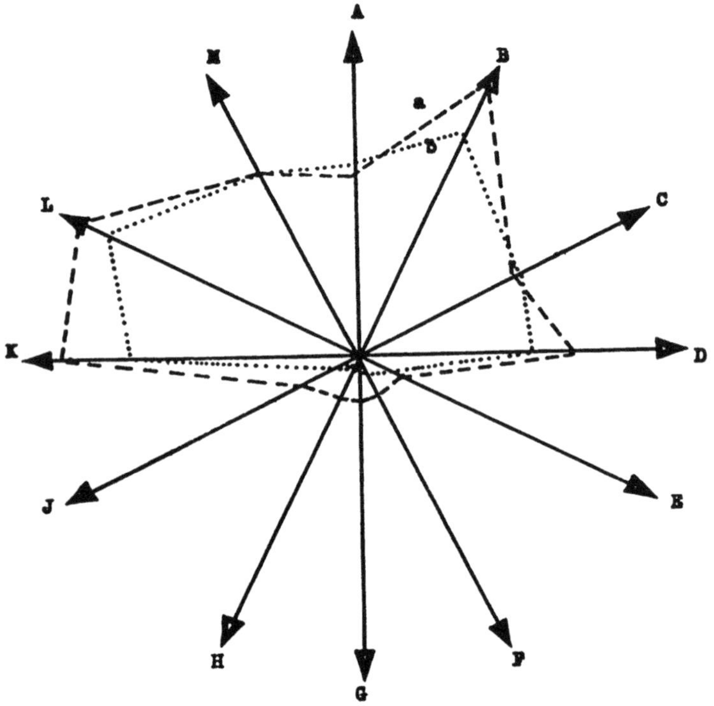

Fig. 5. Comparison of ratings with use of linear graphical scales and of a starlike diagram on the example of an appetizer. Descriptors: A = sweet, B = bitter, C = astringent, D = woody, E = root-like, F = leaf-like, G = floral, H = sharp, J = soapy, K = vanilla, L = cinnamon, M = lemon; in the centre: zero intensity, at the extreme = maximum intensity; a = direct evaluation using the star, b = transferred from linear scales

CONCLUSIONS

From our experience and from the examples shown it may be concluded:

(1) Careful selection of descriptors of flavour notes is necessary so that those appropriate to the task are selected, and their number is not not high (best between 8 - 16).

(2) The descriptors selected should be tested whether their meaning is sufficiently clear to the assessor.

(3) Scales used for the transformation of intensities of flavour notes should be carefully selected to fit the type of sample, qualifications of the assessors,

and their task; this is often neglected.

(4) All profiles in an experiment should be determined in exactly the same way, otherwise, the correct comparison is not possible.

(5) As words are generally used in sensory profiling, the uncertain meaning of most words should be taken into account when preparing the experiments, and when evaluating the results.

(6) Numbers are usually considered as well defined but in human mind the meaning of numbers (especially of large numbers) is also often not clear and precise, which is more pronounced in case of numbers which are not daily used by the assessors.

(7) The determination of sensory profiles is time-consuming and expensive, therefore, great care should be paid to the proper selection of grading scales, and to the proper selection of descriptors of flavour notes, in order to obtain the optimum results.

REFERENCES

[1] Dravnieks, A.: Evaluation of human body odors: Methods and interpretation. J. Soc. Cosm. Chem.,26 (1975), 551 - 571.
[2] Dravnieks, A., F. C. Bock, J. J. Powers, M. Tibbetts and M. Ford: Comparison of odors directly and through profiling. Chem. Senses Flavor 3 (1978), 191 - 225.
[3] Harper, R., E. C. Bate Smith, D. G. Land and N. M. Griffiths: A glossary of odour stimuli and their qualities. Perf. Essent. Oil Rec. 59 (1968), 1 - 16.
[4] ISO 4121: Sensory analysis - Methodology - Tests used for grading of food products. International Organization for Standardization, Genève, 1985.
[5] ISO 6564: Sensory analysis - Methodology - Flavour profile methods. International Organization for Standardization, Genève, 1985.
[6] Kramer, A., G. Kahan, D. Cooper and A. Papaversilieu: A non-parametric ranking method for the statistical evaluation of sensory data. Chem. Senses Flavor 1 (1974), 1, 121 - 133.
[7] Lai, K. G., J. Pokorný, J. Davídek and H. Valentová: Effect of caffeine content on the bitterness of tea infusions. Sb. VŠCHT Praze E, (1988), in press.
[8] McBride, R. L.: Psychophysics: Could Fechner's assumption be correct? Austral. J. Psychol. 35 (1983), 1, 85 - 88.
[9] McBride, R. L.: Taste psychophysics and the Beidler equation. Chem. Senses 12, (1987), 2, 323 - 332.
[10] Molnár, P.: Sensory evaluation of food. 1. Scoring methods. Eur. Org. for Quality Control, Food Section, Budapest, 1982.
[11] Powers, J. J.: Current practices and application of descriptive methods, in: Sensory analysis of foods (J. R. Piggot, ed.). Elsevier Applied Science Publishers, London, New York, 1984.
[12] Schiffman, S. S.: Contributions to the physicochemical dimensions of odor: A psychophysical approach. Ann. N. Y. Acad. Sci. 237 (1974), 164 - 183.
[13] Schiffman, S., D. E. Robinson and R. P. Erickson: Multidimensional scaling of odorants: Examination of psychological and physicochemical dimensions. Chem. Senses Flavor 2 (1977), 375 - 390.
[14] Štern, P., J. Davídek, J. Čmolík and J. Pokorný: Vztahy mezi reologickými a organoleptickými vlastnostmi šlehaných tuků. Sb. ÚVTIZ - Potrav. Vědy 5 (1987), 263 - 278.
[15] Stevens, S. S., and E. H. Galanter: Ratio scales and category scales for a dozen perceptual continua. J. Exp. Psychol. 54 (1957), 377 - 411.

THEORETICAL AND METHODICAL PROBLEMS OF ABSOLUTE THRESHOLD DETERMINATION

K. Hoppe and W. Rödel[1)]

Absolute threshold values characterizing the effectiveness of smell and aroma substances play an important role in aroma research. The concentration transformation of smell substances into the scale of aroma values is important in this connection. With this transformation an effect relationship is established as a result of giving the real concentration in relation to the threshold value. The threshold value which is a measure of the effect can be also used for the determination of structure-effect-relationship. In this work comparable investigations are described concerning the sweetness of sucrose and the partial substitution of cocoa powder by a Maillard product using paired test (2-AFC), triangle test (3-AFC), and categoric test. Medians respectively quartils of the cumulated distributions are used as the criterion of the threshold. Data of literature interpreted by these criteria are involved into the considerations.

However, as a result of methodical approaches concerning experimental procedure and interview of panel members as well as estimation of the results very different threshold values are obtained when evaluating the same substance. In particular non-standardized threshold criteria are supposed to be the cause of this state.

From the stochastic point of view limiting probabilities of zero and one do not have any significance because of the different variability. In sensory analysis the significance of a difference test often serves as threshold criterion. However, this statement which is actually valuable depends on the number of repetitions because the variance of the mean is taken for statistical examination. In the well-known tables of Roessler, Baker and Amerine (1956) which are intended for paired comparison and triangular testing procedures therefore the number of repetitions as characteristic parameter have been taken into account. Even by 1860 Fechner had defined the transition from one category of answer to another as the threshold criterion. That means that there is a threshold if 50 % of the judgements passed have been given in favour of the alternatives. This is true for the nominal answer categories yes/no as well as for the ordinal categories less than/greater than or less than/equal to/greater than. Thus, the threshold value is calculat-

1)Zentralinstitut für Ernährung Potsdam-Rehbrücke, Akademie der Wissenschaften der DDR, Arthur-Scheunert-Allee 114/116, Bergholz-Rehbrücke, DDR-1505

ed using the safest value if the stochastic distribution function of which the median is. The ordinal measuring model is presented in Fig. 1.

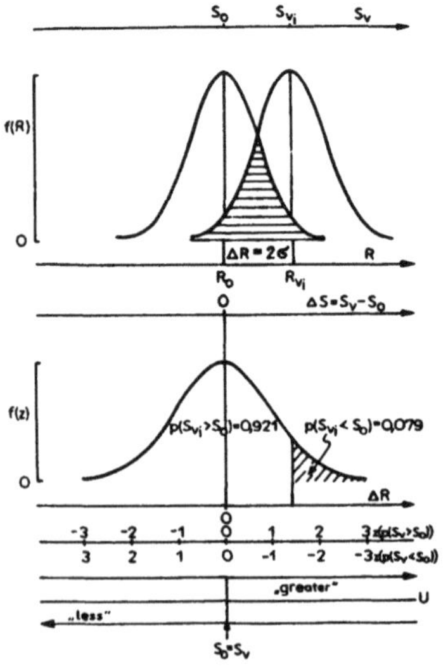

Fig. 1. The ordinal measuring model (paired comparison)

The stimulus continuum (abscissa) is compared with the stimuli S_o and S_{Vi}. We suppose that each stimulus is observed at the sensation continuum with the average sensation R_o or R_V. The sensation is expressed by the judgements 'greater than' or 'less than' as the lower part of the figure shows. When the comparisom has been repeated every sensation reveals normal distribution. The law of comparative judgement according to Thurstone (1927) indicates that under the conditions of independent distribution and the same variance the distribution of judgements is characterized by a standard deviation which is higher by factor $\sqrt{2}$. The x-axis of distribution shown in the lower part is proportional to the differences in sensation and stimulus. When used in connection with the standard normal distribution it is called standard measure z. z = 0 is the mean and the integer values are differences given in the unit of measurement of the standard deviation. The sensation difference of 2 σ shown in the figure results in a probability of p = 0.921 at the resulting distribution if S_{Vi} is greater than S_o.

In the upper part of Fig. 2 the cumulated standard normal distribution dependence of stimulus difference and standard measure z is presented. The lower part makes clear the proportionality between standard measure z and stimulus difference. It is considered as the measuring technical basis for the following data generated. The probabilities p found by experiment were transferred into the corresponding standard measures z with the help of tables and then plotted against the stimulus differ-

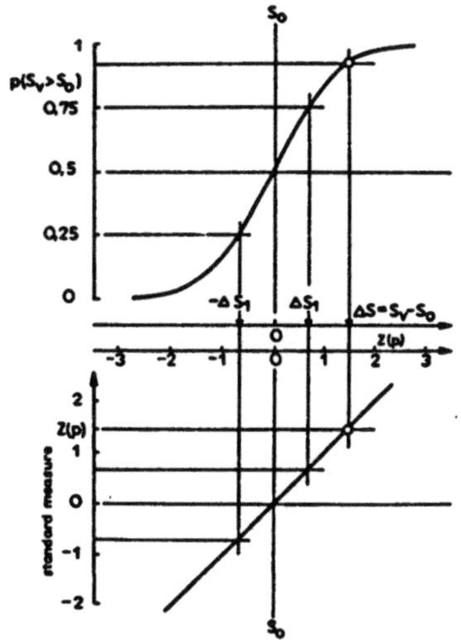

Fig. 2. Cumulated standard normal deviation and standard measure of the ordinal measuring model

ence. The quartiles 0.25 and 0.75 which are shown in the upper part of the figure are supposed to be the criteria of difference threshold ΔS_1. In paired comparison they present the transition from the category less than to equal to or from equal to greater than. Fig. 3 demonstrates the cumulated distributions along with the category 'equal to'. At the top stimulus level in the middle the judgement level and the corresponding standard measures z can be recognized. At the bottom of

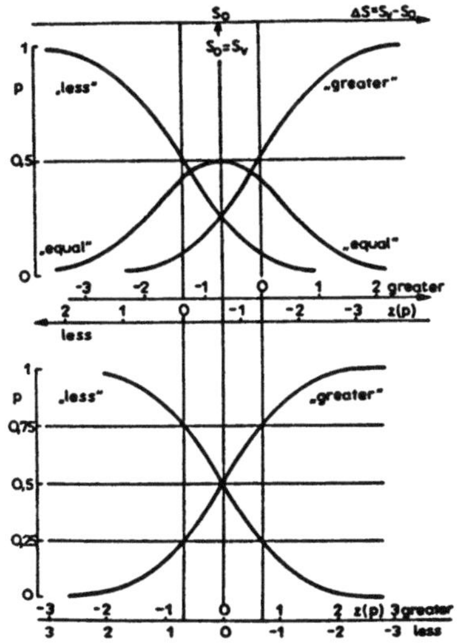

Fig. 3. Paired comparison with and without the judgements 'equal'

Fig. 3 the reduction to the simple alternative is shown. It becomes apparent that threshold values first lie at the distribution means (p = 0.5), and after reduction to the extreme judgements at the quartile.

Under the condition that the measuring model is strictly valid a steady increase in sensation difference expressed by standard measure z is expected to be determined with increasing stimulus difference. Using as an example the comparison between water and sugar solution Filipello (1956) described a significant break with concentrations of about 1 g/l (Fig. 4). This break is particularly obvious when using triangular testing. The zero-positions are found to be in triangular testing at an error probability of 1/3 and at a difference threshold of 2/3. The break in distribution occurs approximately when reaching a difference threshold. Therefore, the concentration differences are given corresponding to half the threshold. They amount to 0.61 and 0.96 g/l, respectively. Please take into account these values when comparing the following experiments. Filipello used 48 repetitions for each difference. The concentrations show relatively greater distances (0.5 g/l). Therefore, in paired comparison (crosses) the break cannot be determined.

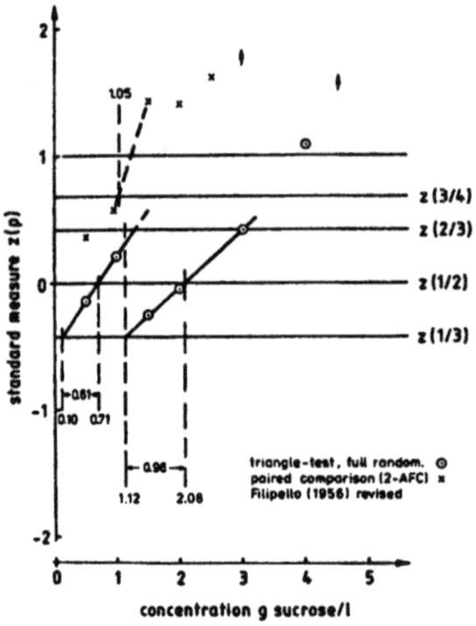

Fig. 4. Triangle test and paired comparison sucrose-water by Filipello (1956)

In carrying out our own experiments the distance between the concentrations has been lowered to 0.125 g/l in critical ranges. The number of repetitions was fixed at 240 for the first distribution, 180 for the following and 100 for the last distribution to guarantee statistical significance. The concentrations are stochastically independent (Fig. 5). There is a break in 4 distributions. The first break is again at about 1 g/l. Two parallel main distributions are clearly revealed. Their zero-positions are at 0.55 and 1.52 g/l. The upper quartile of the first main distribution is between 1.30 and 1.52 g/l which are the zero-positions of the last two distributions. According to the definition this is supposed to be the stimulus threshold. Standard deviation

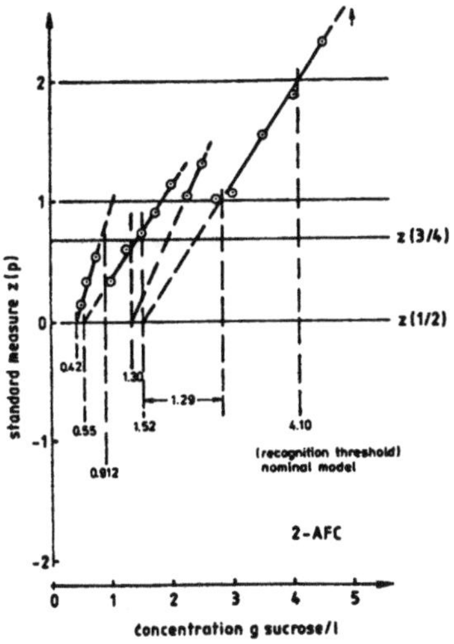

Fig. 5. Paired comparison sucrose-water

of the parallel main distributions amounts to 1.20 g/l. At 5.0 g/l complete discrimination is reached. It is very interesting to see that the recognition threshold of sucrose is about 4 g/l (Lemberger 1908; Richter and Campbell 1940; Herrmann 1972; and Paulus and Reisch 1980). Thus it corresponds to the concentration of two standard deviations above zero. Fig. 6 shows the results of an experiment performed by 3-AFC (three-alternative-forced-choice) technique. In each triangle water is contained twice as much. Distributions are found to be almost parallel. The first zero-position is at concentration zero and the second at 0.54 g/l. The distribution break is at 0.94 g/l. It corresponds to the upper quartile of the first distribution which has not been reached. The upper quartile of the last distribution is at 1.57 g/l. This concentration corresponds to the stimulus threshold detected by the first experiment. Here the characteristics of the 3-AFC technique compared with the full randomized triangle test become appearant. They are expressed by decreasing standard deviation, that is, by a higher test intensity. For each difference 96 triangles were used, therefore, even the first point is significantly different from the error probability 1/3.

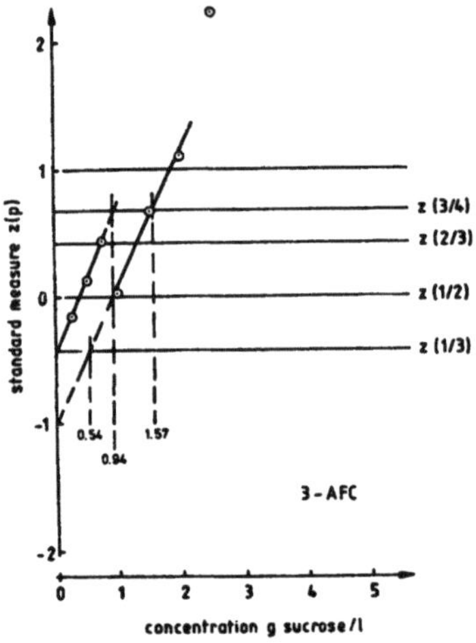

Fig. 6. Three-alternative-forced choice (3-AFC) task for comparison sucrose-water

On the basis of these results generated by the method of constant stimuli an experiment will be presented which was performed by Gregson (1962). 77 test persons were given a series of sucrose solutions with ascending concentrations which had been increased geometrically. 7 categories were permitted to estimate the solutions:

same as water	Likert value 1
doubtful if pure water	Likert value 2
a very faint taste, can't say what	Likert value 3
a very faint sweet taste	Likert value 4
a faint sweet taste	Likert value 5
a weak sweet taste	Likert value 6
a clear sweet taste	Likert value 7

The Likert values were used in analysis but never shown to subjects.

The transition from 3 to 4 was selected by Gregson to be the criterion for the absolute threshold. Therefore an alternative transformation was performed by summation of all judgements greater than and equal to 4, followed by transition into standard measure z (Fig. 7).

As the judgement cases are put down to the nominal measuring model with the categories yes/no, the standard normal distribution is used symmetrically. Based on the content of the categories 3 and 4 the threshold concentration at the medium has to be called the recognition threshold. This is at 1.56 g/l and is equal to the stimulus threshold determined before by using 2-AFC and 3-AFC-technique. The lower quartile as well as the standard deviation correspond to the concentration of 0.93 g/l which is generally found to be the break.

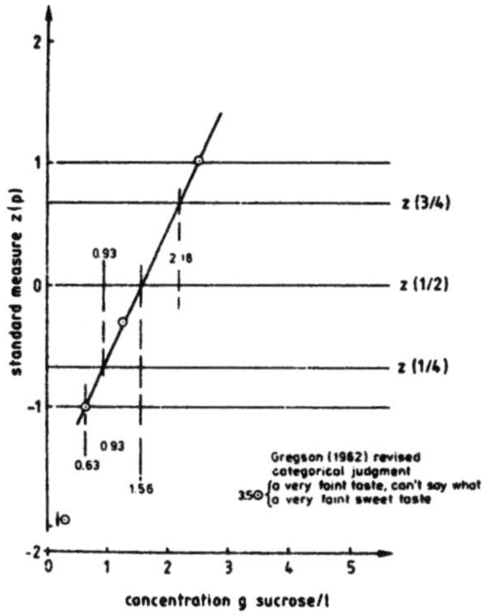

Fig. 7. Alternative transformation of categorical taste judgements for sucrose by Gregson (1962)

From the aspect of experimental expenditure categorical determination of the threshold value is obviously the most favourable method. If necessary, position determination within the measurement model can be performed by means of additional 2-AFC or 3-AFC testing near the threshold. The low test power of a full randomized triangle test expressed by the relatively high standard deviation has been theoretically supported by Ura (1960) and Frijters (1982).

Finally, the question is raised as to whether analogous threshold determinations can also be made in foods with a more complicated structure and whether those distribution breaks can be observed, too.

The abscissa of Fig. 8 shows the substitution rate of cocoa in a cocoa milk drink with a substitute. The full randomized triangle test containing 40 repetitions on each level provides an affirmative answer. The break has been observed near the probability 2/3, as it was also shown by Filipello. The break occurs when a substitution proportion of 12 % has been reached. The fact that with $p = 0.50$ a substitution proportion of 7.5 % as well as of 20.6 % can be used is of great practical importance. To make clear the magnitude of $p = 0.50$ concerning the difference it is to remark that at least 54 repetitions are necessary in order to provide statistical significance on the 1 % level.

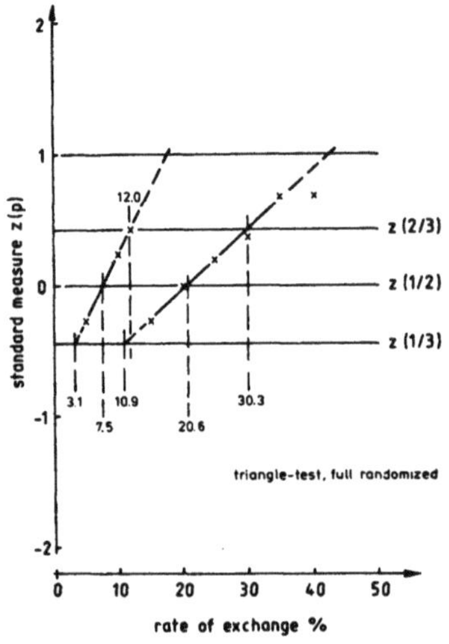

Fig. 8. Triangle-test of cocoa milk drinks with a substitute. Standard measure of difference influenced by the rate of exchange

Fig. 9 shows the categorical investigation into the same subject. Instead of ascending series haphazard presentation of stimuli was performed. 40 repetitions were carried out for each rate. The standard has been permanently available. According to Gregson the categories have been given as follows:

standard	Likert value 1
doubtful if standard	Likert value 2
a very faint difference, can't say what	Likert value 3
a very faint characteristic difference	Likert value 4
a faint characteristic difference	Likert value 5 and so on

The open circles show the main distribution with the transition from 3 to 4. The threshold is 22.6 %. This value corresponds to the value of p = 0.5 obtained by triangle difference test. The medians of the accompanying limits of categories are found to be at the quartiles of this main distribution as was demanded in the multiple category model. However, it is interesting that the first threshold of the transition from 2 to 3 results in a differentiation probability of 2/3 in the triangle. But this is already significantly detectable when making 15 repetitions at the 1 % level.

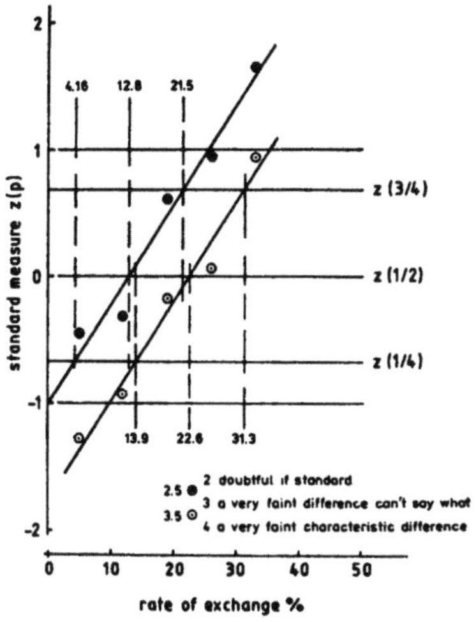

Fig. 9. Alternative transformation of categorical taste judgements for substituted cocoa milk drinks. Standard measure of categorical limens influenced by the rate of exchange

Strategic consequences can be deduced from this interpretation. The advantages of categorical examination are seen in the fact that by using 5 stimulus levels the whole interesting range of substitution can be recorded. The support of these thresholds by elementary difference tests is recommended.

References

[1] Fechner, G. T.: Elemente der Psychophysik, Breitkopf u. Härtel, Leipzig 1860

[2] Filipello, F.: A critical comparison of the two-sample and triangular binomial designs, Food Res. 21 (1956), 235 - 241

[3] Frijters, J. E. R.: An olfactory investigation of the compatibility of oddity instructions with the design of a 3-AFC signal detection task, Acta psychol. 49 (1981), 1 - 16

[4] Gregson, R. A. M.: A rating-scale method for determining absolute taste thresholds, J. Food Sci. 27 (1962), 376 - 380

[5] Herrmann, J.: Neue Methoden der Be- und Auswertung sensorischer Eigenschaften von Lebensmitteln und der Berechnung ihrer Veränderungen. 4. Mitt.: Sensorische Skalen für Geruchs- und Geschmackskonzentrationen, Nahrung 16 (1972), 205 - 213

[6] Lemberger, F.: Psychophysische Untersuchungen über den Geschmack von Saccharin und Zucker, Pflügers Arch. ges. Physiol. 123 (1908), 293 - 311

[7] Paulus, K. and A. M. Reisch: The influence of temperature on the threshold values of primary tastes, Chem. Senses 5 (1980), 11 - 21

[8] Richter, C. P. and K. H. Campbell: Sucrose taste thresholds of rats and humans, Amer. J. Physiol. 128 (1940), 291 - 297

[9] Roessler, E. B., Baker, G. A. and M. A. Amerine: One-tailed and two-tailed tests in organoleptic comparisons, Food Res. 21 (1956), 117 - 121

[10] Thurstone, L. L.: A law of comparative judgement, Psychol. Rev. 34 (1927), 273 - 286

[11] Ura, S.: Pair, triangle and duo-trio test, Rep. Stat. Appl. Res., JUSE 7 (1960), 107 - 119

INTERNATIONAL TRENDS IN STANDARDIZATION OF SENSORY METHODS

R. Neumann[1)]

In the last three decades sensory food analysis was developed in such a way that it may be discussed as a separate discipline within food analytics requiring its own methodology. In this situation standardization efforts as a part of research and development should be focussed to a systematization of the existing knowledge.

Standardized regulations, methods and terms can support an extensive practical application of sensory testing and evaluation of food. The same is true for the necessary raw materials, auxiliary agents and additives including flavouring substances. This should be realized by use of actual methods and results in sensory analytics. Such activities touch research and development as well as process and quality control in food production and in the field of trade.

Finally it is necessary to develop and introduce such methods which provide highly significant results. Large attention must be paid to all factors having a special effect on the significance of the results of sensory analyses; these factors are

. the competence and skill of the 'assessors' who in a metaphorical sense exercise their function as 'measuring instruments', i. e. anthropometers,

- the planning of analytical work including the selection of sensory analytical methods which seem to be most suitable for the task to be solved as well as the mathematical and statistical evaluation of results,

- the thorough technical and organizational preparation and implementation of sensory analytical activities so as to avoid serious errors and mistakes from the very beginning.

Standardization in this field should follow these fundamental lines.

In the CMEA[2)] countries much attention is attached to the standardization efforts in food production since they have to contribute for getting objective data about the development of quality. This becomes evident by the fact that in the GDR a total of 2.002 standards is

---

1) Amt für Standardisierung, Meßwesen und Warenprüfung (ASMW), Fachabt. Nahrungsgüter, Fürstenwalder Damm 388, Berlin, DDR-1162

2) Council of Mutual Economic Assistance

valid for food production with 293 of them for sensory food analytics.

In the following we shall estimate the present situation reached in standardization of sensory food analytics and discuss actual tendencies of development.

Standards covering the field of sensory food analytics can be formally classified into three groups at a national and international level as well.

I. Requirements on panel members (experts) and test rooms as well as on the preparation, presentation and evaluation of sensory analyses

II. Methods of sensory testing and evaluation

III. Terms and definitions

Ad group I.

There are standards covering the selection and qualification of panel members (experts) either in form of national standards or CMEA standards applied in all countries of the CMEA. They include the following aspects:

1. Methods used for determining sensory primary skills

At present these methods cover such skills used for

- detection of the four primary tastes
- description of gustation effects (threshold value)
- detection of just distinguishable differences in the intensity of primary tastes by means of the paired difference, duo-trio and triangel tests
- detection of just distinguishable differences in the intensity of colouring solutions according to the ranking test
- detection of odour qualities.

If the national standards of the CMEA countries are compared there is the tendency of agreement in their methodologies as well as in the parameters concerning sensory minimum requirements. The same applies if this comparison is extended to the ISO[1] standards 3972, 4120 and 5495. Meanwhile the ST-CMEA 5216-85 'Methods used for determining the sensory skills of panel members and assessors' which will be put into force by a number of countries at an international and national level

[1] International Standardization Organisation

in 1987 was harmonized within the CMEA. As GDR Standard TGL 45 121 it will substitute the former regulations in DAMW-VW 656-660 in the GDR.

The standard includes regulations concerning the principle pf the method, the instruments, reagents and material to be used, the sample preparation and realization of tests as well as the evaluation of the data. The last-mentioned item defines the requirements to be met by a panel member for carrying out a test successfully.

The practical application of these standards is expected to deliver new information and findings in the near future. This particularly applies to questions concerning the requirements on panel members. A method used for determining the sensitivity in texture testing of food was not the subject of this standard. This gap has to be filled by the next revision since texture attributes are observed to get increased significance for consumers.

2. Requirements on panel members

In 1984 the methodological material entitled 'Sensory quality testing: requirements on panel members' was harmonized within the CMEA. In the first chapter criteria of evaluation in the field of physiology and psychology are defined. A special section deals with the activities of the leaders of panel sessions since their manager work decides on the validity and reliability of the analytical results of the panels.

In the second chapter of the methodological material the requirements on the qualification of panel members are given. It also deals with the question of how the members have to prepare themselves directly before the sensory food evaluation is performed. It can be stated that this material reflects a state of the art which will only insignificantly develop in the next few years.

In the GDR this methodological material will be put into force as Standard TGL 45 120 in 1987 and will replace DAMW-VW 654.

3. General conditions of testing

Standards of Bulgaria, Hungary, Poland, the GDR and Czechoslovakia impose general requirements on test rooms as well as on the preparation and realization of sensory analyses. The ISO also defines fundamental requirements in the standard 6658-1985 (E) 'Methodology - general rules'.

The ST-CMEA 4710-84 'General conditions' approved in 1984 which also deals with the conduct of sensory tests defines requirements concerning

- sampling, transport, storage of samples
- sample preparation
- rooms
- auxiliary equipment to perform tests
- conduct of tests
- evaluation of test results.

This ST-CMEA was imtroduced in six CMEA countries within the framework of contractual relations from January 1, 1986. It was put into force as GDR standard under TGL No. 43 664 since February 1, 1987. In this case it is also possible to state that it represents a technical level which is fully in line with the international state of the art.

Ad group II

At an international level standardization of sensory testing methods was concentrated on difference comparison testing methods in the past. However, standardized sensory evaluation procedures existed in some CMEA countries such as the GDR, Czechoslovakia, and Poland since more than twenty years, too. In the USSR, Bulgaria, Hungary and the Republic of Cuba standardized evaluation procedures are applied or prepared.

Even in the past 5-year plan period the permanent working group of standardization/quality control of food of the Permanent Commission of Food Industry of the CMEA compiled a list of methods providing a harmonized orientation in this field. It covers 12 methods used for sensory evaluation.

The five types of methods are

1) descriptive tests
2) rating tests
3) difference comparison tests
4) sensitivity tests
5) hedonic tests.

The 12 methods included in these five groups are described in this list according to the following aspects:

1) characterization of the method
2) scope
3) fundamental requirements
4) preparation of testing

5) conduct of testing
6) mathematical and statistical evaluation

This list of methods represents the point of view of the CMEA countries and also takes account of the results of the standardization effort by the ISO. These methods to be applied to routine quality control in enterprises and official state bodies as well as in the development of new processes and new food favourably should be improved and actualized by an international exchange of experiences.

Scalar scoring os of special importance as to quality evaluation of food. Therefore the permament working group of standardization/quality control of food of the Permament Commission of Food Industry of the CMEA has focussed the work to this method when elaborating methodological material entitled 'Fundamental principles for sensory quality evaluation according to scalar scoring'. This material is to ensure harmonized procedures in the cooperation for process and product development in the field of food as well as in the drafting of specific testing standards for food which is subject of international trade.

If standardization efforts oriented on the needs of practical work are to be performed, all evaluating procedures using a scale involve major problems such as [1]

1) definition of decisive quality characteristics and their unambiguous description
2) the number of distinguishable rating classes as a function of the average skill of the'assessors' (experts)
3) definition of the percentage of the individual sensory attributes (odour, flavour, consistency/texture, appearance)
4) mathematical and statistical evaluation of results
5) derivation of quality classes on the basis of scalar scoring.

There are experiences in a number of CMEA countries such as the GDR where this method has been standardized more than 20 years ago being successfully applied to all kinds of food in practical work. In Bulgaria and Hungary a basic method for developing specific product standards is proposed which considers the five mentioned major aspects. In the GDR efforts for improving the actual 5-score scale and preparation of a new edition of ASMW-VW 1149 are under work.

In 1982 the EOQC[1)] published methodological material on this subject which is almost identical with that successfully applied in the countries mentioned. The ISO is also preparing a methodology to be used for food evaluation by means of scales and categories which is to be published as ISO standard 4121.

The same trend can be found in the FRG, too. In the past years efforts favoured and supported by the Deutsche Landwirtschafts-Gesellschaft (German Agricultural Society) as the managing body resulted in the creation of test schemes which use a 5-score scale for a number of different kinds of food such as bakery goods [2], wine [3], meat products [4] and beer [5].

At an international and national level as well there is still a need to develop standards for two types of methods being of large interest: ranking test and flavour profile analysis.

Ranking tests are important for solving a number of practical tasks, e. g. when a large number of samples has to be tested for sensory differences or for correspondence or when samples are to be classified according to quality, intensity of a sensory attribute or popularity [6].

The flavour profile analysis is of large use [7] since it is best suited to detect variations in quality under defined conditions. Simultaneously it can also be applied for the preparation of standards for sensory quality evaluation according to scalar scoring.

The ISO is now preparing standards for both methods which are to be published as ISO 8587 and ISO 6564.

Ad group III

A uniform terminology is a decisive prerequisite for harmonized activities. In the CMEA countries there exists a national standard on this subject, published in Poland under the title 'Expressions and definitions' (PN - 71/A-04025) including 25 terms. The terminology standard which has been valid in Hungary since 1981 covers 90 terms (MSZ 7304/8-80). There exists a factory standard of the Bulgarplod company (Plovdiv) 'Concepts, terms and definitions' in Bulgaria. In the period from 1977 to 1985 the ISO drew up a document entitled 'Vocabulary' (ISO 5492/1-6) containing 134 terms. By making use of national exper-

---

[1)] European Organization for Quality Control

iences and ISO documents the ST-CMEA 5215-85 'Terms and definitions' with 46 standardized terms could be approved within the CMEA. It will be applied in 4 CMEA countries including the USSR and the GDR from 1987 and is intended to create better conditions for the promotion of cooperation in the field of research and development and to fix even more clearly the regulations governing the international trade of food in standards.

The GDR standard 'Terms and definitions' drafted on the basis of ST-CMEA 5215-85 will be put into force as TGL 45 193 in 1987. It will cover a total of 66 terms. In the drafting and harmonization of this standard special allowance has been made of new findings resulting from texture analytics. It will be necessary to intensify standardization efforts in this field in the next few years.

A second essential aspect are the nationally standardized 'principles of sensory evaluation'. Good solutions for generalization in the standardization of terms of sensory attributes were developed in the fruit and vegetable-processing industries [8].

If existing standardized evaluation schemes are analyzed, the terms used in them can be classified into 4 groups (table 1). It often are such terms which can be characterized by associations (such as sweet, fishy). General ambiguous terms (such as specific, aromatic) which are also often used are of limited value. They lead to reliable results only when they are defined by means of reference substances (such as flavouring or aromatic substances).

Terms of a hedonically evaluation character (such as stimulating, palatable) should be avoided in standards.It is useful to prefer terms which assiciatively describe sensations of taste and odour or chemical terms. This requires standardized reference substances which are submitted for testing as purified chemical compounds mostly in a neutral solvent. Defined terms are an essential prerequisite to develop a uniform terminology among the panel members responsible for a group of products and thus implement a uniform quality standard in the panel. For standardization of terms all descriptions used in practical work should be systematized by means of a nomenclature.

Table 1: Terms used to characterize attributes of food

Kind of term	Examples
General ambiguous terms	
a) conditionally definable	specific, typical, aromatic, fragrant, fruity, florid, ester-like, smoky, mouldy, earthy, musty, resinous, oxidized
b) restrictedly definable, partly with a hedonically evaluating character	pure, impure, old, young, ripe, balanced, well-balanced, harmonic, mild, void, full, saturate, rich, great, stimulating, stale, dry, refreshing, appealing, savoury, delicious, sweet-scented, foul, smelly
Terms limitable by associations, which can be referred to gustatoric or olfactory sensations; conditionally definable	salty, sour, bitter, sweet, soapy, rancid, metallic, sulphide-like, grassy, strawy, malty, cramely, fishy, barmy, etheric, butter-aromatic, goaty, cheesy, all kinds of fruit and spices, citrus-like, apple-like, banana-like, peppery, anise-like, caraway-like, nut-like, vinous
Terms which can be referred to haptic sensations; conditionally definable	oily, dry, fresh, tickling, flat, fat, astrigent, warming, cooling, pungent, reedy
Terms unambiguously defined by chemical terms	hydrogen sulphidic, sulphitic, ammoniacal, phenolic, acetonic, acetaldehydic, butyric acidic, vinegary, isovaleric acidic

<u>Summary</u>

Standardization within the CMEA has the aim to get to a true contribution towards the development of socialist economic integration.

As a general principle the tasks of standardization are governed by the needs of cooperation in the field of development of food, auxiliary agents and additives and by the planned exchange of goods among the countries belonging to the CMEA.

On the basis of the defined tasks the Permanent Commission of Food Industry of the CMEA therefore has decided to deal with selected themes set for standardization in the field of sensory analytics according to a fixed plan. With the existing ST-CMEA developed for sensory food evaluation bases were created which can be used to select and qualify panel members in a harmonized manner in the CMEA countries and enable them to exercise their functions in line with standardized conditions. This is to ensure a better objectivication of sensory testing and evaluation of food, auxiliary agents and additives which are important in the exchange of goods within the CMEA.

Such a set of standards for sensory food evaluation also provides essential conditions for the mutual recognition of test results.

After all it should be stressed that a higher level of standardization is required for solving even more effectively the joint tasks resulting from the food requirements of the population in the necessary quantity and quality within the member-countries of the CMEA.

<u>References</u>

[1] Neumann, R. et al.: Zur Standaridisierung der sensorischen Erzeugnisprüfung. Lebensmittelind. <u>21</u> (1974) 8, 352

[2] Seibel, W. et al.: Prüfschemata zur Bewertung der sensorischen Qualitätsmerkmale von Backwaren nach einer 5-Punkte-Skala. Manuskript zum 7. Welt-Getreide- und Brotkongreß·1982, Prague

[3] Anonym.: Neues Fünf-Punkte-Schema zur Weinbeurteilung. Weinwirtschaft 1982, No. 10, 254

[4] Sinell, H. J. et al.: Sensorische Qualitätsmerkmale von Fleischerzeugnissen. Fleischwirtschaft <u>64</u> (1) (1984), 29

[5] Koch, J.: Handbuch der Lebensmitteltechnologie, Getränkebeurteilung. Verlag Eugen Ulmer Stuttgart 1986, 392

[6] Neumann, R. et al.: Sensorische Lebensmitteluntersuchung. Fachbuchverlag Leipzig 1983, 120

[7] Kochan, A. et al.: Die sensorische Profilanalyse, Lebensmittelind. <u>32</u> (1985) 6, 255

[8] TGL 32 336/01. Beurteilung von Erzeugnissen der obst- und gemüseverarbeitenden Industrie. Übersicht und Begriffe

Author Index

Subject Index

List of participants

ABEL, Karl
Akademie-Verlag Berlin/GDR

ADDA, Jacques, Prof.
Institut National de la Recherche Agronomique, Laboratoire de Recherches sur les Arômes, Dijon/France

ANDERSON, Matthias, Dr.
Zentralinstitut für Ernährung Potsdam-Rehbrücke/GDR

BARYLKO-PIKIELNA, Nina, Prof.
Institute of Food and Nutrition Warsaw/Poland

BAUER, I.
VE Forschungsinstitut für Obst- und Gemüseverarbeitung. Magdeburg/GDR

BAYER, Rüdiger
VEB Rotkäppchen-Sektkellerei Freyburg/GDR

BÖHM, Hartmut, Dr. sc.
Institut für Biochemie der Pflanzen Halle/GDR

BRATOVANOVA, Penka, Dr.
University of National Economy 'D. Blagoev' Varna/Bulgaria

BRUNKE, Ernst-Joachim, Dr.
DRAGOCO Holzminden/FRG

CÄMMERER, Bettina, Dr.
Sektion Nahrungsgüterwirtschaft und Lebensmitteltechnologie der Humboldt-Universität zu Berlin/GDR

CIEMINSKI, Bernd, Dr.
VEB Kombinat Spirituosen, Wein und Sekt Berlin/GDR

DAVIDEK, Jiri, Prof.
Institute of Chemical Technology Prague/Czechoslovakia

DENECKE, Paul, Dr.
VEB Kombinat Öl und Margarine Magdeburg/GDR

DOBRANYI, I., Ing.
SZILASMENTI Cooperative Kerepestarcsa/Hungary

ENGST, Wolfram, Dr.
Zentralinstitut für Ernährung Potsdam-Rehbrücke/GDR

FABIAN, Gerd, Dr.
Forschungszentrum für Biotechnologie Berlin/GDR

FARKAS, Pavel, Dr.
Food Research Institute Bratislava/Czechoslovakia

FEDERER, Josef
Firmenich Ges. m. b. H. Wien/Austria

GEBHARDT, Erich, Dr.
VEB Kombinat Nahrungsmittel und Kaffee, Institut für Getreideverarbeitung Bergholz-Rehbrücke/GDR

GOLOVNJA, Rima V., Prof.
A. N. Nesmejanov Institute of Organo-Element Compounds Moscow/USSR

GREN, Andrei Ivanovich, Prof.
A. V. Bogatsky Physiko-Chemical Institute Odessa/USSR

GRIGORYEVA, D. N., Dr.
A. N. Nesmejanov Institute of Organo-Element Compounds Moscow/USSR

GROSCH, Werner, Prof.
Deutsche Forschungsanstalt für Lebensmittelchemie (DFA) Garching/FRG

GRÜB, Helmut, Dr.
Fa. Haarmann & Reimer GmbH Holzminden/FRG

GRUCHAŁA, Leonarda
Institute of Food Technology Poznań/Poland

GRUNERT, Siegfried, Dr.
Sektion Chemie, WB Lebensmittelchemie der Humboldt-Universität zu Berlin/GDR

HAENEL, Helmut, Prof.
Zentralinstitut für Ernährung Potsdam-Rehbrücke/GDR

HEINIG, Rüdiger
VEB Chemisches Werk Miltitz/GDR

HEINIG, Wolfgang
VEB Nordbrand Nordhausen/GDR

HERMERSDÖRFER, Heino, Dr.
Zentralinstitut für Ernährung Potsdam-Rehbrücke/GDR

HERRMANN, Heinrich, Dr.
VEB Geflügelwirtschaft Dresden/GDR

HOPPE, Konrad, Dr. sc.
Zentralinstitut für Ernährung Potsdam-Rehbrücke/GDR

HORN, Annerose, Dr.
VEB Kombinat Nahrungsmittel und Kaffee, VEB Gewürzmühle Leipzig/GDR

JENSEN, Brigitte
VEB Getränkekombinat Dresden/GDR

KAMINSKI, Edward, Prof.
Institute of Food Technology of Plant Origin Poznań/Poland

KARWOWSKA, Krystyna, Prof.
University of Agriculture Warsaw, Department of Medicinal and Spices Plants, Warsaw/Poland

KERENYI, Zoltan, Dr.
Research Institute of Viticulture Kecskemet/Hungary

KIM-SUM-IK, Dr.
Institute of Basic Research Phjoengjang/Corea

KOCHAN, Anita, Prof.
Technische Universität Dresden, Industrie-Institut, Dresden/GDR

KONJA, Gordana, Dr.
University of Zagreb, Institute of Food Biotechnology, Zagreb/Yugoslavia

KORNELSON, Carla
Zentralinstitut für Ernährung Potsdam-Rehbrücke/GDR

KOSCHEMANN, Uwe
VEB Kombinat Milchwirtschaft Dresden/GDR

KOSTRZEWA, Ewa, Dr.
Institute of Fermentation Industry Warsaw/Poland

KRAUSE, Wolfgang, Dr. sc.
Sektion Chemie, WB Lebensmittelchemie und techn. Biochemie der Technischen Universität Dresden/GDR

KRUSE, Hans-Peter, Dr.
Zentralinstitut für Ernährung Potsdam-Rehbrücke/GDR

KUNZ, Artur
Kammer der Technik, Fachausschuß "Alkoholfreie Erfrischungsgetränke", Burkhardswalde/GDR

LINSSEN, Josef Peter Hubertus
Agricultural University Wageningen/The Netherlands

ŁOBASZEWSKI, Edward, Dr.
Institute of Human Nutrition Warsaw/Poland

LUDWIG, Eberhard, Prof.
Sektion Chemie, WB Lebensmittelchemie und techn. Biochemie der Technischen Universität Dresden/GDR

MAARSE, Henk, Dr.
TNO CIVO Food Analysis Institute Zeist/The Netherlands

MACHOLZ, Rainer, Prof.
Zentralinstitut für Ernährung Potsdam-Rehbrücke/GDR

MISHARINA, T. A., Dr.
A. N. Nesmejanov Institute of Organo-Element Compounds Moscow/USSR

MITEVA, Ecaterina, Dr.
Institute of Meat Industry Sofia/Bulgaria

MOR, Juan-R., Dr.
Givaudan Research Co. Ltd. Dübendorf/Switzerland

MÜLLNER, Franz-Xaver, Dr.
Firmenich SA Genf/Switzerland

NEUMANN, Ralph, Dr.
Amt für Standardisierung, Meßwesen und Warenprüfung Berlin/GDR

NYKÄNEN, Lalli, Dr.
Research Laboratories of the Finnish State Alcohol Company, Alko Ltd. Helsinki/Finland

OFFENHEIMER, Julio
Coprima S. A., Internationales Handelszentrum Berlin/GDR

ÖRSI, Ferenc, Prof.
Technical University Budapest/Hungary

OSUCHA, Andrzej
Institute of Food and Nutrition Warsaw/Poland

PAK-ZANG-DZUN, Dr.
Institute of Basic Research Phjoengjang/Corea

PAULENZ, Horst, Dr.
Ministerium für Gesundheitswesen, Staatliche Hygieneinspektion, Berlin/GDR

PETRO-TURZA, Marta, Dr.
Central Food Research Institute Budapest/Hungary

PIETZ, Rüdiger
VEB Rotkäppchen-Sektkellerei Freyburg/GDR

POKORNY, Jan, Prof.
Institute of Chemical Technology Prague/Czechoslovakia

QUEHL, Adelheid, Dr.
Zentralinstitut für Ernährung Potsdam-Rehbrücke/GDR

RÖDEL, Wolfgang, Dr.
Zentralinstitut für Ernährung Potsdam-Rehbrücke/GDR

ROHNS, Günter, Dr.
H. Bahlsens Keksfabrik KG Hannover/FRG

ROOZEN, Jacobus Petrus, Dr.
Agricultural University Wageningen/The Netherlands

ROSCHE, Rolf
Bezirkshygiene-Inspektion und -Institut, Insp. Lebensmittel- und Ernährungshygiene Leipzig/GDR

ROTHE, Manfred, Dr. sc.
Zentralinstitut für Ernährung Potsdam-Rehbrücke/GDR

RUDOLPH, Reiner
VEB Forschung und Rationalisierung, Betrieb des VEB Kombinat Süßwaren Leipzig/GDR

RUTTLOFF, Heinz, Prof.
Zentralinstitut für Ernährung Potsdam-Rehbrücke/GDR

SASS-KISS, Agnes, Dr.
Central Food Research Institute Budapest/Hungary

SCHLEGEL, Bernd, Dr.
Bezirkshygiene-Inspektipn, Hygiene-Institut Zwickau/GDR

SCHLIEMANN, Jürgen, Dr.
Zentralinstitut für Ernährung Potsdam-Rehbrücke/GDR

SCHREIER, Peter, Prof.
Institut für Pharmazie und Lebensmittelchemie der Universität Würzburg/FRG

SCHRÖDTER, Rüdiger, Dr.
Zentralinstitut für Ernährung Potsdam-Rehbrücke/GDR

SCHULZE, Marita
VE Forschungsinstitut für Obst- und Gemüseverarbeitung Magdeburg/GDR

SPECHT, Manfred
Kreispoliklinik Teltow/GDR

STEČENKO, Alexandra Vasiljevna, Dr.
NPO 'Masloshirprom' Leningrad/USSR

STRÖHMER, Gundolf, Dr.
Sektion Nahrungsgüterwirtschaft und Lebensmitteltechnologie der Humboldt-Universität zu Berlin/GDR

STRÜBER, Franz-Josef, Dr.
VEB Chemisches Werk Miltitz/GDR

SUCHORSKA, Krystyna, Dr.
University of Agriculture Warsaw, Department of Medicinal and Spices Plants, Warsaw/Poland

SULEIMANOVA, Elmira Tejmburovna, Dr.
Institute of Petrochemical Processes Baku/USSR

SZABO-FORRAS, Zsuzsanna, Dr.
Central Food Research Institute Budapest/Hungary

SZENTE, Lajos, Dr.
Cyclodextrin Research Lab. of CHINOIN Budapest/Hungary

TABAJDI, Gábor
SZILASMENTI Cooperative Kerepestarcsa/Hungary

TELEKY-VAMOSSY, Gyönggi, Dr.
Central Food Research Institute Budapest/Hungary

TERENINA, M. B., Dr.
A. N. Nesmejanov Institute of Organo-Element Compounds Moscow/USSR

THOMANN, Ralph, Dr.
VEB Kombinat Nahrungsmittel und Kaffee, Institut für Getreideverarbeitung Bergholz-Rehbrücke/GDR

TIETZ, Uta
VEB Kombinat Nahrungsmittel und Kaffee, Institut für Getreideverarbeitung Bergholz-Rehbrücke/GDR

TSCHAKAROFF, Marin, Dr.
Institute of Meat Industry Sofia/Bulgaria

VALDEIG, Klaus
VEB Forschung und Rationalisierung, Betrieb des VEB Kombinat Süßwaren Leipzig/GDR

VENSKUTONIS, P. R., Ass. prof., cand. sc.
Kaunas Politechnic Institute, Kaunas/USSR

WALPUSKI, Jörg, Dr.
Zentralinstitut für Ernährung Potsdam-Rehbrücke/GDR

WASOWICZ, Erwin, Dr.
Institute of Food Technology of Plant Origin Poznań/Poland

WESTPHAL, Günter, Prof.
Sektion Nahrungsgüterwirtschaft und Lebensmitteltechnologie der Humboldt-Universität zu Berlin/GDR

WÖLM, Gerhard, Dr.
Zentralinstitut für Ernährung Potsdam-Rehbrücke/GDR

ZEISE, Simone
Sektion Nahrungsgüterwirtschaft und Lebensmitteltechnologie der Humboldt-Universität zu Berlin/GDR

ZHURAVLEVA, Irina L., Dr.
A. N. Nesmejanov Institute of Organo-Element Compounds Moscow/USSR

ZIEGLEDER, Gottfried, Dr.
Fraunhofer-Institut für Lebensmitteltechnologie und Verpackung München/FRG

ZILAI, Tibor
Research Institute of Viticulture Kecskemet/Hungary

ZIMINSKA, Halina, Dr.
Politechnika Gdanska, Dept. of Food Preservation, Gdańsk/Poland

ZINTERHOF, Herbert, Dipl.-Ing.
DRAGOCO Ges. m. b. H. Wien/Austria